Natural Resource Sovereignty and the Right to Development in Africa

This book explores the nexus between natural resources ownership and the right to development in Africa.

The right to sovereignty over natural resources and the right to development are recognised and protected in an extensive framework of international, regional and domestic instruments. They guarantee people's entitlement to fully and freely utilise their natural resources as a means of subsistence and for economic, social and cultural development. Yet, despite the abundance of natural resources in Africa a majority of the people on the continent remain largely impoverished. This book articulates the central argument that to achieve the right to development in Africa requires appropriate governance of the continent's natural resources to which the people of Africa are guaranteed sovereign ownership. With case study illustrations from Zimbabwe, Ghana, Ethiopia and the Democratic Republic of Congo, chapters explore the normative measures, specific guarantees, and community entitlements to natural resources for the realisation of the right to development.

The book will be an invaluable guide to scholars and postgraduate students of Natural Resources, Development and African studies as well as policy makers and practitioners in these areas.

Carol Chi Ngang is a Researcher at the Free State Centre for Human Rights, University of the Free State, South Africa and Senior Lecturer at the National University of Lesotho, and is also the editor of *Insights into Policies and Practices on the Right to Development* (2020)

Serges Djoyou Kamga is a Professor of Law at the Thabo Mbeki African School of Public and International Affairs, University of South Africa and is the author of *The Right to Development in the African Human Rights System* (Routledge, 2018)

Earthscan Studies in Natural Resource Management

Interdisciplinary Collaboration for Water Diplomacy
A Principled and Pragmatic Approach
Edited by Shafiqul Islam and Kevin M. Smith

Peatlands
Ecology, Conservation and Heritage
Ian D. Rotherham

The Moroccan Argan Trade
Producer Networks and Human Bio-Geographies
Daniel F. Robinson

Fair and Equitable Benefit-Sharing in Agriculture
Reinventing Agrarian Justice
Elsa Tsioumani

Participatory Governance of UNESCO Biosphere Reserves in Canada and Israel
Resolving Natural Resource Conflicts
Natasha Donevska

Balancing the Commons in Switzerland
Institutional Transformations and Sustainable Innovations
Edited by Tobias Haller, Karina Liechti, Martin Stuber, François-Xavier Viallon and Rahel Wunderli

Natural Resource Sovereignty and the Right to Development in Africa
Edited by Carol Chi Ngang and Serges Djoyou Kamga

For more information about this series, please visit: www.routledge.com/books/series/ECNRM/

Natural Resource Sovereignty and the Right to Development in Africa

Edited by
Carol Chi Ngang and
Serges Djoyou Kamga

Routledge
Taylor & Francis Group
LONDON AND NEW YORK

earthscan
from Routledge

First published 2022
by Routledge
2 Park Square, Milton Park, Abingdon, Oxon OX14 4RN

and by Routledge
605 Third Avenue, New York, NY 10158

Routledge is an imprint of the Taylor & Francis Group, an informa business

British Library Cataloguing-in-Publication Data
A catalogue record for this book is available from the British Library

Library of Congress Cataloging-in-Publication Data
Names: Ngang, Carol Chi, editor. | Kamga, Serges Djoyou (Serges Alain Djoyou), editor.
Title: Natural resource sovereignty and the right to development in Africa / edited by Carol Chi Ngang and Serges Djoyou Kamga.
Other titles: Earthscan studies in natural resource management.
Description: Abingdon, Oxon ; New York, NY : Routledge, 2022. | Series: Earthscan studies in natural resource management | Includes bibliographical references and index. |
Identifiers: LCCN 2021011212 (print) | LCCN 2021011213 (ebook) |
Subjects: LCSH: Natural resources--Africa--Management. | Law and economic development--Africa. | Sustainable development--Africa.
Classification: LCC HC800.Z65 N384 2022 (print) | LCC HC800.Z65 (ebook) | DDC 333.7096--dc23
LC record available at https://lccn.loc.gov/2021011212
LC ebook record available at https://lccn.loc.gov/2021011213

ISBN: 978-1-032-04906-9 (hbk)
ISBN: 978-1-032-04907-6 (pbk)
ISBN: 978-1-003-19511-5 (ebk)

Typeset in Bembo
by Deanta Global Publishing Services, Chennai, India

Contents

Illustrations

Figures

Tables

Editors and Contributors

Editors

Carol Chi Ngang (LLD, LLM, IDHA, SUSTLAW and LLB) is a Researcher at the Free State Centre for Human Rights, University of the Free State, South Africa and Senior Lecturer at the National University of Lesotho. His areas of expertise include international human rights law, public international law, sustainable development law and constitutional law. His research interest combines human rights and development with a niche focus on the right to development and socio-economic and cultural rights.

Serges Djoyou Kamga is a Professor of Human rights law at the Thabo Mbeki African School of Public and International Affairs, University of South Africa. He is a member of the Building Committee of the Cross-Cultural Human Rights Centre based at the Free University of Amsterdam. His areas of interest include the right to development, African renaissance and disability rights. He publishes in these areas.

Contributors

Aaron Olaniyi Salau holds a PhD in Public Law from the University of Cape Town, South Africa and an LLB and LLM from Obafemi Awolowo University, Ile Ife, Nigeria. He is a Senior Lecturer and current HOD at the Department of Jurisprudence and International Law and is a Coordinator of Clinical Legal Education at the Faculty of Law, Olabisi Onabanjo University, Ago Iwoye, Nigeria. His research interests include African human rights law, freedom of expression, access to information, constitutional law and public international law.

Annelie de Man holds an LLD (UFS), a European master's degree in Human Rights and Democratisation (EIUC), an LLM (UFS), an LLB (UFS) and a Bcom Law (UFS). She has previously worked as Intern at Human Rights Watch, Brussels, Legal Researcher at the South African Supreme Court of Appeal and Coordinator of the Human Rights Desk of the Institute for

Reconciliation and Social Justice, University of the Free State. She is currently the Coordinator of the Advocacy Division since 2016 and Coordinator of the master`s module in Human Rights and Development at the Free State Centre for Human Rights.

Bridget Kafui Anthonio-Apedzi is a Senior State Attorney at the Office of the Attorney General and Ministry of Justice in Ghana. Having worked previously in various capacities as the State's principal legal adviser, she has experience in international arbitration, advocacy and legal drafting. She is an ardent advocate for environmental justice, indigenous rights and sustainable natural resource exploitation. She holds a PhD in Law from University of Birmingham and an LLM in International Environmental Law from University of Nottingham. She also attended Ghana School of Law, University of Ghana and Kwame Nkrumah University of Science and Technology.

Cephas Mandirahwe is a Lecturer in the Department of Development Studies at Midlands States University Gweru, Zimbabwe. He has passion and interest in researching cross cutting issues of rural development, natural resources management, agro-biotechnology and small-scale banana farming. Cephas has presented papers at different international conferences in Zimbabwe and abroad. He has also worked as a Research Supervisor with Mercy Corps in collaboration with the United National World Food Programme. He holds a Master of Arts in Development Studies, a Bachelor of Arts in Development Studies and a National Certificate in Environmental Health.

Daphine Kabagambe Agaba is a Postdoctoral Research Fellow at Thabo Mbeki African School of Public and International Affairs (TM-School) affiliated to the University of South Africa (UNISA). She holds a Doctorate in Law (PhD) from the University of the Western Cape, obtained in August 2017, a Master of Laws (LLM) in Human Rights and Democratisation in Africa from the Centre for Human Rights, University of Pretoria. Her current areas of research include de-coloniality, political leadership and governance, human rights, health law and policy and human rights.

Ebun Abolarin is the Managing Solicitor at ELI-AB Associates in Lagos, Nigeria. Her area of specific interest covers sustainable development, natural resources, project finance and the rights of indigenous peoples. Her passion and vision for human rights and development led to her research for an LLD in Jurisprudence at the University of South Africa. She holds an LLM (International Commercial and Business Law) University of East Anglia, Norwich, and an LLB from the Lagos State University, Ojo. She was called to the Nigerian Bar and is a Notary Public of the Supreme Court of Nigeria.

Esther E. Njieassam is a Postdoctoral Research Fellow at the National Research Foundation Chair in International Law, Faculty of Law, University of Johannesburg. She holds a Doctor of Philosophy (PhD) and a Master of Laws (LLM) from North-West University, South Africa and a Bachelor of Laws (LLB) from the University of Buea, Cameroon. Her research interests include public law, human rights, environmental law, social justice, labour law and socio-economic rights.

Faith Kabata is a Lecturer in Law at Kenyatta University School of Law in Nairobi, Kenya, where she specialises in human rights law, public international law and international human rights law. Her main areas of legal practice are human rights and the rule of law. She has previously served as an Adviser on the rule of law and human rights in the Presidency in Kenya. She holds a Doctor of Laws (LLD) from the University of Pretoria, South Africa; a Master of Laws (LLM) from the University of Notre Dame, in the USA, and a Bachelor of Laws (LLB) from the University of Nairobi, Kenya.

Furaha Joy Sekai Saungweme is a Lawyer and founder of Africa End Sexual Harassment Initiative (AESHI), a law reform and social movement project addressing sexual harassment in Africa. She is a fellow of Women Leaders for the World in California, which promotes women in leadership. She is a member of the Berkeley University Centre on Comparative Equality and Anti-Discrimination Law and sits on its editorial board and is part of the Berkeley COVID-19 Equality Law Working Group. She has presented research papers at academic conferences in different institutions across Africa, Europe and the USA.

Gedion G. Jalata is CEO of the Centre of Excellence International Consult in Addis Ababa, Ethiopia. He has previously been the Programme Manager of the Africa–China Dialogue Platform at Oxfam International Pan Africa Office in Addis Ababa, Ethiopia from January 2016 to January 2018 and Research Consultant with the United Nations Economic Commission for Africa on policy issues related to economic development and governance, capacity development, public administration, anti-corruption, and illicit financial flows and China/Brazil, Russia, India, China and South Africa (BRICS) relations with Africa from 2008 to 2015. He is also a vetted Senior Adviser on South–South Cooperation for the United Nations Development Programme (UNDP) Regional Office in Africa since 2018.

Isaac Shai is an Associate Professor of Law in the Department of Public, Constitutional and International Law at the University of South Africa. He earned his LLD from the University of Pretoria focusing on legal philosophy, constitutional theory and the systems theoretical approach to law. His research interests include post-apartheid jurisprudence, South African legal culture and postcolonial and decolonial approaches to law. His current projects include modernity, colonialism and constitutionalism, legal culture and legal reasoning, law and development, constitutionalism and radical transformation in South Africa.

Jean Claude N. Ashukem holds an LLD and LLM from the North-West University in South Africa and Maitrise en Droit, DEUG and Licence en Droit from the University of Yaoundé II. He is currently an Independent Researcher in environmental rights and natural resource governance. The chapter contribution in this book was done during his tenure as Postdoctoral Researcher at the University of the Free State, Bloemfontein.

Mofihli Teleki is a PhD candidate at the School of Public Leadership, Stellenbosch University, South Africa. His research interests include governance, cultural affairs, development, human rights and public policy. He has published a number of book chapters and academic journal articles on the Right to Development and Public Policy.

Muyamba Mangu has served at the Belgium Technical Cooperation (Enabel) where he occupied various positions such as Programme Officer, National Director and Provincial Coordinator. He also formerly served as acting Director of the Department of Development at the Council of Christian Churches in the Democratic Republic of Congo. He holds a Master of Laws (LLM) in Multidisciplinary Human Rights from the Centre for Human Rights, University of Pretoria. He is currently a doctoral candidate in Law at the University of South Africa. His field of expertise comprises gender, community development and human rights.

Osuntogun Abiodun Jacob is a Senior Lecturer at the Faculty of Law, University of Ibadan, Nigeria. He served previously as the acting Head at the Department of Commercial and Industrial Law of the University. He holds a PhD from the University of the Witwatersrand, a Master of Laws (LLM) from Obafemi Awolowo University, Ile Ife, Nigeria, a master's degree (LLM) in International Trade and Investment Law in Africa, from University of Pretoria and a Master of Arts (MA) in Peace and Conflict Studies from the University of Ibadan, Nigeria. He has authored papers on human rights law, trade and commercial transactions.

Tsion Tadele Dechassa is Deputy CEO at the Centre of Excellence International Consult (CEIC) in Addis Ababa, Ethiopia since January 2018. Prior to that, she has also previously served as Research Consultant at the United Nations Economic Commission for Africa (UNECA) Social Policy Division.

Foreword

I am very happy to have been asked to write the Foreword to this exciting new book. Our institution, the Free State Centre for Human Rights has from the outset been involved, with our partners, the Thabo Mbeki African School of Public and International Affairs, University of South Africa and the Centre for Human Rights, University of Pretoria in arranging the annual International Conference on the Right to Development, the third edition of which, presented at our Centre in Bloemfontein, gave rise to this book. This relationship has over the past three years proven to be one of our most active and productive research collaborations, not only in terms of the production of research such as this book, but also in establishing and deepening academic community. Importantly, the collaboration, the conferences and the various publications that have arisen from them reflect the approach to research and to human rights that we espouse at our Centre: one that is critical in the sense of being politically and ideologically aware; contextually engaged, in the sense of concerning itself with the continent in which we live; and interdisciplinary; and that values human rights for their potential to transform society, not their tendency to preserve the status quo.

It is also important to express my appreciation of and support for this book because, as I write this, the global competition for access to sufficient quantities of COVID-19 vaccines again illustrates the importance of the right to development, which

> incorporates the right to self-determination and permanent sovereignty over natural wealth and resources is understood universally as a composite entitlement that provides the framework for the realisation of all other human rights; requiring every human person and all peoples to participate in, contribute to, and equitably benefit from the processes of socio-economic and cultural as well as political development.[1]

Again, there is the danger that African countries such as my own, despite contributing mightily to the natural resource needs of the rest of the world (and in some cases to the development of the vaccines) will be pushed to the back of the vaccination line, having to feed off the scraps of those countries who consume the bulk of our natural resources.

Although written and completed before the COVID-19 crisis took hold of our world, the chapters in this book, in their engagements in different ways with the question of control of and self-determination with respect to our natural resources will appear when their message is particularly apt.

My congratulations and appreciation to the editors, Dr Ngang and Prof Kamga and to each and every author for their gift to us.

Prof Danie Brand
Bloemfontein, South Africa

Note

1 See Ngang and Kamga in Chapter 1.

Introduction

1 The natural resource and right to development dilemma

Serges Djoyou Kamga and Carol Chi Ngang

The growing dynamism in the demand for raw materials and natural resources resulting from pressures exerted by dominant market economies has been on a dramatic increase.[1] While this may look like a global phenomenon, Africa appears to bear the pressure a bit more disproportionately compared to other regions of the world. Despite the series of international treaties, resolutions and declarations that recognise and protect a peoples' permanent sovereignty over their natural resources, the race for Africa's resources, dating several centuries back remains unprecedented.[2] Africa is indeed, abundantly endowed with an avalanche of natural resources, which in spite of the rate at which they are abusively exploited, more and more continue to be discovered in increasing quantities. As the principal drivers to economic growth and development, Africa should by virtue of its natural resource endowments, stand out prominently as a development hub in the world but unfortunately, the natural resources extracted from the continent have largely only been shipped abroad to fuel the pace of economic growth and technological advancement in more developed and industrialised parts of the world, leaving the peoples on the continent (unfortunately) only with a long-term aspiration to '[t]ransform, grow and industrialise our economies through beneficiation and value addition of natural resources' by 2063.[3]

In the broadcast of the programme *Africa Debate* on 15 October 2012, the British Broadcasting Corporation noted that

> [w]hether Africa will ever benefit from its natural resources is a question that is more relevant now than ever, as new discoveries of coal, oil and gas across East Africa look set to transform global energy markets and—people hope—the economies of those countries.[4]

Although the academic literature and other commentaries on the relevance and need to more productively rationalise Africa's natural resources continue to grow, policy direction appears glaringly lacking. With illustrations from countries in Latin America where the exploitation of natural resources has had a positive impact on development in the long term, Jeffrey Sachs and Andrew Warner write that 'If low-income equilibrium traps are important in economic development—so that a big push is necessary for development—then natural resource booms can be potentially important catalysts for growth and development'.[5] Envisaging the abundance of natural resources to accelerate development in Africa, the reversal of it creates a scenario that remains unideal with regard to the extremely low standards of living that the populations are subjected to. In order that natural resources effectively contribute to the realisation of the right to development on the continent of Africa, the scenario demands more comprehensive ways of shifting the dynamics. The question is how?

Natural resource sovereignty and the right to development under international law

The law provides an interesting angle to redressing the natural resource and right to development dilemma in Africa which, shifting from the controversies that have shaped the debates on the subjects; both need to be explored more in-depth as the chapters in this book provide direction to. Sovereignty over natural resources as it is sanctioned by law for the achievement of human rights in general and the right to development in particular, cannot be over emphasised. This has been acknowledged over the years by the international community, which has put in place a robust legal normative framework on the right to self-determination and permanent sovereignty over natural resources and anevolving framework on the right to development. Sovereignty over natural resources is principally defined by United Nations (UN) General Assembly Resolution 1803 (XVII) on Permanent Sovereignty over Natural Resources adopted on 14 December 1962, which though not legally binding, provides a solid basis for absolute autonomy in asserting natural resource ownership rights, which should be exercised in the interest of national development and of the well-being of the peoples concerned. Sovereignty over natural resources is also codified in the twin covenants on human rights, which guarantee that all peoples may freely dispose of their natural resources, which they may in no instance be deprived of, considering that the resources constitute a means of subsistence and thus, forbid the interpretation of any provision in the covenants as impairing the right to utilise and enjoy the benefits obtained from the exploitation of natural resources.[6]

These provisions cascade down to regional human rights instruments. For instance, the principle of permanent sovereignty over natural resources is infused in article 21 of the American Convention on Human Rights under the right to property as vindicated in the *Mayagna (Sumo) Awas Tingni Community v Nicaragua* case.[7] In Africa, 'the right to free disposal of wealth and natural resources' is unambiguously enshrined in article 21 and reiterated in article 22 of the African Charter as interconnected to the enjoyment of the right to development.[8] On the basis of these provisions, both rights have concurrently been recognised in case law (discussed in some of the chapters) wherein, the African Commission[9] and the African Court on Human and Peoples' Rights found the defendant states in contravention of the substantive rights or of the obligation to protect the interest of the peoples involved.[10] The direct connection between the right to natural resources and the right to development is contained in the UN Declaration on the Right to Development, formulated in these terms:

> The human right to development also implies the full realization of the right of peoples to self-determination, which includes, subject to the relevant provisions of both International Covenants on Human Rights, the exercise of their inalienable right to full sovereignty over all their natural wealth and resources.[11]

This is to say that to deprive a people of their natural resources would amount to a violation of their right to development. Therefore, achieving the right to development requires a sustainable governance of and management of natural resources, entailing as Nicolaas Schrijver posits, 'imposing accountability at the national and international levels'[12] to ensure that the exploitation of natural resources effectively translate into the enjoyment of the right to development. As Schrijver further observes, '[t]he challenge is how to inject these established principles of self-determination of peoples and sovereignty over natural wealth and resources into the basic tenets of the right to development'.[13]

Conceptual framing

The discourses in this book are based on the interconnected conceptual framing of the international law principle of permanent sovereignty over natural resources and the concept of the right to development, which are both envisioned to equip the peoples in developing countries, particularly in Africa against dominant political stakeholders and the exploitative practices of the extractive industry. The right to development, which incorporates the right to self-determination and permanent sovereignty over natural wealth and resources is understood universally as a composite entitlement that provides the framework for the realisation of all other human rights; requiring every human person and all peoples to participate in, contribute to and equitably benefit from the processes of socio-economic and cultural as well as political development.[14] It implies the right to own, control and utilise natural resources not only for purposes of economic growth or technological advancement but importantly, as situational exigencies may necessitate, for survival, well-being and livelihood sustainability. It is stated in common article 1(2) of the international covenants on human rights and in article 21(1) of the African Charter that all peoples may, for their own purposes, freely dispose of their natural resources, conceived as a means of subsistence, which they may not be deprived of.[15]

While in the context within which the book is written, some of the chapters focus on particular resources, natural resources are conceived herein, in the broadest sense as comprising of the lands, minerals, gas and fossil fuels, fauna and flora and aquatic resources among others, which are recognised as a collective entitlement belonging to the peoples of Africa and envisaged to facilitate realisation of their right to development. Use of the description 'common heritage' in the African Charter denotes the fact that natural resources are a collective possession, which must be shared equitably to the exclusive collective benefit all of the peoples on the continent. Although the pressure on natural resources in Africa also emanates from an ever-increasing population and the failing resilience to confront environmental hazards and climate risk factors, this book overlooks these natural phenomena and focuses on the human-induced factors that dispossess the peoples of Africa of the right to sovereignty over their natural resources as a means to achieve the right to development.

Literature on the principle of permanent sovereignty over natural resources is inexhaustive, and although characterised by an unsettled polarised debate on whether sovereignty pertains to peoples or to states, is fundamentally undergirded by the universal guarantee that every state is obligated to exercise absolute legitimate control over its natural resources to the ultimate exclusive benefit of their peoples.[16] Indeed, international law has evolved in this direction to recognise a number of substantive and procedural rights for local communities and indigenous peoples, including ownership rights over their natural resources; the right to participate in decision-making and to free prior informed consent and the sharing of benefits arising from the commercial exploitation and extraction of the natural resources on their territories.

As opposed to economic growth models that promote the exploitation of natural resources on an unfettered scale, often without consideration of the adverse impact on human livelihood and sustenance, the contributions in this book explore from a rights-based point of view with emphasises on the human aspects of development and hence, peoples' sovereign ownership rights over their natural resources as a determining factor for the attainment of development objectives. It is of essence for the purpose of clarity, to highlight that the right to permanent sovereignty over natural resources is by nature, just like the derivative right to self-determination and the right to development, recognised fundamentally as a collective entitlement. Seen in this light, the book is centred on the overarching argument that sovereignty over natural resources pertains to peoples as a collective and therefore, they are entitled to reap exclusive benefits from the free disposal or exploitation of the resources.

The interconnectedness between natural resource sovereignty and the right to development as it is articulated in this book, is to the effect that their combined realisation should culminate into improved standards of living, where the benefits obtained from the development process are shared equitably and all human rights are fulfilled in the process. While Africa is yet to demonstrate genuine commitment to the right to development, with regard to natural resource sovereignty, the prevailing trend across the continent is that natural resources are not only exploited almost on a gratis basis without commensurate benefit to local populations, but in most instances, the latter are forcibly displaced and their livelihoods disrupted. The arguments advanced in this regard are to the effect that a reversal of the trend is envisaged to significantly transform the development landscape on the continent.

Approach and scope of the book

Taking a shift away from the controversy over the conceptual nature of the right to development and the discourse that portrays the overabundance of natural resources in Africa as a curse, this book adopts a pragmatic approach in exploring the nexus between the international law principle of permanent sovereignty over natural resources and the concept of the right to development, which are both envisioned to equip the peoples in developing countries, particularly in Africa against dominant political stakeholders and the exploitative practices of the extractive industry and thus, guarantee sustainable socio-economic and cultural development. These are the thought patterns that formed the core of the discussions at the 3rd International Conference on the Right to Development hosted by the University of the Free State in Bloemfontein, South Africa from 25 to 27 September 2019 under the auspices of the Free State Centre for Human Rights, University of the Free State; Centre for Human Rights, University of Pretoria and Thabo Mbeki African Leadership Institute (now Thabo Mbeki African School of Public and International Affairs) and the University of South Africa on the theme 'Natural Resources Ownership and the Right to Development in Africa'. The chapters in this book were selected from the papers presented at the conference and rigorously peer reviewed.

While three of the contributions are based on empirical findings, the rest are more analytical in approach wherein, the reading of legal texts, cases and related literature are comprehensively juxtaposed to illustrate how entitlement to natural resource sovereignty could contribute to actualising the right to development in Africa. The purpose of accentuating the law and the legal dimensions to natural resource sovereignty and the right to development are to the effect that when the law is contravened, remedy could be sought and indeed, as the African jurisprudence illustrates, has happened in a couple of instances with positive rulings in favour of dispossessed communities. Even as the central theme is anchored on the law, the limits of law in redressing the political questions relating to natural resource sovereignty and sustainable development is acknowledged, which is reflected in the book's multidisciplinary blend with chapter contributions from authors with diverse academic inclinations and thus, is intended to serve a broad academic audience among others who may equally find the book resourceful.

While the book zooms in on Africa as a whole, specific cases and illustrations are drawn from Nigeria, Angola, South Sudan, South Africa, Kenya, Ghana, the Democratic Republic of Congo (DRC), Ethiopia, Zanzibar, The Gambia, Botswana, Uganda and Zimbabwe. Greater attention is also directed more at indigenous peoples, discussed in some of the chapters, rather than on other groups of peoples in Africa. The contributions in Part 1 and Part 2 are mainly theoretical, based on interpretations of relevant texts of the law, policies, jurisprudence and related literature and commentaries. Three of the contributions in Part 3 are empirical in nature—involving case studies of specific communities in Addis Ababa in Ethiopia; Kenyasi No 2 in Ghana and Mutasa District in Zimbabwe. The three other chapters in Part 3 are more socio-legal in the analysis relating to benefit sharing in communities hosting mining operations in the DRC and South Africa.

Overview of the chapters

The book is divided into three main parts. The first part explores the normative considerations on ownership rights over natural resources and the right to development in Africa. It combines a reading of the law with theoretical reasoning and practical understanding of the entitlement to natural resource sovereignty and the right to development. The second part delves into more technical the matic issues relating to natural resource sovereignty guaranteed to the peoples of Africa, which however, is impacted by the rush for the resources located in their communities. Part 3 looks at specific case studies on collective benefit sharing of natural resources in sustaining the right to development.

PartI—Normative considerations on natural resource sovereignty and the right to development

This section consists of six chapters. Carol Ngang begins with a discussion on sustainable right to development governance of natural resources in Africa. He explores how the right to development governance as a model for development could, in combination with natural resource ownership, be explored more practically to support efforts to achieve structural transformation and sustainable development in Africa. He contends that while Africa is known to be the natural resource supply base that drives the economies of other regions and countries around the world, development on the continent or the lack there of, has been theorised to be hampered by scarcity or over abundance. As a point of departure, he refutes the natural resource curse and resource scarcity theories, which he says have frequently been used to justify why Africa remains poor and under developed. He advances the argument that the actual problem Africa is confronted with is not scarcity or excess but rather the lack of a sustainable governance framework to enable the conversion and judicious use of natural resources to the benefit of the populations across the continent. In his view, both the right to natural resource ownership and the right to development are legitimate guarantees, which the peoples of Africa are entitled to claim, including through legal processes.

Isaac Shai explores the notion of theright to development and ownership of natural resources by drawing from fictional literature. He posits that literature has the ability, through the re-presentation of life, to disrupt historicist accounts of life as presented by certain legal and jurisprudential accounts. His chapter therefore attempts to reorient the idea and story of ownership of natural resources and the right to development as hither to a story of loss, anguish and postponement—and refusal too. Njabulo Ndebele's *The Cry of Winnie Mandela* and Kafka's *Before the Law* are taken not only as disruptive to historicist accounts of life but also disruptive to the meta narrative and teleological accounts of law. The limits of law and its ability to create 'possible impossibilities' are poignant in Kafka's *Before the Law* and are taken to also afflict the notions of the right to development and ownership of natural resources in Africa.

Daphine Agaba proceeds to explain why Africa should be able to assert legitimate ownership over its natural resources. She argues that Africa's historical positioning, which manifests as a policy taker rather than a policy maker and financial dependence on the Western world hinders effective and independent management of its resources. This translates into the inability to transform the enormous natural resource endowments on the continent into sustainable development. She critiques the 'good governance' narrative advanced by 'Western nations' as a solution to Africa's 'resource curse' with the counter argument that itis only used to back up foreign extractive companies' rent-seeking motives to maximise profits from exploitation of the resources. The chapter presents the lack of transformational and visionary leadership as a contributing factor to the failure of resource-rich countries in sustaining themselves. The author finds that without transformational leadership and critical strategies aimed at dismantling the power imbalances that abound in the extractive sector, policies adopted will continuously be 'modified',with little or no success in changing the *status quo*.

For Aaron Salau, the questions of self-determination, sovereignty over natural resources and the right to developmentneed to be explored from a 'resource constitutionalism' point of view. He argues that to actualise Africa's developmental imperatives, petrostates must embrace 'resource constitutionalism' for the efficient management of their petroleum resources as a means to escape the resource curse syndrome. He points out that after independence, oil-rich African states nationalised petroleum ownership, supposedly to speed up economic development in accordance with the principle of permanent sovereignty over natural resources. He asserts that states' exclusive ownership of petroleum provokes power struggles by rent-seeking elites, corruption and underdevelopment, disregard for sustainable development and suppression of oil communities' rights to free, prior and informed consent to oil extraction operations in those communities. He further explains that recourse to 'resource nationalism' by African petrostates in the 1990s through Production Sharing Contracts has proved insufficient to wrest de facto control of petroleum resources from transnational companies. Through a comparative study of petroleum industries, legislative and policy frameworks in Nigeria, Angola and South Sudan, he suggests that resource constitutionalism can enhance the goals of economic diversification for the reason that it has the potential to moderate state ownership through community participation in petroleum production which enables equitable access to socio-economic development.

Jacob Abiodun looks at 'peoples' right to natural resources in Africa and state sovereignty'. He argues that this right should be viewed as a 'quest for equilibrium'. He explains that most countries in Africa are exposed to protracted conflicts related to the ownership of natural resources, for which he illustrates the African Commission and the African Court on Human and Peoples' Rights have developed an impressive jurisprudence to protect the rights of indigenous peoples on the continent. He says, the jurisprudence lays down the rule that a section of a population can legitimately claim ownership of natural resources within their territory and in the event of dispossession, restitution can be ordered. He opines that jurisprudenceis indispensable in actualising the right to development, which he interrogates in determining compatibility with domestic, regional and international law with argument to the effect that jurisprudence conflicts not only with state sovereignty and the spirit of communalism in Africa but most often also, with the fundamental law of the land.

De Man and Ngang delve into the extraction of natural resources during colonialism and the impact it has had on the right to development in Africa today. They argue that while underdevelopment in Africa may be attributed to other factors, the wanton extraction of natural resources during the colonial period has contributed majorly to the development challenges that Africa is confronted with and thus, impairs prospects for actualising the right to development on the continent. The chapter highlights the contradictions of the civilising mission on the basis of which European colonisers justified their invasion of the African continent. It further examines the postcolonial anticipations of natural resource sovereignty in achieving the right to development, which is hampered by unprecedented tensions over natural resource ownership rights, which is largely still informed by colonial models of dispossession of local populations sanctioned by states. The chapter concludes with a proposition in favour of African peoples' entitlement inasserting sovereignty over their natural resources as a means of exercising their right to development.

Part 2—Permanent sovereignty over natural resources

This second part is made of four chapters. Focusing on issues of natural resource sovereignty in Africa, Faith Kabata argues that the right to socio-economic and cultural self-determination imposes a limitation on states' sovereignty and grounds a claim to peoples' collective ownership rights over natural resources. The chapter begins by tracing the evolution of the doctrine of permanent sovereignty over natural resources and then turns to look at how the doctrine could be reconciled with the right to self-determination under international human rights law. Turning to prescriptions on how peoples within a state can assert the right to development by claiming sovereignty over natural resources, the chapter draws from the achievements in the rights of indigenous peoples. In closing, the author suggests the need to infuse the right to socio-economic and cultural self-determination with the right to development to place people within the state at the centre of national development processes.

Esther Njieassam draws attention to natural resource ownership and the realisation of the right to development for indigenous communities in Africa. She estimates that more than half of the world's natural resources are located in Africa and predominantly in indigenous territories, which constitutes for them the only means of survival. She argues that the over exploitation of these resources triggers agitations over ownership and control rights between indigenous populations and national governments. For her, government policies that vest resource ownership rights with the state hinder indigenous peoples from actualising their right to development in the sense that they do not get to benefit from the proceeds that accrue from the exploitation of the resources in their territories. She suggests as a remedy to the problem, policy measures to protect indigenous peoples' rights to own, control and manage the natural resources located in their territories as a means, not only to enhance sustainability of the resources but also to facilitate development within the indigenous communities.

Ebun Abolarin equally looks at natural resource sovereignty and the right to development for Africa's indigenous peoples. She notes that under international human rights law, the right to natural resources is a collective right of peoples. She highlights common article 1 of the twin international covenants on human rights, which guarantees to all peoples the right to freely dispose of their natural resources and that, 'in no case may a people be deprived of its own means of subsistence'. She explains that Africa's indigenous peoples have with the backing of international and regional human rights law, resisted and continue to challenge practices that dispossess them of their lands and natural resources. The Rio Declaration, as she points out, portrays human beings as the central focus for sustainable development, necessitating the fulfilment of their right to development so as to equitably meet the developmental and environmental needs of present and future generations. With regard to engaging indigenous peoples, she makes reference to the Sustainable Development Goals and Agenda 2063, which provide insights into the relationship between governments, indigenous peoples and the sustainable management of their natural resources.

With a thematic focus on water, Joy Saungweme examines the policies and practices within the tourism industry and how they impact on the right to development in selected African countries. She interrogates the ways in which the tourism industry impacts the right to water as a basic human right relevant to the right to development. The arguments she advances are extrapolated from national, regional and international instruments that affirm water as a socio-economic right, fundamental to human life and dignity and therefore, cannot be construed just as periphery but indeed, is central to the realisation of the right to development. By looking at court decisions, UN resolutions, policy papers and academic opinions, her analysis draws attention to water use within the tourism sector, which she argues must be conceived as a basic right, requiring private and state actors to adopt pro-poor policies in facilitating water use, ownership, and access within the broader construct of the right to development.

Part 3—Collective benefit sharing of natural resources in sustaining the right to development

Made of six chapters, the contributions in this section focus mainly on specific illustrations on collective benefit sharing of natural resources in sustaining the right to development. Jean-Claude Ashukem's reflections on the Sustainable Development Goals (SDG) and the bumpy road to 2030 in Ethiopia draw from the challenges posed by land grabbing. Three years after the SDGs were adopted, he notes that it is doubtful that their objectives will be achieved by 2030. He hypothesises that the negative impact of the phenomenon of land grabbing could pre-empt the achievement of the SDGs. His chapter critiques and analyses the negative impact of land grabbing to illustrate the potential difficulty of achieving the SDGs. He makes the argument that actions directed towards achieving any of the goals could, in the face of persistent land grabbing, have a considerable unintended adverse impact on other goals and the notion of sustainability. He makes the point that the phenomenon of land grabbing constitutes a huge impediment to meeting the SDGs targets.

Building on the idea of freedom from want and the constitutional right to development in Ethiopia, Dechassa and Jalata's empirical findings are obtained from the Urban Productive Safety Net Programme (UPSNP) developed by the Ethiopian government and other stakeholders to enhance food security in the country. They indicate that the UPSNP's main purpose is to ensure freedom from want, which is associated with the right to development, measured by the degree to which citizens enjoy improved living standards as indicated in article 43 of the Ethiopian Constitution. The chapter assesses the implementation, challenges and determinants of food security with respect to the experiences and prospects of the UPSNP beneficiaries in Woreda 08, Arada sub-city and Addis Ababa. Based on the survey conducted and the nature of the dataset, the Tobit regression model is employed to identify the determinants of households' food insecurity experiences, which is then also analysed using the binary logic regression model. They identify as significant challenges, selection problems and challenges associated with institutional arrangements, political intervention and lack of transparency, insufficient coordination among partners, lack of awareness about the project and delayed payments, which they say, need to be addressed as a means to increase prospects of enhancing food security in Ethiopia. They recommend regular monitoring, evaluation, learning, and accountability of the UPSNP, which necessitates strengthening its support and service components.

With an eye on the mining sector, Mangu and Kamga explore the role of corporate social responsibility (CSR)in achieving the right to development for local communities in the DRC. They assert that the traditional approach to human rights is generally to guarantee protection against the might of the state. Yet, in recent times transnational companies (TNCs) have become stronger than some states, especially in developing countries where they operate. Within this backdrop, the chapter explores the role of TNCs in fostering the right to development with focuses on extractive companies in the DRC, selected because it is one of the wealthiest countries globally in terms of mineral resources that attract TNCs. While TNCs resort to CSR in discharging their human rights obligations, the authors contend that the voluntary nature of CSR compromises its potential contribution to realising the right to development. They recommend for the state to establish a regulatory framework that would make CSR obligatory, through which TNCs can be compelled to comply with human rights standards. With regard to the state's human rights obligations, they further deconstruct corporate responsibility in light of the Ruggie Principles on Business and Human Rights and in doing so, proffer suggestions on how to engage TNCs in advancing the right to development.

Anthonio-Apedzi examines laws on natural resources ownership in Ghana. She observes that the concept of states' sovereignty over natural resources is entrenched both in international and domestic law, which guarantees the right to explore and exploit natural resources for development and the well-being of their citizens. However, some countries are in constant confrontation with their citizens who are indigenes of where the natural resources are found. Much of the problem as she explains, results from the domestic arrangements regarding benefit sharing of the proceeds from natural resources to the people. She illustrates that in Ghana, the law is not clear on ownership of natural resources, which are deemed state property held by the President in trust for the citizens, resulting in conflicts between local communities, traditional leaders, state institutions and exploration companies. Her contribution is the outcome of an empirical study conducted in the gold mining community of Kenyasi No 2 wherein, the findings reveal that members of the community have no problem with the current arrangement of trusteeship over their natural resources, but the manner in which proceeds from exploitation of the resources are distributed inhibits community development. She recommends that the laws on the management of natural resources be reviewed to be more inclusive and accessible as a means of realising the right to development.

In looking at the policy implications of the right to development, Mofihli Teleki portrays the South African Mining Charter as a regulatory instrument that provides guidelines for 'effective symbiotic growth and mutual transformation'. He points out that while the Mining Charter highlights aspects that are obtained from article 8 of the Declaration on the Right to Development, including housing, adequate living conditions, health care and education, intended to cater for communities hosting mining activities, these aspects have rather been the subject of contestation. For him, the guidelines in the Mining Charter ought to provide direction to mining companies and the South African government in ensuring that the socio-economic aspects contained therein are catered for and thus, needs to be interrogated for its potential to inform policy formulation. He argues that besides the Mining Charter, which is barely a regulatory instrument, a policy framework on social responsibility for mining companies needs to be put in place to regulate and ensure best practices within the mining sector for the development of mining communities. This chapter taps into the transposition of the guidelines of the Mining Charter into a policy framework that can effectively regulate mining companies and the actions of government in relation to the right to development.

With a similar focus on the mining sector, Cephas Mandirahwe explores the impact of artisanal and small-scale gold mining in Mutasa District, Zimbabwe. He traces the history of mining in Zimbabwe, involving trade in gold between the Shona and Portuguese, which he says dates back to the 15th century and is today a major driver of the country's economy. He highlights the debate on the impact of artisanal and small-scale gold mining on livelihoods in the local communities, which he says leaves scholars divided on the understanding of artisanal and small-scale gold mining. Situated within the scholarship on the relevance of natural resources for development, Mandirahwe examines the extent to which the activities of artisanal and small-scale gold mining impact on community livelihood and contribute to the realisation of the right to development in Mutasa District. His empirical study is conducted using a qualitative research method through convenience sampling of selected key small-scale miners and data collected through in-depth interviews and focus group discussions. Based on the findings, the author recommends for the government of Zimbabwe to formulate a mining policy, which he opines, would formalise the operations of artisanal and small-scale mining.

The concluding chapter makes a projection into how the free disposal and exploitation of natural resources and equitable (re)distribution of the benefits resulting therefrom could ensure socio-economic and cultural development and livelihood sustainability in Africa, especially looking at the unprecedented public health emergency fall-out from the COVID-19 pandemic, which has exposed the weaknesses and coping capacity of state governments in dealing with eventualities of the sort. It is suggested that the continual roll back in development gains and in living standards across Africa obligates state governments to seriously rethink the sources of development financing, which could indeed sufficiently be generated from the natural resources that abound on the continent .

Notes

1 Gilbert, Jérémie. 2018. *Natural Resources and Human Rights*. Oxford: Oxford University Press, 1.

2 Lawal, A.B. 2020. *The Big Rush for the Control of African Natural Resources.* A.B. Prominent Publisher; Scholvin, Sören. (Ed). 2015. *A New Scramble for Africa?: The Rush for Energy Resources in Sub-Saharan Africa.* New York: Routledge; Okereke, Chris. 2008. 'The Second Scramble for Africa's Oil and Mineral Resources: Blessing or Curse? *The International Lawyer*, 193–209.

3 AU Commission. 2015. 'Agenda 2063: The Africa We Want'. *African Union*, para. 72(d).

4 BBC. 2012. 'Africa Debate: Will Africa Ever Benefit from Its Natural Resources?' *BBC Africa*, 15 October <www.bbc.com/news/world-africa-19926886> accessed 24 December 2020.

5 Sacks, Jeffrey and Warner, Andrew. 1999. 'The Big Push, Natural Resource Booms and Growth'. *Journal of Development Economics* 59: 43–46.

6 International Covenant on Economic, Social and Cultural Rights, General Assembly, Resolution 2200A(XXI) of 16 December 1966, arts 1(2) and 25; International Covenant on Civil and Political Rights, General Assembly Resolution 2200A(XXI) of 16 December 1966, arts 1(2) and 47.

7 *Inter-American Court of Human Rights, Mayagna (Sumo) Awas Tingni Community v Nicaragua*, judgment of 31 August 2001.

8 Ngang, Carol. 2020. 'Right to Development in Africa and the Common Heritage Entitlement'. *Journal for Juridical Sciences* 45(1): 41; Kamga, Serges. 2011. 'The Right to Development in the African Human Rights System: The *Endorois* case'. *De Jure* 44(2): 381–391.

9 *African Commission on Human and Peoples' Rights (Ogiek) v Republic of Kenya* (2017) Appl No 006/2017.

10 *Centre for Minority Rights Development (Kenya) and Minority Rights Group International on Behalf of Endorois Welfare Council v Kenya* Comm 276/2003 (2009) AHRLR 75 (ACHPR 2009); *Democratic Republic of Congo v Burundi, Rwanda and Uganda* (2009) AHRLR 9 (ACHPR 2009); *Social and Economic Rights Action Centre and the Centre for Economic and Social Rights v Nigeria*, Comm No 155/96 (2001).

11 Declaration on the Right to Development, General Assembly Resolution A/RES/41/128 of 4 December 1986, art 1(2).

12 Schrijver, Nicolaas. 2013. 'Self-Determination of Peoples and Sovereignty over Natural Wealth and Resources'. In *Realizing the Right to Development: Essays in Commemoration of 25 Years of the United Nations Declaration on the Right to Development*, edited by UN Human Rights, 95–102. Geneva: United Nations Publication, 101.

13 *Ibid*, 102.

14 Declaration on the Right to Development (DRTD), note11, art 1(1) and (2); African Charter on Human and Peoples' Rights, adopted by the Organisation of African Unity (OAU) in Nairobi Kenya on 27 June 1981, art 22; UN Declaration on the Rights of Indigenous Peoples, Gen Ass ResolutionA/RES/61/295 of 13 September 2007, arts 25 and 26.

15 ICCPR, note 6; ICESCR, note 6; African Charter, note 14.

16 Ng'ambi, Samgwani. 2015. 'Permanent Sovereignty over Natural Resources and the Sanctity of Contracts, from the Angle of Lucrum Cessans'. *Loyola University Chicago International Law Review* 12(2/3): 154; Chekera, Yolanda and Nmehielle, Vincent. 2013. 'The International Law Principle of Permanent Sovereignty over Natural Resources as an Instrument for Development: The Case of Zimbabwean Diamonds'. *African Journal of Legal Studies* 6(1): 69–101; Atsebua, Lawrence. 1993. 'Principle of Permanent Sovereignty over Natural Resources and its Contribution to Modern Petroleum Development Agreements'. *Journal of the Indian Law Institute* 35(1/2): 115–126.

References

Atsebua, Lawrence. 1993. 'Principle of Permanent Sovereignty over Natural Resources and its Contribution to Modern Petroleum Development Agreements'. *Journal of the Indian Law Institute* 35(1/2): 115–126.

Chekera, Yolanda and Nmehielle, Vincent. 2013. 'The International Law Principle of Permanent Sovereignty over Natural Resources as an Instrument for Development: The Case of Zimbabwean Diamonds'. *African Journal of Legal Studies* 6(1): 69–101.

Gilbert, Jérémie. 2018. *Natural Resources and Human Rights*. Oxford: Oxford University Press.

Kamga, Serges. 2011. 'The Right to Development in the African Human Rights System: The *Endorois* Case'. *De Jure* 44(2): 381–391.

Lawal, A.B. 2020. *The Big Rush for the Control of African Natural Resources*. A.B. Prominent Publisher.

Ng'ambi, Samgwani. 2015. 'Permanent Sovereignty over Natural Resources and the Sanctity of Contracts, from the Angle of Lucrum Cessans'. *Loyola University Chicago International Law Review* 12(2/3): 153–172.

Ngang, Carol. 2020. 'Right to Development in Africa and the Common Heritage Entitlement'. *Journal for Juridical Sciences* 45(1): 28–50.

Okereke, Chris. 2008. 'The Second Scramble for Africa's Oil and Mineral Resources: Blessing or Curse? *The International Lawyer* 193–209.

Sacks, Jeffrey and Warner, Andrew. 1999. 'The Big Push, Natural Resource Booms and Growth'. *Journal of Development Economics* 59: 43–76.

Scholvin, Sören. (Ed). 2015. *A New Scramble for Africa?: The Rush for Energy Resources in Sub-Saharan Africa*. New York: Routledge.

Schrijver, Nicolaas. 2013. 'Self-Determination of Peoples and Sovereignty over Natural Wealth and Resources'. In *Realizing the Right to Development: Essays in Commemoration of 25 Years of the United Nations Declaration on the Right to Development*, edited by UN Human Rights. Geneva: UN Publication, 95–102.

Part 1

Normative considerations on natural resource sovereignty and the right to development

2 Sustainable right to development governance of natural resources in Africa

Carol Chi Ngang

Introduction

In this chapter, I explore how the right to development governance as a model for development in combination with the entitlement to natural resource ownership, could in more practical ways accelerate efforts to achieve socio-economic and cultural transformation and sustainable development in Africa. While Africa is known to be the supply base of natural resources that drives the economies of other regions and countries around the world, the governance thereof for development on the continent has wrongly been theorised. As a point of departure, I refute the natural resource curse and resource scarcity theories, which have frequently been used to justify why Africa remains poor and underdeveloped.[1] I advance the argument that the actual problem Africa is confronted with is not scarcity or a curse but rather the lack of a sustainable governance framework to enable the judicious use of natural resources to the benefit of its peoples. The right to natural resource ownership and the right to development are guaranteed to the peoples of Africa as legitimate entitlements, claimable through political processes, requiring the crafting of appropriate policies or through judicial processes when these rights are contravened.

Prior to decolonisation and in the period there after, a range of United Nations (UN) resolutions and declarations were adopted recognising that in order to develop and guarantee better living standards, decolonised territories merited sovereign ownership rights over their natural resources, which incorporate the right to dispose of same in accordance with each state's development imperatives and (governance).[2] However, because natural resources are often not evenly distributed, the state, as the guarantor of human rights is imposed with the duty to harness and effectively coordinate in a fair (equitable and redistributive) manner, the disposal and exploitation of the reservoir of resources as a prerequisite for their socio-economic and cultural development.

Even though the law entitles peoples with ownership rights, several factors, especially those that expose vulnerable populations to loss of benefits or unlawful dispossession lend justification to states, as juristic persons under the law and as guarantor of rights, to exercise sovereignty in the governance of natural resources on behalf of and to the exclusive benefit of their peoples. While effective governance of natural resources is estimated to accelerate socio-economic and cultural development and improve human well-being, the UN Economic Commission for Africa notes that 'Africa has been slow to convert its natural resources endowments to tangible development outcomes because of weaknesses in governance and the wider capacities of the African state'.[3]

African state governments indeed exhibit attributes that create uncertainty over whether they take cognisance of the duty to maximise exploitation of the natural resources on the continent to meet the socio-economic and cultural development expectations of the populations. In the face of competing priorities between the pursuit of economic growth objectives and the compelling commitment to meet livelihood exigencies, the situational realities necessitate exploring how the African commonwealth could be harnessed to accelerate socio-economic and cultural development gains.

The chapter is structured in the following manner: In the section 'Natural resource sovereignty', I outline the context for natural resource governance in Africa, where in I look at the intertwined concepts of permanent sovereignty and the right to development on the one hand, and the underlining principles with respect to utilising natural resources as a principal driver for development on the other hand. In the section 'Sustainable governance of natural resources for development', I deduce from the treaty instruments under consideration, commitments that necessitate putting into place regulatory protocols for the sustainable governance of natural resources to accelerate people-centred development in Africa. In the concluding section, I submit that African state governments have a duty to exercise territorial sovereignty in ensuring a sustainable regime of governance of the African common heritage for the collective substantive benefit of the populations across the continent.

Natural resource sovereignty

African patrimony

Fola Adeleke points out that '[t]he socio-economic development of African states is heavily dependent on how effectively governments are able to manage their natural resources'.[4] The law requires for the peoples of Africa to equitably benefit from the natural resources that make up the Africa patrimony, referred to in the African Charter as a common heritage, defined within context as composing of'the ensemble of natural wealth and resources that are considered a shared communal possession, to which the peoples of Africa are legitimately entitled by virtue of their African ancestry'.[5] The right to development enshrined in the African Charter thus envisages a commonwealth mechanism, which African state governments are compelled to establish to ensure subsistence and sustainable livelihood.

Quantified in actual terms, the African commonwealth can be demarcated to consist of: about 30 percent of the global supply of minerals; 10 percent of fossil fuel and 8 percent of the world's stock of oil reserves; the largest cobalt, diamonds, platinum and uranium reserves; 70 percent of columbite—tantalite deposits, 63 river basins covering approximately 64 percent of the continent's land area and 93 percent of the total surface water resources, a fair share of the world's fertile lands and forests, which unfortunately, are yet to translate into development.[6] Globally, South Africa alone ranks fifth in the production of gold and remains a cornucopia of other minerals; the world's largest producer of chrome, manganese, platinum, vanadium and vermiculite; the second largest producer of ilmenite, palladium, rutile and zirconium and the third largest producer and exporter of coal and iron ore.[7] These are just some illustrations of the extent of Africa's wealth of natural resources.

The African Convention on the Conservation of Nature and Natural Resources affirms that the 'natural resources with which Africa is endowed are an irreplaceable part of the African heritage and constitute a capital of vital important to the continent'.[8] The Convention enshrines the principle that obligates state parties to 'individually and collectively ensure the enjoyment of the right to development', which of course, reiterates the commitment under article 22 of the African Charter. Natural resources are accordingly perceived in economic terms as constituting 'capital of vital importance' and as a major source of national income.[9] The Organisation of African Unity (OAU) took cognisance of this reality and expressed the need to harness the continent's natural resources for the advancement of the African peoples [10] This formulates a duty for states, which according to Ng'ambi, entails regulating the operations and conduct of non-state actors in exploiting and more so, sustainably utilising the resources to achieve national development objectives.[11]

Although the African Charter emphasises the interrelatedness of the rights enshrined therein, there is indeed a visible grouping of rights wherein the right to natural resources and the right to development are categorised as collective entitlements, which can only be claimed by groups of peoples and not by individuals. This is affirmed in the International Labour Organisation (ILO) Convention No 169 on Indigenous and Tribal Peoples that recognises the collective right of indigenous and tribal peoples to ownership and control over their natural resources.[12] The African Commission equally notesthat indigenous peoples (with specific reference to the Endorois) enjoy certain explicit rights, such as the right to property in a distinctly collective manner.[13]

Both the right to natural resource ownership and the right to development originate from and are an extension of the right to self-determination, which failed to be achieved in full measure when the colonised territories of Africa gained political freedom from colonial rule without equally acquiring socio-economic and cultural autonomy. Sanginga, Ochola and Bekalo state that 'natural resources contribute significantly to development in different ways, including as an economic activity and source of growth and as a source of livelihood'.[14] The right to natural resources and the right to development thus have an identical purpose aimed at activating self-sufficiency as stipulated in the following wording:

> The human right to development also implies the full realization of the right of peoples to self-determination, which includes, subject to the relevant provisions of both International Covenants on Human Rights, the exercise of their inalienable right to full sovereignty over all their natural wealth and resources.[15]

Means of livelihood and subsistence

Natural resources are conceived in legal terms as an indispensable source of livelihood security, which peoples cannot be denied access to or deprived of. Common article 1(2) of the twin international covenants on human rights highlights the guarantee that '[i]n no case may a people be deprived of its own means of subsistence'. The African Charter equally states in article 21(1) that '[i]n no case shall a people be deprived of [their wealth and natural resources]'. Qualifying natural resources as a 'means of subsistence' makes it directly connected to human livelihood and well-being and in essence, defines the fundamental purpose of development, which aims primarily at promoting social progress and the highest attainable standard of living in larger freedoms (from the fears and wants that derive from deprivation and impoverishment).[16]

The Convention on the Conservation of Nature and Natural Resources reiterates the duty of states to conserve, protect and utilise their natural resources in a sustainable manner so as to meet their objectives for development for the present and future generations of the African peoples,[17] which as underscored in the Convention andthe African Charter, should focus more on economic, social and cultural aspects.[18] Interestingly, these measures have largely remained abstract entitlements in the sense that states have either ignored or outright contravened the provisions that compel them to harness their natural resources for the betterment of living conditions. The Kenyan government for example, has persistently demonstrated that as far as it is concerned the conservation of nature and natural resources for the well-being of local and indigenous communities does not count, which of course became the subject of litigation in the *Endorois* and *Ogiek Community* cases.[19]

Put side by side with the obligation to harness the natural resources on the continent for the advancement of the African peoples and the obligation to ensure equitable redistribution of the commonwealth as a determining factor to actualising socio-economic and cultural development, with the perceived lapses, it is essential to deduce from the applicable instruments the principles that may suitably guide sustainable governance of the natural resources on the African continent as part of the commitment to give practical effect to the right to development.

Sustainable governance of natural resources for development

Contrary to the provisions under human rights law that attribute permanent sovereignty over natural resources to peoples, states have increasingly arrogated that right, understandably on the basis of exercising their territorial sovereignty, which incorporates the duty to respect, protect and enable their peoples to fulfil their right to freely dispose of their natural resources and to enjoy the benefits of doing so.[20] Rightly, the state is only mandated to exercise custodianship in the governance of natural resources on behalf of and to the exclusive interest of the peoples to whom ownership rights are attributed.[21] The trouble, as indicated earlier, is the absence of a sustainable model by which to curb abusive exploitation, prevent illicit financial outflows and ensure that the benefits and proceeds from exploitation are retained at base to serve the domestic purpose. To construct such a governance model entails a comprehensive understanding of the principles discussed in the proceeding section.

Natural resource governance principles

Free disposal

Even though African state governments generally exercise territorial sovereignty rights over the natural resources within their geographic space, the principle of free disposal allocates the right of ultimate decision-making to the peoples rather than the state. Natural resources are acknowledged to constitute an important source of livelihood and subsistence and, therefore, should not be disposed of unless under circumstances that ensure proceeds are retained at source. The law makes provision to the effect that the choice to dispose of natural resources must be born out of the peoples' free resolve to do so. Common article 1(2) of the international human rights covenants stating that '[a]ll peoples may, *for their own ends*, freely dispose of their natural wealth and resources' (emphasis added), highlight the principle that peoples are ultimately empowered to determine whether to dispose of their resources, obviously in accordance with their livelihood exigencies.

Local and indigenous communities in Africa have historically preserved in a sustainable manner the natural resources on which their identities, spirituality, lifestyle and livelihood inseparably depend.[22] While the conservation of natural resources is presented as a matter of law with the drafting of conventions to that end, it is worth noting that local and indigenous communities have traditionally done so not because they are constrained by law but because of the importance of sustainably utilising the resources to the benefit of present and future generations. Curiously, conservation has rather come to mean empowering the state to coerce local communities with arbitrary evictions or forcible displacement like it has happened with the Endorois and the Ogiek in Kenya and many other communities across Africa. Free disposal as the Inter-American Court pointed out in the *Saramaka* case, entails effective consultation with respect to the traditions and customs of the peoples, who the Court held, have the right to either give or withhold consent regarding any development or investment that may adversely affect their livelihood, and in the instance where such development or investment is inevitable, for the community to share in the benefits.[23]

Jérémie Gilbert affirms that the principle of free disposal imposes limitations on the extent to which a state may exercise its sovereignty over natural resources.[24] This is intended to mean that in the instance where a people chose not to dispose of the natural resources they are entitled to, the state would be restricted from encroaching into that protected domain. When it becomes imperative for a community to dispose of its natural resources, the decision to do so must be taken freely and without external pressures either by the state or a third party. As Gilbert further notes, state practice has not always followed the principles of the law with respect to their entitlement to dispose of natural resources.[25] The free disposal principle, which is interpreted to also imply prior informed consent, equips the peoples of Africa with sufficient armoury against exploitation.

The law does not allow states to gratuitously deprive peoples of their natural resources without their free consent. The requirement of prior informed consent as a qualifying feature of the principle of free disposal is justified by the fact that Africa's natural resources constitute what Okoth-Ogendo defines as 'ontologically organised land and associated resources available exclusively to specific communities, lineages or families operating as corporate entities'.[26] Natural resourcesare a collective entitlement, requiring that consent for their disposal be obtained through broad-based consultations on the basis of communal values and principles that reflect the collective interest of the entire community. It implies also, as established in the *Endorois* and *Ogiek* cases that representation cannot substitute consent in the disposal of natural resources.[27] In the event of spoliation, article 21(2) grants the right to lawful recovery and a claim to adequate compensation for damages incurred.

Exclusive interest

The exclusive interest principle is enshrined in article 21(2) of the African Charter. It represents the understanding that the right to freely dispose of natural resources shall only be exercised to the extent that the exclusive interest of the peoples of Africa is prioritised. I define exclusive interest as the livelihood sustainability exigencies recognised as such by norms and value attributes for which natural resources are acknowledged to constitute a primary source of subsistence, which peoples may under no instance be deprived of. Exclusive interest equally conveys the idea of context-specificity; meaning that it is not generic but can only selectively apply on the basis of specific needs and entitlements. The exclusive interest of indigenous peoples for example, whose cultural and spiritual lifestyle patterns are closely associated with the natural environment, would differ from the exclusive interest of other groups of peoples who live a more urban lifestyle.

The exclusive interest principle enshrined in the right to natural resource ownership in the African Charter is quite radical; embodying the promise that in the instance where a people undertake to freely dispose of their natural resources, their livelihood sustainability exigencies must take centre stage and accordingly, inform the operational modalities for the exploitation and extraction of the natural resources within their communities. The principle further defines a mutual relationship and adherence to guarantees that natural resources can gainfully be exploited without compromising human livelihood and survival. African jurisprudence has, in this regard, ascertained that initiatives and practices that deprive indigenous peoples and local communities, or destroy the ecosystems on which they depend for their survival and spirituality, constitute not only a violation of the right to natural resources ownership but indeed, also a violation of the right to development.[28]

African state governments are accordingly obligated by their undertakings under the African Charter to individually and collectively ensure as indicated in article 22(2) that the exclusive interest is infused into the exploration and exploitation of natural resources to the benefit of indigenous and local communities and other disadvantaged groups. For exclusive interest to be achieved necessitates a fair measure of redistributive justice.

Equitable redistributive justice

It would seem that because the right to development incorporates the right to full sovereignty over natural resources, resource-rich countries have greater propensity to deliver on the right to development much more than resource-deficient countries. That is not the case. While the Democratic Republic of Congo for instance, is enormously endowed with some ofthe most valuable minerals, the wealth generated therefrom does not translate into enjoyment of the right to development for the Congolese who remain among the most impoverished globally. Equally, mineral resources have been the principal driver of the South African economy, but as Jessie Duarte opines, the mineral wealth has largely only benefited a few mining companies, leaving the rest of the population unjustly impoverished.[29] Even as the cost of mining in South Africa has been borne by host communities in terms of harm to community life for over a century, it was not until 2010 that a 'profit-based royalty' regime was instituted to allocate some benefits to mining host communities.[30]

From a right to development point of view, it does not count as development that Nigeria and Angola are the largest producers of crude oil on the African continent, yet are host to exceedingly large numbers of impoverished people; 61 percent (approximately 100 million) in Nigeria[31] and 94 percent of rural households in Angola are categorised as extremely poor.[32] Meanwhile, standards of living in other oil-producing countries like Saudi Arabia are remarkably high, ranking 36th out of 189 on the human development index.[33] These examples illustrate that the variation in the enjoyment of the right to development lies in how efficiently the gains from natural resource exploitation are distributed and more so, whether they are targeted at maximising living standards for the often-neglected local communities.

Because the right to natural resources is guaranteed as a legal entitlement, complicity in ensuring appropriate governance of same, would in terms of the applicable provisions of the African Charter, constitute a violation of the right to development.

The African Commission affirms in the *SERAC* case that article 21 of the African Charter was designed to reverse the trend of deprivation of the 'birth right' of ownership over the natural resources, which has principally benefited external stakeholders rather than the peoples of Africa.[34] Articles 2(3) and 4 of the Declaration on the Right to Development (DRTD) reiterate the obligation on states to formulate suitable policies to ensure the sustainable use of natural resources for socio-economic and cultural development. It requires a framework for redistributive justice and a commitment to sustainable governance of natural resources, which as the Economic Commission for Africa observes, is generally constrained by conflicting ownership rights.[35] As a matter of law, if the peoples of Africa are to fully enjoy the right to development, redistribution must inevitably apply.

Historically, local and indigenous African communities have been the most affected by the acts of deprivation that alienate them from the benefits that derive from the exploitation of their natural resources. Equitable redistributive justice requires in this instance, as stated in article 21(2), that in the event of spoliation, the dispossessed people are entitled to lawful recovery and/or adequate compensation. The Declaration on the Rights of Indigenous Peoples, in a number of provisions reminiscent of article 21(2) of the African Charter, provides that indigenous peoples deprived of the lands, territories and natural resources, which they have traditionally owned or otherwise occupied or used as a means of subsistence, (by lands being confiscated, taken, occupied, used or damaged without their free, prior and informed consent) are entitled to just and fair redress.[36]

Policy obligations

Deriving from the right to self-determination, which incorporates the right to full sovereignty over natural wealth and resources, the right to development obligates states to adopt appropriate national development policies to ensure constant improvement in living standards for their peoples.[37] This is summed up in the concept of the right to development governance which connotes among other things, the 'advancement of human capabilities for the sustainable management of Africa's resources [...] within a legal framework that guarantees genuine accountability and equitable redistribution for the improved collective well-being of the African peoples'.[38] Its implementation demands sensitivity to the rules and processes embodied in the instruments of law, which African state governments are obligated to adhere to in harnessing the African common heritage to advance productive capabilities and the socio-economic and cultural development exigencies of their peoples as I proceed to explain.

Natural resource governance responsiveness

The governance of natural resources requires states to exercise the power to control, monitor and regulate, in a sustainable manner, how the exploitation of natural resources is carried out so that the practice does not become deleterious to livelihood. The African Charter obligates states parties to eliminate all forms of foreign economic exploitation.[39] Although extractive industry multinationals are known for excesses in the exploitation of natural resources in Africa, they are generally insulated from accountability and often only required to demonstrate moral responsibility in respecting human rights in the course of their operations, which they frequently ignore with impunity. Excesses in the exploitation of natural resource refer to the malpractices committed in the communities where extraction takes place, which by and large, inflict pain and misery on human life, damage to means of subsistence or arbitrary displacements.

The right to permanent sovereignty over natural resources granted to states is justified to the extent that the state government scan effectively regulate the sector to the benefit of the peoples of Africa. Article 21(3) of the African Charter, in stating that the free disposal of natural resources shall not prejudice the obligation to promote international economic cooperation, in principle allows state parties to grant concessions to foreign investors to exploit the resources on the continent. In the *Saramaka* judgment, the Inter-American Court of Human Rights held that article 21 of the Inter-American Convention cannot be interpreted in a way that prevents the state from granting concessions for the exploration and extraction of the natural resources within the Saramaka territory.[40]

However, beyond doing so, states have the primary responsibility and a higher obligation of accountability to put in place appropriate measures and mechanisms to curb excesses in the exploitation of natural resources. Read in its entirety, section 21 of the African Charter does not allow state parties the liberty to loosely grant concessions without constraints and regulation. Stephan Hobe argues in this regard that international development cooperation can only aim to further the national development priorities of developing countries and must be defined by respect for the peoples' sovereignty over their natural wealth and resources.[41]

With respect to the provision that forbids depriving a people of their natural resources, assurance is granted that in the event of spoliation, a claim can be made for recovery and/or adequate compensation. The DRTD, although inherently non-binding, provides as part of the requirement to exercise the right to full sovereignty over natural wealth and resources, for states to assume the duty to formulate suitable national development policies aimed at improving living standards for the populations.[42] It emphasises states' regulatory duty with respect to the sustainable governance of natural resources to ensure equitable distribution of the benefits resulting there from. The regulatory standard provides the basis on which the peoples of Africa may not only have to make a claim when their right to development is contravened but in essence, to be able to also demand of their governments to take positive action in ensuring sustainable governance of natural resources for the common good of the peoples of Africa as prescribed in relevant provisions of the law.

Taking a look at domestic law, the Ethiopian Constitution for instance, provides with respect to the right to development that '[a]ll international agreements and relations concluded, established or conducted by the State shall protect and ensure Ethiopia's right to sustainable development'.[43] It means that international agreements entered into for the extraction of natural resources in Ethiopia must inevitably comply with article 43(3) and accordingly show evidence of contribution to realising the right to sustainable development in the country. Whether Ethiopia does in fact ensure that this constitutional obligation is adhered to in the international agreements and relations that the government enters into with foreign partners can only be ascertained to the extent that the state has put in place the necessary regulatory measures and enforcement mechanisms to effectuate compliance.

The increasing rate at which natural resources are extracted from the continent, often through shady deals without corresponding benefits to the local populations, causing harm in some instances, reveals that the regulatory framework in Africa is either too rudimentary or completely lacking. The scenario provides leeway for absolutism by foreign investors and multinationals, many of which wield excessive economic power that dwarfs the sovereignty of most African state governments in regulating their operations.[44] While land remains an invaluable natural resource for socio-economic and cultural development, empirical findings of research conducted in a number of African countries shows an escalation in land grabbing by foreign investors with the complicity of host governments, notwithstanding the attendant consequences on livelihood and the food security needs of displaced local communities.[45]

It does not suffice for state governments to make commitments to protect the rights of their people. Reality demands that the undertakings to protect may not only be presumed to be done but crucially, must be seen to be done. Controversially, of the range of existing and predominantly non-binding mechanisms for regulating the activities of extractive industry multinationals, Ekhator observes that the Kimberly Process appears to be the only African initiative,[46] which also, only regulates trade in conflict diamonds. He argues that in spite of the range of regulatory paradigms conceptualised by scholars, the African Union (AU) has not been able to explore the scholarly knowledge to develop a normative and institutional framework for regulating the activities of multinationals operating in Africa, which appears not to be a major or pressing priority for the AU.[47] The Economic Commission for Africa notes that

> Good natural resource governance requires that institutions—formal and informal—be mandated and capacitated to manage resources efficiently and to formulate, implement and enforce sound policies and regulations.

> Institutions must be held accountable for their decisions and must ensure that decision-making processes are transparent and widely participatory.[48]

By this suggestion, African state governments are required to translate the commitments undertaken under various treaties at the international and continental levels as well as the constitutional obligations undertaken at the domestic level into enforceable policies and regulations. The regulatory framework needs to be backed by functional institutions that can respond adequately to local socio-economic and cultural development exigencies and be held accountable when they fail to do so. African state governments are accordingly required to have monitoring systems in place to ensure that extractive industry actors stringently adhere to the laws and policies in force.

The need to establish formal and informal institutions, as the Economic Commission insinuates, probably implies, as a matter of transparency and accountability, curbing states' unfettered natural resource governance responsibilities, which in a good number of cases, have been abused to the detriment of the peoples whose exclusive interest they are supposed to protect. In effect, the regulatory framework is supposed to set the rules in ensuring that the disposal of natural resources is defined by the underlining purpose to ameliorate rather than cause a regression in living standards for the populations across the continent. It requires, in addition, a laid-down strategy for mobilising revenue from the exploitation of natural resources for development financing.

Development financing

Socio-economic and cultural development requires extensive resource mobilisation to be able to drive poverty eradication and essential service delivery in the areas of education, healthcare, housing, household energy, water and sanitation, food security, social assistance and livelihood security among others on a sustainable basis. These are arduous challenges for poor countries to surmount, which of course has forced most, if not all, African countries into the bad habit of begging and borrowing from the rest of the world. Contrary to the prescriptions of international law that promote systemic dependency on development assistance, the realities in Africa compel state governments to explore the option of generating development financing from the continent's natural resources, which indeed constitutes 'capital of vital importance'.[49] The Economic Commission for Africa affirms that

> Africa is endowed with huge untapped renewable and non-renewable natural resources that account for about 30 per cent of the global supply of mineral resources and 8 per cent of the world's stock of oil reserves. [...] Today, more than 70 per cent of total African exports derive from the oil, gas and mineral sectors, which account for about half of Africa's gross domestic product and contribute significantly to government revenues.

> Resource revenues represent on average about 40 per cent of domestic public revenues in Africa.[50]

Acknowledging that the 'most sustainable path to economic development and transformation is domestic revenue mobilization',[51] African state governments' fiduciary obligation in the governance of the natural resources on the continent necessitates ensuring that the proceeds derived from their exploitation are effectively utilised in financing socio-economic and cultural development on the continent. Schrivjer affirms that the Declaration on Permanent Sovereignty over Natural Resources underscores the principle that 'profits derived must be shared in the proportions freely agreed upon' with due regard for the state's sovereignty over its natural resources.[52]

There is no justification why Africa cannot develop the capacity to convert its natural resource endowments, which are predominantly only exported as unprocessed raw materials into consumable finished products. It is affirmed in Agenda 2063 that Africa has the capacity to generate development financing from its natural resources to accelerate growth and transformation.[53] Botswana for example, where the main economic growth driver remains the extraction and domestic processing of diamonds—accounting for 88 percent of total exports in 2016—has, as a result of discreet management of the country's natural resources, experienced rapid growth in per capit a income, from USD80 at independence in 1966 to USD6924 in 2016. [54] Distinguished by the remarkable decline in poverty and an impressive positioning on the Human Development Index,[55] the Botswana experience illustrates that the mobilisation of development financing from natural resources is not unattainable for the rest of Africa.

However, as the Executive Secretary of the Economic Commission for Africa opines, 'many natural resource-rich African countries have not managed their natural resources to increase government revenue and foster inclusive and transformative growth'.[56] Africa's natural resources are indeed an invaluable capital asset and a prerequisite for development, which as indicated, are not only insufficiently explored but are also increasingly compromised by debt entrapment resulting from unsustainable burrowing that amounts to mortgaging the resources to repay debts. Botswana's strategy in prioritising community-based natural resource management, which has supported the country's major development initiatives for over two decades[57] provides best practice lessons on how other African countries could generate development financing through a sustainable governance of their natural resources.

A major challenge that Africa would have to deal with in the extractive sector is the persistent lack of accountability, especially by foreign multinationals whose operations are expanding exponentially across the continent and more so, posing a threat to the environment and to livelihood for local populations.[58] The absence of accountability within the extractive sector marked by high levels of corruption, tax evasion and market abuse (trade mispricing and misinvoicing) among other malpractices, is reported to cost Africa an estimated USD100 billion in gross domestic product losses annually through illicit financial outflows.[59] While commodity prices have continued to soar, especially in the last decade, putting extractive industry actors at an advantages in terms of profit margins, the average royalty rate that these companies pay remains at a nominal 3.5 percent,[60] amounting to lost revenue, which on average should contribute to financing development in the host countries.

Conclusion

Against the backdrop of the continuous extraction of natural resources from Africa to feed the economic growth needs and elevated standards of living in other parts of the world while the peoples of Africa remain impoverished, the analysis and arguments advanced in this chapter have aimed to shift the frame of reasoning to looking at how sustainable governance of those resources could be maximised to yield substantive socio-economic and cultural development benefits on the continent. I have endeavoured to do so by looking into the provisions of the law that entitle the peoples of Africa to permanent sovereign ownership over natural resources and the right to development. With reference to the range of international and continental instruments that enshrine these exclusive collective rights, which the peoples of Africa are entitled to explore in their pursuit of better standards of living, I submit that African state governments have a duty to exercise territorial sovereignty in ensuring a sustainable regime of governance of the African common heritage for the collective substantive benefit of the populations across the continent.

Notes

1 Ngang, Carol. 2020. 'Guest Editor's Note: The Right to Development in Africa and Natural Resources Ownership'. *Journal for Juridical Science* 45(1): v.

2 The instruments that guarantee permanent sovereignty over natural resources include United Nations General Assembly (UNGA) Resolution 626(VII) on the Right to Exploit Freely Natural Wealth and Resources adopted on 21 December 1952; UNGA Resolution 1803(XVII) of 14 December 1962, on Permanent Sovereignty over Natural Resources para 1; common art 1(2) of the International Covenant on Economic, Social and Cultural Rights (ICESCR) adopted by the UN Gen Ass, Resolution 2200A(XXI) 1966 and the International Covenant on Civil and Political Rights (ICCPR) adopted by the UNGen Ass, Resolution 2200A(XXI) 1966; Resolutions 2158(XXI), 2386(XXIII) and 2692(XXV) adopted in 1966, 1968 and 1970, respectively; Declaration on the Right to Development Resolution A/RES/41/128 adopted by the UNGA on 4 December 1986, art 1(2).

3 The United Nations Economic Commission for Africa (UNECA). 2018. 'African Governance Report V: Natural Resource Governance and Domestic Revenue Mobilization for Structural Transformation'. Addis Ababa: UN Economic Commission for Africa, 13.

4 Adeleke, Fola. 2017. 'The Extractive Industries Transparency Initiative and Africa's Mineral Governance Regime'. *South African Institute of International Affairs–Occasional Paper 266*, 3.

5 Ngang, Carol. 2020. 'Right to Development in Africa and the Common Heritage Entitlement'. *Journal for Juridical Science* 45(1): 62; African Charter on Human and Peoples' Rights adopted in Nairobi, Kenya on 27 June 1981.OAU Doc CAB/LEG/67/3 Rev.5(1981): preamble para 8 and art 22(1).

6 Kumeh, Mensah. 2017. 'Natural Resource Governance in Africa'. *Tana High-Level Forum on Security in Africa* 1–2.

7 Duarte, Jessie. 2017. 'Opinion: Is South Africa Benefiting Enough from Extraction of its Resources?' *IOL Business Report*, 18 September <www.iol.co.za/business-report/opinion-is-south-africa-benefiting-enough-from-extraction-of-its-resources-11263735>.

8 African Convention on the Conservation of Nature and Natural Resources (Revised) adopted in Maputo, Mozambique on 11 July 2003, preamble para 1.

9 UNECA, note 3, 14.

10 Charter of the Organisation of African Unity adopted in Addis Ababa, Ethiopia on 25 May 1963, preamble para 3.

11 Ng'ambi, Sangwani. 2015. 'Permanent Sovereignty over Natural Resources and the Sanctity of Contracts, From the Angle of Lucrum Cessans'. *Loyola University of Chicago International Law Review* 12(2): 153–172.

12 International Labour Organisation. 1989. Indigenous and Tribal Peoples Convention No 169, art 15(1).

13 *Centre for Minority Rights Development (Kenya) and Minority Rights Group International on behalf of Endorois Welfare Council v Kenya* Comm 276/2003 (2009) AHRLR 75 (ACHPR 2009), para 161.

14 Sanginga, Pascal; Ochola, Washington and Bekalo, Isaac. 2010. 'Natural Resource Management and Development Nexus in Africa'. In *Managing Natural Resources for Development in Africa: A Resource Book*, edited by Ochola, Washington; Sanginga, Pascal and Bekalo, Isaac. Nairobi: University of Nairobi Press, 25.

15 DRTD, note 2, art 1(2).

16 UN Charter adopted in San Francisco on 26 June 1945, and Universal Declaration of Human Rights adopted by the UNGA Resolution 217 A(III) of 10 December 1948, preamble.

17 African Convention on the Conservation of Nature, note 8, arts 2 and 4.

18 African Charter, note 5, preamble; Convention on the Conservation of Nature and Natural Resources, note 9, art 14(b).

19 *Endorois*, note 13; *African Commission on Human and Peoples' Rights (Ogiek) v Republic of Kenya* (2017) Appl No 006/2017.

20 Cambou, Dorothée and Smis, Stefaan. 2013. 'Permanent Sovereignty over Natural Resources from a Human Rights Perspective: Natural Resources Exploitation and Indigenous Peoples' Rights in the Arctic'. *Michigan State International Law Review* 22(1): 350.

21 Ngang, note 5, 117.

22 Ngang, Carol. 2015. 'Indigenous Peoples' Right to Sustainable Development and the Green Economy Agenda'. *Africa Insight* 44(4): 33–34; Abate, Randall and Warner, Elizabeth. 2013. 'Commonality among Unique Indigenous Communities: An Introduction to Climate Change and its Impacts on Indigenous Peoples'. *Tulane Environmental Law Journal* 26(2): 179–195.

23 *Inter-American Court of Human Rights, Saramaka People v Suriname* (Judgment of 28 November 2007), para. 95.

24 Gilbert, Jérémie. 2013. 'The Right to Freely Dispose of Natural Resources: Utopia or Forgotten Right?' *Netherlands Quarterly of Human Rights* 13(3): 319.

25 *Ibid*, 326–327.

26 Okoth-Ogendo, Hastings. 2003. *The Tragic African Commons: A Century of Expropriation, Suppression and Subversion*. Cape Town: Programme for Land and Agrarian Studies, 2.

27 *Endorois*, note 13, paras 281–282, 290–291 and 297–298; *Ogiek Community*, note 19, para 210.

28 *Social and Economic Rights Action Centre et al v Nigeria (Ogoni)* Comm 155/96 (2001) AHRLR 60 (ACHPR 2001), *Endorois*, note 13; *Ogiek Community*, note 19.

29 Duarte, note 7.

30 Gajigo, Ousman; Mutambatsere, Emelly and Ndiaye, Guirane. 2012. 'Royalty Rates in African Mining Revisited: Evidence from Gold Mining'. *AfDB Africa Economic Brief* 3(6): 1.
31 BBC. 2012. 'Nigerians Living in Poverty Rise to Nearly 61%'. *British Broadcasting Corporation*, 13 February <www.bbc.com/news/world-africa-17015873>.
32 Kraus, Kimber. 2016. 'Poverty in Angola: Causes, Updates and Statistics'. *The Borgen Project*, 19 February <https://borgenproject.org/poverty-angola-causes-updates-statistics/>.
33 UNDP. No date. 'Saudi Arabia: Human Development Indicators'. UNDP Human Development Reports <http://hdr.undp.org/en/countries/profiles/SAU>; Arab News. 2018. 'Saudi Arabia Rises in Human Development Index' *Arab News*, 9 March <www.arabnews.com/node/1613221/saudi-arabia>.
34 *Ogoni*, note 28, para 56.
35 UNECA, note 3, 13.
36 Declaration on the Rights of Indigenous Peoples Resolution 61/295 adopted by UN General Assembly on 13 September 2007, arts 10, 20(2) and 28(1); Schrivjer, Nicolaas. 2013. 'Self-Determination of Peoples and Sovereignty over Natural Wealth and Resources'. In *Realizing the Right to Development: Essays in Commemoration of 25 Years of the United Nations Declaration on the Right to Development*, edited by UN Human Rights. Geneva: UN Publication, 98–99.
37 DRTD, note 2, art 1(2) and 2(3).
38 Ngang, note 5, 115.
39 African Charter, note 18, art 21(5).
40 Inter-American Court of Human Rights, *Saramaka People v Suriname* (Judgment of 28 November 2007), para 214(8).
41 Hobe, Stephan. 2015. 'Evolution of the Principle on Permanent Sovereignty over Natural Resources'. In *Permanent Sovereignty over Natural Resources*, edited by Bungenberg, Marc and Hobe, Stephan. Geneva: Springer International Publishing, 6.
42 DRTD, note 2, arts 1(2) and 2(3).
43 Constitution of the Federal Democratic Republic of Ethiopia adopted on 8 December 1994, art 43(3).
44 *Ogoni*, note 28.
45 Kachika, Tinyade. 2010. 'Land Grabbing in Africa: A Review of the Impacts and Possible Policy Responses'. *Oxfam International*, 8–13.
46 Ekhator, Eghosa. 2016. 'Regulating the Activities of Oil Multinationals in Nigeria: A Case for Self-Regulation?' *Journal of African Law* 60(1):2–12.
47 Ekhator, Eghosa. 2018. 'Regulating the Activities of Multinational Corporations in Nigeria: A Case for the African Union?' *International Community Law Review* 20(1): 30.
48 UNECA, note 3, 13.
49 African Convention on the Conservation of Nature, note 8, preamble.
50 UNECA, note 3, 8.
51 *Ibid*, 12.
52 Schrivjer, not 36, 97.
53 AU Commission. 2015. *Agenda 2063: The Africa We Want*. Addis Ababa: African Union, paras 60 and 74(b).
54 Honde, George. 2018. 'Botswana: 2018 African Economic Outlook Country Note'. *African Development Bank* 1; UNECA, note 3, 11.
55 Honde, note 54, 8.
56 UNECA, note 3, 8.
57 Honde, note 54, 7.
58 Gedicks, Al. 2015. 'Transnational Mining Corporations, the Environment, and Indigenous Communities'. *Brown Journal of World Affairs* 22(1): 129.
59 Forstarter, Maya. 2018. 'Illicit Financial Flows, Trade Misinvoicing, and Multinational Tax Avoidance: The Same or Different?' *Centre for Global Development Policy Paper 123*, 1–33;

Kar, Dev and Spanjers, Joseph. 2015. 'Illicit Financial Flows from Developing Countries: 2004-2013'. *Global Financial Integrity*, 10–12; Cobham, Alex. 2014. 'The Impacts of Illicit Financial Flows on Peace and Security in Africa'. *Tana High-Level Forum on Security in Africa*, 1–3.

60 Gajigo, Mutambatsere and Ndiaye, note 30, 1–3.

References

Abate, Randall and Warner, Elizabeth. 2013. 'Commonality among Unique Indigenous Communities: An Introduction to Climate Change and its Impacts on Indigenous Peoples'. *Tulane Environmental Law Journal* 26(2): 179–195.

Adeleke, Fola. 2017. 'The Extractive Industries Transparency Initiative and Africa's Mineral Governance Regime'. South African Institute of International Affairs—Occasional Paper 266.

Cambou, Dorothée and Smis, Stefaan. 2013. 'Permanent Sovereignty over Natural Resources from a Human Rights Perspective: Natural Resources Exploitation and Indigenous Peoples' Rights in the Arctic'. *Michigan State International Law Review* 22(1): 347–376.

Cobham, Alex. 2014. 'The Impacts of Illicit Financial Flows on Peace and Security in Africa'. High-Level Forum on Security in Africa, 1–36.

Ekhator, Eghosa. 2016. 'Regulating the Activities of Oil Multinationals in Nigeria: A Case for Self-Regulation?' *Journal of African Law* 60(1): 1–28.

Ekhator, Eghosa. 2018. 'Regulating the Activities of Multinational Corporations in Nigeria: A Case for the African Union?' *International Community Law Review* 20(1): 30–68.

Forstarter, Maya. 2018. 'Illicit Financial Flows, Trade Misinvoicing, and Multinational Tax Avoidance: The Same or Different?' Centre for Global Development Policy Paper 123, 1–33.

Gajigo, Ousman; Mutambatsere, Emelly and Ndiaye, Guirane. 2012. 'Royalty Rates in African Mining Revisited: Evidence from Gold Mining'. *AfDB Africa Economic Brief* 3(6): 1–12.

Gedicks, Al. 2015. 'Transnational Mining Corporations, the Environment, and Indigenous Communities'. *Brown Journal of World Affairs* 22(1): 129–152.

Gilbert, Jérémie. 2013. 'The Right to Freely Dispose of Natural Resources: Utopia or Forgotten Right?' *Netherlands Quarterly of Human Rights* 13(3): 314–341.

Hobe, Stephan. 2015. 'Evolution of the Principle on Permanent Sovereignty over Natural Resources'. In *Permanent Sovereignty over Natural Resources*, edited by Bungenberg, Marc and Hobe, Stephan. Geneva: Springer International Publishing, 1–11.

Honde, George. 2018. 'Botswana: 2018 African Economic Outlook Country Note'. African Development Bank.

Kachika, Tinyade. 2010. 'Land Grabbing in Africa: A Review of the Impacts and Possible Policy Responses'. Oxfam International, 1–69.

Kar, Dev and Spanjers, Joseph. 2015. 'Illicit Financial Flows from Developing Countries: 2004–2013'. *Global Financial Integrity*, 1–72.

Kumeh, Mensah. 2017. 'Natural Resource Governance in Africa'. Tana High-Level Forum on Security in Africa, 1–10.

Ng'ambi, Sangwani. 2015. 'Permanent Sovereignty over Natural Resources and the Sanctity of Contracts, From the Angle of Lucrum Cessans'. *Loyola University of Chicago International Law Review* 12(2): 153–172.

Ngang, Carol. 2015. 'Indigenous Peoples' Right to Sustainable Development and the Green Economy Agenda'. *Africa Insight* 44(4): 31–46.

Ngang, Carol. 2020. 'Guest Editor's Note: The Right to Development in Africa and Natural Resources Ownership'. *Journal for Juridical Science* 45(1): i–xii.

Ngang, Carol. 2020. 'Right to Development in Africa and the Common Heritage Entitlement'. *Journal for Juridical Science* 45(1): 28–50.

Okoth-Ogendo, Hastings. 2003. *The Tragic African Commons: A Century of Expropriation, Suppression and Subversion*. Cape Town: Programme for Land and Agrarian Studies.

Sanginga, Pascal; Ochola, Washington and Bekalo, Isaac. 2010. 'Natural Resource Management and Development Nexus in Africa'. In *Managing Natural Resources for Development in Africa: A Resource Book*, edited by Ochola, Washington; Sanginga, Pascal and Bekalo, Isaac. Nairobi: University of Nairobi Press.

Schrivjer, Nicolaas. 2013. 'Self-Determination of Peoples and Sovereignty over Natural Wealth and Resources'. In *Realizing the Right to Development: Essays in Commemoration of 25 Years of the United Nations Declaration on the Right to Development*, edited by UN Human Rights. Geneva: UN Publication, 95–102.

3 Loss, anguish and postponement

The story of African development, natural resource ownership and the paradox of waiting

Isaac Shai

Introduction

In this chapter I explore the notions of the right to development and ownership of natural resources by drawing from literary genres, in particular fictional literature. According to Stuart Motha,

> the common perception is that literature and law are of a different order. After all, what of the world of law, what of its judgments, pain and death-dealing can literature accomplish, other than the re-presentation of the 'real'? At best, literature stands in through a cultural resemblance through metaphor or allegory.[1]

My suggestion is that 'literature should be placed at the very heart of law and jurisprudence'.[2] As Motha puts it, '[a]t the heart of the argument is that literature has the same terminal potential as law. Like law, literature has the capacity to put the question of being (including its own being) into question'.[3] For Stuart Motha, 'it is in literature, in the work of art, that the complexity of a political crisis, as well as the urgency and limits of critical responses, can be discerned'.

Literature also has a disruptive effect. For instance, Stuart Motha tells of J.M. Coetzee's novel titled *Disgrace* written during South Africa's transitional period. The novel was greeted with anger, precisely because it had the effect of disrupting the dominant constitutional narrative that South Africa belongs to all who live in it. Motha correctly states that 'far from a "rainbow nation" or "post-racial nation", this novel depicts relations between people that remain explicitly racially divided. It also explicitly tackles the problem of living with sexual violence'.[4] In this sense, the novel has the ability, through the re-presentation of life, to disrupt historicist accounts of life as presented by certain legal and jurisprudential accounts.

It is in this context that I recall Njabulo Ndebele's novel titled *The Cry of Winnie Mandela* and Franz Kafka's short story called *Before the Law*. Central to these works of fiction are the notions of waiting and refusal.[5] I take the notions of waiting and refusal as central to the discussion on the right to development and ownership of natural resources precisely because they are suggestive of the idea of decolonisation that is still to be and also the ability of the law to create conditions of possibility that have so far been impossible. Interpreted radically, these two works of fiction serve to provide counter histories and counter-narratives and can further serve to instil the belief in the art of the possible. I will further suggest that contained in the vortex of waiting and seeming impossibility lies the emancipatory notion of refusal—refusal as a psychic insurrectionary attitude.[6]

The 'sheer allegorical force'[7] of the notion of waiting is pungently captured by these two works of fiction. In *The Cry of Winnie Mandela*, Njabulo Ndebele tells the story of four South African women who are forced to wait for their husbands to return. Ndebele calls these women the 'descendants of Penelope'. In Homer's Odyssey, Penelope is the wife of Odysseus, and she is reputed for having waited for her husband for years. In Ndebele's *The Cry of Winnie Mandela*, the women's waiting is characterised by continuously shifting moments of hope and despair that are mediated by the permanence of anguish, but with hope ever present. As Ndebele states, all these women, 'they all seem to be struggling to wriggle out of the cocoon of thought, seeking to emerge as fully fledged being'.[8] They are continuously caught in the 'Gramscian' liminal moment where the old is refusing to die and the new is struggling to be born. In the end, the women do not have the strength to wait anymore. Nor do they have 'the strength to start a new life'.[9]

As they wait, the future keeps mocking them and the past haunting them. Patience, the second descendant, 'is never released from waiting'.[10] If in Ndebele's *The Cry of Winnie Mandela* the women are not really released from waiting for the husbands, in Kafka's *Before the Law*, the man from the country waits until the end of his living days to get entry into the law. Kafka's *Before the Law* is a parable about the man from the country who tries to get through a gateway that allows him to enter into the law. The gate into the law is open but the gatekeeper does not let the man in. The man asks if he can be allowed to get into the law later and the gatekeeper tells the man that it is possible but does not allow the man to get inside the law. The man waits by the gate, offering the gatekeeper his belongings and the gatekeeper takes the man's belongings and says: 'I am taking this only so that you do not think you have failed to do anything'. The man from the country waits at the gate until he is old and deaf and when he realises that he is about to die, he decides to ask the gatekeeper one last question: He asks,

> everyone strives after the law [...] so how is it that in these many years no one except me has requested entry? The gate keeper then replies: Here no one else can gain entry because this entrance was assigned only to you and I am now going to close it.

The man from the country dies and the gatekeeper closes the entrance.

Ndebele's *The Cry of Winnie of Mandela* and Kafka's *Before the Law* can be taken as allegories of the oppressor and the oppressed, of the oppressed waiting for liberation, of the unhomed who although at home, are nonetheless outside of home and are perpetually negotiating their place in the world. Kafka's *Before the Law* is particularly a strong reminder of the ability of law to mock by creating possibilities that are sometimes made impossible by the very same law. Njabulo Ndebele's 'women in waiting' and Franz Kafka's man who waits for the law are also ways of retelling the tragedy of colonialism and the colonial dispossession, and what seem to be an endless waiting of African peoples, which is now and then disrupted by the inevitability of death.

I have already indicated that it is this notion of waiting, its causes and implications that is central to this chapter. I therefore wish to re-orient the idea and story of ownership of natural resources and the right to development as a story of anguish, loss, postponement and waiting—in short as a story of not knowing whether and when. This approach, I suggest, allows us to descend from high-level abstractions to the concreteness of the everyday lives of the African multitudes living a life of loss that is nevertheless sustained by hope, the sort of hope that can only be in waiting—waiting in hope and hope in waiting. In engaging the notion and story of waiting, I necessarily draw from the past and the future 'as objects of memory and expectation'[11] because central to waiting is memory and expectation, both located in the annals of the past and the future.

Ndebele's women in waiting and Kafka's man who waits for the law poignantly bring forth the unfulfilled promises and dreams deferred, dreams that could have been realised but are betrayed by those entrusted with the role of making them true. This approach allows us to rethink the notions of natural resource ownership and the right to development from a phenomenological prism. However, as Ndebele says, 'the life of waiting is a minefield of ambiguities'.[12] As such, the story of the women in waiting is also the story of refusal. I introduce the notion of refusal to wait within the context of refusal to be 'unhomed' or put differently, the refusal to continue to be 'outside a home', of 'not being at home in the home'.[13] In this sense, the question of Africans and ownership of natural resources, particularly in South Africa, is akin to 'not being at home in the home'. However, I begin with where it all started, which is, colonialism as the beginning of 'unhoming'. I then contextualise the notion of waiting by anchoring it within modernity and the politics of time and how law is used to produce time. This is followed by a detailed analysis of Ndebele's *The Cry of Winnie Mandela* and Kafka's *Before the Law* and the meaning of waiting. I then proceed to give a counter-hegemonic narrative about the right to development and ownership of natural resources within the context of the limits of law.

Colonisation as 'unhoming'

Hommi Bhabha describes the condition of being unhomed, not as a question of being 'homeless' but rather a question of being outside of home, 'of being forced to renegotiate one's place in the world'.[14] In this chapter, I suggest that the question of unhoming as deprivation and dispossession should start with the European spatial view of the world. Looked at from this prism, the question of development and ownership of natural resources should therefore, firstly, contend with the following question: To whom does the Earth belong? Despite its deceptively simple posture, the centrality of this question inheres in that it is this very question that has fundamentally shaped the past, brought about the present and defines the future.

The Earth, according to the English enlightenment philosopher John Locke, belonged to the Europeans because 'the savage' 'was a wanderer or related to land in an indefinite communal way, not sufficiently 'removed from the common state nature placed it in'.[15] According to Peter Fitzpatrick, enlightenment philosophers generally agreed that 'the savage had no sufficiently fixed relation to things to support a legal right to them because property was the basis of law'.[16] The 'African savage' is characterised by the European philosophers of modernity as living in a state of nature and therefore, characterised by a lack of a fixed permanent relationship with land and as a result cannot possess any legal rights.[17] The lack of a fixed permanent relationship with land, which was ostensibly a result of the absence of reason among the 'African savages', justified the European to acquire sovereignty over the lands and the natural resources. For John Locke, 'the absence of a fixed, cultivating relation to land accounted for the lack of reason itself'.[18] This is precisely why for Achille Mbembe, the question of whom the Earth belongs to

> was the key question underlying colonial conquest and imperial expansion since the fifteenth century. With the partition of Africa in the nineteenth century, European powers have decided that the earth in its entirety belonged to them. They were its true owners, and they could occupy lands that were populated by foreign people. They could exploit these lands as well as the people who had always inhabited them, thereby carving out spheres of influence each of them had control over.[19]

Modernity, waiting and the politics of time

Kathleen Davis states that the techniques of colonialism continue to this day and that our duty is to discover these techniques and deconstruct them.[20] The imposition of a particular view of historical time has had a deep impact on how we look at the past and the future and consequently, on why and how the dispossessed and the excluded wait for the final moment when they also cease to be excluded and marginalised. In this section, I suggest that the often commonsensical periodisation of historical time and law's productive relationship with periodisation of historical time have been instrumental in how the notions of the right to development and ownership of natural resources are predominantly conceived.

In demonstrating law's productive relationship with periodisation of historical time and colonialism, Peter Fitzpatrick puts forth a particular conception of 'ancien law': 'ancient law is written backwards'[21] to enable it to form a seamless synergy with the notion of 'stages of social development envisaged [as] the march of the social unit from primitive family group to the modern territorial state'.[22] In this sense, law is described and describes itself in evolutionary terms, where for instance 'civil law develops simultaneously with private property out of the disintegration of the natural community'.[23] This is how law becomes an instrument of periodisation of time. The link with periodisation and colonialism is in the sense that law is 'brought' to Africa through colonisation in order to, as per Lord Lugard, 'raise the mass of the people of Africa to a higher plane of civilisation'.[24]

According to Davis, the notion of time has up to now been tethered to the sacred. This approach has, according to Davis, created 'conceptual limit' as the tendency has been to naturally divide historical time into a religious Middle Ages and secular modernity.[25] In this sense, historical change is almost always an uninterrupted linear movement from bad to good and from evil to virtuous.

Davis suggests that historical periodisation derives its'explanatory power from economic models of periodization by aligning a transition from ecclesiastical to secular society with a transition from medieval, rural agrarian economy to a modern, urban, commercial economy, most typically expressed as the transition from feudalism to capitalism'.[26] For Davis, the notion of dividing time and history into segments is nothing but a 'political technique'[27] that, I argue, engenders a partial and ethnocentric way of looking at the past and the future. Moreover, this division of time and history between the ostensibly medieval and modern periods has implications on our understanding of the notions of sovereignty, the state and the meaning of politics—and as I suggest, our understanding of the notions of development and ownership of natural resources and most importantly, what it means to wait. This historicist understanding of time leads to the justification of colonialism as having been necessary for the 'development' and 'civilisation' of Africa.

The law has been central to the maintenance of periodisation of historical time and modernity's triumphant teleology. Historically, those regarded as not possessing Western structures and notions of law were and are still regarded as pre-modern. In this sense, law comes to occupy a central role in cementing the modern/non-modern binary and becomes imbricated in temporal ordering. This is why Davis calls periodisation a fundamentally juridical function precisely because, for instance, the feudal narrative is constructed and developed 'within the larger project of revising and interpreting law'[28]—hence the reference to periodisation as fundamentally a juridical construction. The rhetoric of law becomes a 'political technique' in the sense that development must be in accordance with or follow the same route as the West. In this sense, development must be in accordance with the (Western) law or Western understandings of what law is. The ethic of ownership and what it means to own is forced to be in accordance with the Western-centric legal structures.

What is less interrogated is the fact of the relationship between periodisation of historical time and colonialism. According to Davis,

> As medievalists have recently argued, the genealogies of the 'Middle Ages' colonialism are intertwined. The construction of a 'medieval 'period characterised by irrational superstition was fully involved with the identification of the colonial subjects as irrational and superstitious, and this process bore concrete effects upon colonised peoples through the system of rule that it generated and legitimised. The idea of a superstitious Middle Ages, in other words, did not pre-exist the 'superstitious' upon which it became mapped; rather, they emerged together, each simultaneously making possible and verifying the other.[29]

In other words, there is a relationship between periodisation and colonialism in the sense that periodisation enables and justifies colonialism. Davis continues by stating that the notions of a superstitious, feudal Middle Ages did not exist before colonialism and would never have existed but for colonialism. In fact, the 'colonisers could not have mapped and administered foreign lands and bodies as they did without the simultaneous process of imagining the "Middle Ages"'.[30] In this sense, the construction of the Middle Ages becomes central to the construction of the pre-political 'African savage' that requires colonialism in order to be nudged or if need be, violently dragged towards the modern.

I have introduced the notion of periodisation of historical time and law's imbrication to demonstrate and critique the idea of a linear and chronological temporality that modernity has peddled so much as the only way to conceive of the past, the present and the future and how, as a result, other alternatives of conceiving of the world are radically undercut. In this sense, I am calling for what Karin van Marle calls the 'disruption of a chronological and linear conception of time that could contribute to an acceptance of the notion of multiple truths and fluidity of meanings'.[31] Looked at from this prism, it could be argued that the problems of the present are fundamentally due to how the past has been and continues to be constructed and reconstructed and it is precisely this construction of the past that radically undercuts alternative epistemologies.

Drawing largely from Davis, I have so far attempted to expose and critique the notion of periodisation and to demonstrate that periodisation is a political technique with the sole purpose of undercutting alternative epistemologies of the past and the future. It is in this sense that I suggest that to critique the notion of periodisation is at the same time positing the politics of time. The notion of a linear teleological time has implications for the notion of waiting, precisely because of the centrality of time within the notion of waiting. Ben Okri opens his novel *Infinite Riches* with the following:

> Who can be certain where the end begins? said Dad, shortly before he was arrested for the murder of the carpenter. 'Time is growing,' he added.

> 'And our suffering is growing too. When will our suffering bear fruit? One great thought can alter the future of the world. One revelation. One dream. But who will dream that dream? And who will make it real?[32]

The father's words are narrated by Azaro, the spirit child who before he was born had promised his spiritual companions that he will return to the spirit world as soon as he was born. Borrowing from David Scott's words in *Conscripts of Modernity*, the words uttered by the father may denote 'confident anticipation' or 'anxious foreboding' or either of the two but 'always with a sense of fervent expectation and hope'[33]—'one great thought can alter the future of the world. One revelation. One dream'. Most importantly and for our purposes, the words by Azaro's father serve to question a linear chronological time that tends to inhabit the notion of waiting, at least within the context of postcolonial (South) Africa. In a melancholic tone, Azaro's father asks 'who can be certain where the end begins' and by implication who can be certain when the beginning ends. Azaro's father questions the notion of certainty about a particular way that certain things will end precisely because for him as time grows, his suffering grows too. The long wait does not seem to bear any fruit. The deep questioning by Azaro's father of the relationship between time and suffering and the end of the suffering as a teleological process is deeply apposite to my metaphorical appropriation of the notion of waiting and its relationship with the right to development and ownership of natural resources.

As indicated above, a fictional literary approach holds promise as an alternative narrative to better tell, expose and critique the notion of historical time as teleological. I now turn to Njabulo Ndebele's *The Cry of Winnie Mandela* and Franz Kafka's man from the country as, like Azaro's father, literary narrative that question, critique and dispute latent theodicy that predominantly inhabit the notion of waiting.

Njabulo Ndebele's women in waiting, Franz Kafka's man from the country

It could plausibly be argued that with the partition of Africa in the nineteenth century, Africans, in the same way as Penelope's descendants in *The Cry of Winnie Mandela* have been waiting in anguish to be reunited with their partners and transcend the chasm inflicted by racial domination. In Kafka's *Before the Law*, the man from the country does not lose hope of gaining entry into the law until he dies. In Ndebele's *The Cry of Winnie of Mandela*, there is a sense in which, despite many years having passed since the disappearance or absence of the husbands, the women in waiting hold on to the hope of final reunification with their husbands. The man from the country dies without gaining entry into the law. Having waited for their husbands, the first woman's husband never comes back, the second and the third women's husbands come back only to divorce and the fourth woman's husband who has, for all intents and purposes been present but absent at the same time, dies. But they had all dutifully waited, 'now with confident anticipation, now with anxious foreboding, but always with a sense of fervent expectation and hope'.[34] The question therefore is why: Why did the man before the law wait to gain entry into the law until he died and why did the women wait for their husbands for what looked like eternity. In the context of Africa, of international law, of the right to development, and ownership of natural resources, the question is sharply raised by Sundya Pahuja in her book titled *Decolonising International Law*. Pahuja asks the following question:

> Why has international law, from the perspective of the Third World, been so disappointing? What it is about international law that makes it simultaneously full of promise, and yet again and again a contributor to the failure of projects articulated in its name? And in the face of these disappointments, why do so many people from both inside and outside the discipline mount what are often devastating critiques of international law—(such as) its uses by the powerful, its implication in imperialism, its capacity to facilitate exploitation, its other manifold dark sides—only to conclude with a plea for the reinterpretation of international law, or its retrieval for the powers of goodness?[35]

Within the context of South Africa, the following questions can be asked; why is it that the dawn of liberation that South Africans had waited more than 350 years for did not bring about substantial transformation? Alternatively, why is it that despite the promises of the Constitution, there has been, twenty-five years later, failure to resolve the issue of land and the attendant ownership of natural resources to the benefit of African peoples. To put it in Azaro's father's words, why it is that the end has not begun and instead with the growth of time, 'our suffering is growing too'. I suggest that the answer lies in our conception of time and therefore, what it means to wait. Our conception of historical time and what it means to wait in turn influences and is influenced by our construction of the past and the productive relationship that law has with periodisation in the maintenance of our conception of the past, the present and the future. As Siam Beynon-Jones and Emily Grabham state, 'many of the technical means by which law works have the effect of instantiating an orientation to a constructed past, present, or future in linear or teleological form'.[36] Beynon-Jones and Grabham further state that 'law's linearity and forward moving teleology [...] are achieved as a result of one of the archetypal modes of Western law: the common law's reliance on precedent and its own understanding of its capacity to generate certain stances towards history'.[37]

The debate around land in South Africa is a more apposite example of law's productive relationship with periodisation and the implication that this relationship has on the notion of waiting. As indicated in the preface, European land dispossession of the indigenous African peoples started in the middle to late seventeenth century. When Jan van Riebeeck informs Autshumao that the land now belongs to him and he does not recognise the Khoi ownership of land, he is essentially speaking the language and philosophy of enlightenment. This is so because enlightenment thought contrasted law with the savage state.[38] The mythical notion that pre-colonial Africa was inhabited by savages in a state of disorderly stasis because it had no 'assemblies to debate and had no ancestral ordinances' is central to understanding law's productive role in the maintenance of periodisation.[39] As a result, according to Peter Fitzpatrick, 'many mythic origins of modern law are compressed in this description of the savage's disorderliness'.[40]

It is therefore common to identify 'an uncivilised or wild state with the absence of (European) law'.[41] Law in this sense becomes central because it presumably tamed the 'wild, promiscuous, propertyless and lawless' African.[42] It is precisely this deeply entrenched view that galvanised Helen Zille, the leader of the Democratic Alliance party in South Africa, to proudly proclaim on twitter in March 2017 that colonialism gave Africa law and infrastructure. The Hobbesian brutish state is in this sense ascribed to the absence of law. 'From this negation is derived the positive content of the law of the land in its unconditional and unlimited validity'.[43] In other words, the positive content of modern Western law is based on European self-definition, which is simultaneously a negation of the other. In this sense, 'the savage state provides more than the force creating and sustaining law and political society'.[44] The concern of modern law has always been with savagery. Fitzpatrick notes that '[i]t is the savage which in negative terms gives context to the political and gives content to law'.[45]

Fitzpatrick proceeds to state John Austin's views to the effect that the content of law is informed by savagery on the one hand and also by 'the challenge posed by the poor and the ignorant'.[46] However, for the poor, according to Austin, 'if they adjusted their numbers to the demand for their labour, they would share abundantly, with their employers in the blessings of property. Unlike the stupid savage who can only respond to the imperatives of the inexorable'.[47] The primordial and savage societies are defined by the absence of law and therefore civilisation. The absence of law suggests the absence of property because property can only be owned legally. There is a profound relationship between savagery, law, property and the civilisation of the savage. The savage is a savage because it has no law; the savage needs (European) law to lift 'it' up to civilisation; and the savage could not have owned property or land because it had no law and was therefore uncivilised.

Like Njabulo Ndebele's women in waiting and Franz Kafka's man who waits for the law, the advent of the new constitutional dispensation in South Africa was expected to 'bring back the land' to the dispossessed, deliver 'the heaven it promised'.[48] Various legal mechanisms were put into place to ensure that the issue of land is resolved.[49] Nearly 400 years later, the dispossessed indigenous African majority are still waiting for what is now called the resolution of the land question. Thiven Reddy, in his book titled *South Africa, Settler Colonialism and the Failure of Liberal Democracy*, takes it even further by stating that today, 'everywhere we see the unravelling of the dreams of post-Apartheid expectation, general disappointment and political malaise replacing the optimism and enthusiasm of 1994'.[50]

In this, there are echoes of Njabulo Ndebele's women in waiting; Patience, one of the women in waiting says that '[w]omen are the prisoners of the dream of romance'. She may well have been referring to South Africa's propensity to, partially borrowing from David Scott in *Conscripts of Modernity*, 'fade' and 'dissolve' into that 'seductive silence' occasioned by a people used to waiting. In reference to Njabulo Ndebele, Reddy states that 'Njabulo Ndebele offers a nuanced and close reading of South Africa's present conjuncture. After two and a half decades of democracy, both blacks and whites engage dishonestly: 'The whites are pretending it [apartheid] didn't happen; the Blacks are pretending to forgive'. Ndebele could well be saying, like Patience, albeit in different words, that South Africans 'are the prisoners of the dream of romance'. Marara, one of the women in waiting, states as follows in her conversation with Deli, another woman in waiting: 'I think of your husband, Sis' Deli, who returned home only to leave again. To what notion of home did he return? Was it a country as home he returned to, and not the home that was your house? Then home and house are not the same thing'.[51] Ndebele says: 'All these women, they all seem to be struggling to wriggle out of the cocoon of thought, seeking to emerge as fully fledged beings'.[52] In the end, all these women, who to varying extents and at different moments waited 'with confident anticipation', sometimes 'with anxious foreboding' but nonetheless 'always with a sense of fervent expectation and hope', experience the impossibility of Hegelian theodicy and the fact that historical time is not linear.

Returning to Franz Kafka's *Before the Law*, Costas Douzinas and Adam Gearey ask the following, 'how can we read this story justly?'[53] Having asked this question, the two scholars concede to the capacious nature of the parable of *Before the Law* in that 'to be recognised and to achieve the status of literature', a good work of literature must be 'repeated, disseminated, even plagiarised and retold in new versions'.[54] In essence, there are plural ways in which to justly read a story. It is in this sense that I read *Before the Law* as a strong reminder of the ability of law to mock by creating possibilities that are sometimes made impossible by the very same law. Douzinas and Gearey make almost a similar observation when they state that 'the law too [...] is caught in the ambiguities of the textual archive, despite the attempts to present it as a closed textual machine, a self-interpreting or self-executing system'.[55]

In this section, I have sought to read Ndebele's women in waiting and Kafka's man who waits for the law as stories that disrupt Hegelian the odicy wherein life objectively moves from bad to good. In other words, there is always a possibility that waiting itself can turn out to be an eternal duration[56]—'an endlessly fading twilight'.[57] I have attempted to demonstrate law's productive relationship with time wherein law plays a central role in how we view time and consequently how and why (South) African people continue to be subjected to a 'disenchanted world of deadening and meaningless routine'—what Dilip Gaonkar calls 'a Sisyphean world of repetition devoid of a subjectively meaning telos'.[58]

In the context of South Africa, we are getting closer to 400 years of waiting for justice for the dispossessed. The constitutional dispensation in South Africa, itself encumbered by historicist narratives, has so far proven true the fallacy of a linear teleological time. In the next section, I turn to the notion of refusal in the sense of refusing the dominant narrative of what waiting should entail. In other words, I suggest that the answer largely lies in the notion of refusal in the sense that South Africans have thus far refused to refuse. This is because to be deprived of ownership of natural resources is essentially to be unhomed and dislocated. It is to emasculate the essence of home. It is to be at home but outside the home. Refusal as a psychic revolt therefore becomes a solution.

Refusing to wait any longer

Karin van Marle, relying on Julia Kristeva and Hannah Arendt, talks about 'the damage of a society overtaken by law, human rights and constitutional discourse',[59] which results in 'complacent society where political action, thought, eternal questioning and contestation are absent and replaced by an understanding of freedom as mere commercial freedom and thought as calculated and instrumental'.[60] Refusal as 'revolt in a psychic sense refers to a state of permanent questioning, of transformation, change, and endless probing of appearances'.[61] The notion of refusal within the context of (South) Africa and within the context of the right to development and ownership of natural resources should start with, to borrow from Fitzpatrick, appreciating the fact that 'occidental history has itself been revealed as mythic, both in its arrogation of universal history and its specific constructions'.[62] Modernity's greatest success was its mapping of historical time as homogenous.

A deconstruction of the commonsensical notion of periodisation would reveal, according to Davis, that

> there was no such 'superstitious, feudal Middle Ages' prior to colonialism, and doubtless there never would have been such without colonialism; vice versa, colonisers could not have mapped and administered foreign lands and bodies as they did without the simultaneous process of imagining such a 'Middle Ages'.[63]

The invocation of the Middle Ages as a 'temporal marker' enables the unproblematic narration of linearity, civilisation, progress and development—and a particular version of the notion of development as constituting the end of history. It is thus important, as a point of departure, to expose the political trick that works by conceptualising and consolidating the feudal past and then rejecting it as a way of distancing Europe from its others.[64] In short, deconstructing the myth of periodisation, I suggest, should be the basis for refusal in the context of the right to development and ownership of natural resources.

South Africa's liberal democratic era is testimony to problems of periodisation anchored in historicism.[65] Since the 1990s, the approach to analysing South Africa has been predominantly historicist and it is this historicist approach that has constituted the framework wherein 'the experience of Europe and North America serves as the template' and 'democratic politics' is narrowly conceived to mean 'electoral practice'.[66] For Reddy, the predominant approach tends to analyse

> South African politics within a linear narrative of political modernity: colonialism to Apartheid to democracy. Post-World War II modernisation theorists praised capitalist modernity as the preferred model for the Third World, so post-Apartheid South Africa was seen to be joining (or maybe rejoining) the normal path.[67]

In terms of this view, the notion of development is imbricated in the historicist Hegelian theodicy which subscribes to a view of history where freedom is progressively achieved and all that is evil fades away.[68] The problem with a historicist approach is that, within the context of South Africa, it tends to miniaturise the present centrality of colonial apartheid by, at all costs, violently yanking South Africa towards 'the modern'. The past ceases to be the central problematic and the future becomes available only to those able to pull themselves up by their bootstraps. To deconstruct periodisation is to at the same time debunk the notion of a South Africa that is moving towards 'the end of history'. I therefore suggest that it is imperative that in appreciating refusal as 'revolt in a psychic sense' which 'refers to a state of permanent questioning, of transformation, change, and endless probing of appearances', we should critique notions of development and what it means to own so that we avoid, in Mbembe's words, pseudo-radical calls for ownership and return of land that are based on 'mimetic inversions of what was previously showered with malediction'.[69] Refusal in the context of the right to development and ownership of natural resources suggests that Africans should among others, refuse 'traditional ways of thinking and doing law',[70] or put in the language of Kafka's man before the law, refusal suggests that we refuse to wait for the gatekeeper to open entry into the law.

This insurrectionary refusal implores us to perhaps heed critical voices such as Balakrishnan Rajagopal who states as follows within the context of the notion of development:

> Given the centrality of colonialism and development in producing the architecture of modern international law, and their impact on substantive areas of law, a critical approach to international law must begin by closely interrogating the meaning, purpose, goals and means of development itself. This is not what one sees in international and transnational legal scholarship, such as 'law and development', where the meaning of development is rarely, if ever, critically engaged in all its historical complexity. I would suggest

that to be relevant to the third world, a critical approach to development is a key component of international law as a political project.[71]

Conclusion

In this chapter, I have attempted to deal with the two notions of the right to development and ownership of natural resources by firstly, demonstrating the complicity of the Western philosophical tradition from the era of European modernity in the unhoming of Africans. I have argued that periodisation of historical time has been central as a political technique to justify colonialism and that mediating this periodisation was law. In other words, periodisation is nothing but political, hence the notion of the politics of time. I have sought to critique the notions of periodisation of time and law's complicity in its maintenance by relying on fictional literature. In particular, I have recalled Ndebele's *The Cry of Winnie Mandela* to critique what Mark Antaki calls 'superficial understandings' and 'hasty celebrations of a legal modernity'. For Antaki, 'legal forms shaped by ideologies (such as apartheid) tend to survive even when those ideologies have ostensibly died or even been consciously disavowed'.[72] Moreover, law has the ability to mock and create conditions of possibility, which, like in Kafka's *Before the Law*, may only remain possible. In the end, my literary approach is informed by Stuart Motha's prescient observation 'I want to suggest that it is in literature, in the work of art, that the complexity of a political crisis, as well as the urgency and limits of critical responses, can be discerned'.[73]

Notes

1 Motha, Stewart. 2013. 'As if: Constitutional Narratives and "Forms of Life"'. In *Genres of Critique: Law, Aesthetics and Liminality*, edited by Van Marle, Karin and Motha, Stewart. Stellenbosch: Sun Press, 106.
2 *Ibid.*
3 *Ibid.*
4 *Ibid*, 103.
5 I am indebted to van Marle for use of the notion of refusal. Van Marle, Karin. 2009. 'Laughter, Refusal, Friendship—Thoughts on a "Jurisprudence of Generosity"'. In *Refusal, Transition and Post-Apartheid Law*, edited by Van Marle, Karin.Stellenbosch: Sun Media.
6 *Ibid*, 18.
7 I borrow these words from Pavalich, George. 2009.'Dissociative Grammar of Constitutional Critique'. In *Genres of Critique: Law, Aesthetics and Liminality*, edited by Van Marle, Karin and Motha, Stewart. Stellenbosch: Sun Press, 17.
8 Ndebele, Njabulo. 2013.*The Cry of Winnie Mandela*.Johannesburg: Picardo Africa, 65.
9 *Ibid*, 34.
10 *Ibid.*
11 Osborne, Peter. 1995. *The Politics of Time: Modernity and Avant-Garde*. London: Verso.
12 Ndebele, note 8, 65.
13 Bhabha, Homi. 1992. 'The World and the Home'. *Social Text* 31(32): 141–153.
14 Masgrau-Peya, Elisenda. 2004. 'Towards the Poetics of the "Unhomed": The House in Katherine Mansfield's "Prelude" and Barbara Hanrahan's "The Scent of Eucalyptus"'. *Antipode* 18(1): 60–66.

15 Fitzpatrick, Patrick. 1992. *The Mythology of Modern Law*.New York: Routledge, 82.
16 *Ibid.*
17 *Ibid.*
18 *Ibid.*
19 Bangstad, Sindre and Nilsen, Tumyr. 2018. 'Thoughts on the Planetary: An Interview with Achille Mbembe'. *Klasse Kampen*, 30 November <www.newframe.com/thoughts-on-the-planetary-an-interview-with-achille-mbembe/>.
20 Davis, Kathleen. 2008. *Periodisation and Sovereignty: How Ideas of Feudalism and Secularism Govern the Politics of Time*. Philadelphia: University of Pennsylvania Press, 20.
21 Fitzpatrick, note 15, 104.
22 *Ibid*, 103.
23 *Ibid*,105.
24 Fitzpatrick, note 15, 107 quoting Lord Lugard. 1965. *The Dual Mandate in British Tropical Africa*.Abingdon: Frank Cass and Co, 546–547.
25 Davis, note 20, 1.
26 *Ibid.*
27 *Ibid*, 5.
28 Davis, note 20, 23.
29 *Ibid*, 20.
30 *Ibid.*
31 Van Marle, Karin. 2003.'Law's Time, Particularity and Slowness'. *South African Journal of Human Rights* 19(2): 241.
32 Okri, Ben. 1999. *Infinite Riches*. Frankfurt: Phoenix Publishing, 5.
33 Scott, David. 2004. *Conscripts of Modernity: The Tragedy of Colonial Enlightenment*. Durham: Duke University Press, 29.
34 *Ibid*, 29.
35 Pahuja, Sundya. 2011. *Decolonising International Law: Development, Economic Growth and the Politics of Universality*. Cambridge: Cambridge University Press, 1.
36 Beynon-Jones, Seân and Grabham, Emily. 2019. 'Introduction'. In *Law and Time*, edited by Beynon-Jones, Sean and Grabham, Emily. New York: Routledge, 19.
37 *Ibid*, 19.
38 Fitzpatrick, note 15, 72.
39 *Ibid.*
40 *Ibid.*
41 *Ibid.*
42 Fitzpatrick, note 15, 72.
43 *Ibid*, 76.
44 *Ibid.*
45 *Ibid*, 79.
46 *Ibid.*
47 *Ibid*, 80.
48 Reddy, Thiven. 2015. *South Africa, Settler Colonialism and the Failures of Liberal Democracy*. London: Zed Books, 13.
49 Sect 25 of the Constitution of South Africa deals with land and the pillars of land reform in South Africa are redistribution, restitution and tenure.
50 Reddy, note 48, 2.
51 Ndebele, note 8, 99.
52 *Ibid*, 35.
53 Douzinas, Costas and Gearey, Adam. 2005. *Critical Jurisprudence: The Political Philosophy of Justice*. London: Hart Publishing, 356.
54 *Ibid.*
55 Douzians and Gearey, note 53, 357.
56 Gaonker, Parameshwar. 1999.'On Alternative Modernities'. *Public Culture* 11(1): 13.
57 *Ibid*, 13.

58 *Ibid*, 8.
59 Van Marle, note 5, 17.
60 *Ibid*.
61 *Ibid*, 18.
62 Fitzpatrick, note 15, 101.
63 Davis, note 20, 20, 23.
64 *Ibid*, 23.
65 I borrow Reddy's use of historicism as a reference 'to a conception of history drawn from the European experience with the endpoint of modernity defined as Western liberal democracy. In historicism, the Western experience of the modern serves as the universal template. It occupies a space of privilege and functions as the reference point against which other histories are understood. In its aim to demonstrate a universal process, a historicist frame emphasises those attributes familiar to the Eurocentric historical narrative whilst marginalising or excluding features that do not fit this narrative'; Reddy, note 48, 56–57.
66 Reddy, note 48, 41.
67 *Ibid*, 42–43.
68 Pellican, Luciano. 1994. *The Genesis of Capitalism and the Origins of Modernity* (James, C. trl). New York: Telos Press, 13.
69 Mbembe, Achille. 2017. *Critique of Black Reason*. Johannesburg: Wits University Press, 7.
70 Van Marle, note 5.
71 Rajagopal, Balakrishnan. 2006. 'Counter-Hegemonic International Law: Rethinking Human Rights and Development as a Third World Strategy'. *Third World Quarterly* 27(5): 783.
72 Antaki, Mark. 2009. 'From the Bridge to the Book: An Examination of South Africa's Transformative Constitutionalism's Neglected Metaphor'. In *Genres of Critique: Law, Aesthetics and Liminality*, edited by Van Marle, Karin and Motha, Stewart. Stellenbosch: Sun Press, 43.
73 Motha, note 1.

References

Antaki, Mark. 2009. 'From the Bridge to the Book: An Examination of South Africa's Transformative Constitutionalism's Neglected Metaphor'. In *Genres of Critique: Law, Aesthetics and Liminality*, edited by Van Marle, Karin and Motha, Stewart. Stellenbosch: Sun Press, 49–74.

Beynon-Jones, Seân and Grabham, Emily. 2019. 'Introduction'. In *Law and Time*, edited by Beynon-Jones, Sean and Grabham, Emily. New York: Routledge, 1–28.

Bhabha, Homi. 1992. 'The World and the Home'. *Social Text* 31(32): 141–153.

Davis, Kathleen. 2008. *Periodisation and Sovereignty: How Ideas of Feudalism and Secularism Govern the Politics of Time*. Philadelphia: University of Pennsylvania Press.

Douzinas, Costas and Gearey, Adam. 2005. *Critical Jurisprudence: The Political Philosophy of Justice*. London: Hart Publishing.

Fitzpatrick, Patrick. 1992. *The Mythology of Modern Law*. New York: Routledge.

Gaonker, Parameshwar. 1999. 'On Alternative Modernities'. *Public Culture* 11(1): 1–18.

Lord, Lugard. 1965. *The Dual Mandate in British Tropical Africa*. Abingdon: Frank Cass & Co.

Masgrau-Peya, Elisenda. 2004. 'Towards the Poetics of the "Unhomed": The House in Katherine Mansfield's "Prelude" and Barbara Hanrahan's "The Scent of Eucalyptus"'. *Antipode* 18(1): 60–66.

Mbembe, Achille. 2017. *Critique of Black Reason*. Johannesburg: Wits University Press.

Motha, Stewart. 2013. 'As if: Constitutional Narratives and "Forms of Life"'. In *Genres of Critique: Law, Aesthetics and Liminality*, edited by Van Marle, Karin and Motha, Stewart. Stellenbosch: Sun Press, 89–106.

Ndebele, Njabulo. 2013. *The Cry of Winnie Mandela*. Johannesburg: PicardoAfrica.

Okri, Ben. 1999. *Infinite Riches*. Frankfurt: Phoenix Publishing.

Osborne, Peter. 1995. *The Politics of Time: Modernity and Avant-Garde*. London: Verso.

Pahuja, Sundya. 2011. *Decolonising International Law: Development, Economic Growth and the Politics of Universality*. Cambridge: Cambridge University Press.

Pavalich, George. 2009. 'Dissociative Grammar of Constitutional Critique'. In *Genres of Critique: Law, Aesthetics and Liminality*, edited by Van Marle, Karin and Motha, Stewart. Stellenbosch: Sun Press, 31–48.

Pellican, Luciano. 1994. *The Genesis of Capitalism and the Origins of Modernity*. (James, C. trl). New York: Telos Press.

Rajagopal, Balakrishnan. 2006. 'Counter-Hegemonic International Law: Rethinking Human Rights and Development as a Third World Strategy'. *Third World Quarterly* 27(5): 767–783.

Reddy, Thiven. 2015. *South Africa, Settler Colonialism and the Failures of Liberal Democracy*. London: Zed Books.

Scott, David. 2004. *Conscripts of Modernity: The Tragedy of Colonial Enlightenment*. Durham: Duke University Press.

Van Marle, Karin. 2003. 'Law's Time, Particularity and Slowness'. *South African Journal of Human Rights* 19(2): 239–255.

Van Marle, Karin. 2009. 'Laughter, Refusal, Friendship—Thoughts on a "Jurisprudence of Generosity"'. In *Refusal, Transition and Post-Apartheid Law*, edited by Van Marle, Karin. Stellenbosch: Sun Media, 15–28.

4 Rationale for Africa's 'legitimate' ownership of her natural resources

Daphine Kabagambe Agaba

Introduction

Africa is endowed with a multiplicity of natural resources. The continent ranks first or second in quantity as well as production of diamonds, cobalt, manganese, gold, uranium, petroleum, vermiculite, aluminium, phosphate rock, bauxite and platinum group minerals (PGM), to name a few.[1] The actual mineral capacity of Africa is not fully known because large tracts of its land are still unexplored. Yet, with a few exceptions, these resources do not translate into bringing about sustainable development on the continent. A series of reasons have been forwarded for this set of circumstances, including the resource-curse concern and poor governance, among others. On the contrary, this chapter argues that the inability to translate Africa's natural resources into sustainable development is attributable to the failure to fully understand or engage with African communities, the specific contexts within which Africans live, their specific challenges and their ethos (cultures, values, norms). This translates into the blind adoption of policies and measures that do not directly contribute to development within these contexts.

The critical question to be asked where extraction has not yet happened is whether the extraction itself is even important. It is vital from the beginning to critically establish whether the proposed extractive activity is the best economic strategy for the particular area or whether some other economic initiative could be more relevant for that context.[2] This requires undertaking environmental, social and human rights impacts assessments and genuinely negotiating agreements that are beneficial for local communities. Unfortunately, these critical questions are usually not asked as the driving force is usually the economic gain to the political elites and corporations, most times to the detriment of the communities living in the areas of extraction. Several reasons are provided for this set of circumstances including; uncritical acceptance of external development strategies, the role of Africa in the world, which for the most part is dependent on the outside, the powerful influence of the extractive industry, Africa's leaders who are, often, not committed to transformational leadership, Africa's blind adopting governance policies from without rather than from within and the exclusion of local communities from the relevant decision-making processes.

This chapter argues for the need for Africa to fully 'own' her natural resources. Africa's historical positioning as a policy taker rather than as a policy maker in addition to its reliance on the outside world hinders effective and independent management of its resources, which translates into the unfortunate scenario of the inability to transform its natural resource potential into sustainable development. The chapter also critiques the 'good governance' narrative advanced by 'Western nations' as a solution to what is largely regarded as the 'resource curse' concern, a narrative that does nothing but cover up the real intention of foreign extractive companies in seeking to maximise profits accruing from the extraction of natural resources. The chapter posits that without forwards-looking practices or transformational leadership coupled with critical strategies aimed at dismantling the power imbalances that abound in the extractive sector, the policies adopted will continue to be 'modified' replicas of what is available and, thus, achieve little to no success.

Ghosts of the past

Due to its disadvantaged historical positioning, Africa continues to be a bystander in charting its own path. As far back as 1973, Rodney strongly cautioned that the only way Africa could fully develop is by effectively breaking the vicious cycles of dependence and exploitation which are a key characteristic of its relationship with the outside world in collusion with local political interests. Rodney expounded on the fact that Europe's development was and is at the expense of Africa's underdevelopment. Tracing Europe's relationship with Africa back to the late 15th century, Rodney defines it as one that was principally about the transfer of wealth including gold, trading in slaves and ivory from Africa to Europe, thus, enabling development in Europe and the subsequent underdevelopment of Africa.[3] The trade in slaves was instrumental to Europe's development as the slaves were made to work in cotton and sugarcane plantations, build ships and the towns and cities of Europe and America. In the 19th century when slavery was abolished, Europe resorted to the colonisation of territories and enclaves in Africa in search for raw materials ranging from palm products, cotton, rubber and groundnuts among others for its growing economy.

Ocheni and Nwankwo assert that industrial capitalism, which was pioneered in Europe, multiplied the demand on world resources as it fostered Europe to seize for her factories agricultural and mineral resources from other parts of the world.[4] Ironically, this arrangement has persisted to date. There has been and continues to be an increasing call for the industrialisation of Africa. Efforts are being undertaken to see the fruition of this project with some commendable efforts in some countries. Yet, for the most part, Africa still stands in an inferior and disadvantaged economic position, eerily similar to its colonial past of exporting raw materials at very meagre prices and importing finished goods and products at exorbitantly higher rates. It has been re-echoed that despite Africa being the primary producer of minerals such as gold, platinum, diamond and cobalt, most of these materials are still exported in their raw form to developed and emerging economies like China.[5]

Taylor debunks the misguided notion that Africa is on the rise, which relies on measuring Africa's gross domestic product (GDP) without taking into consideration the quality of growth.[6] The emphasis on GDP is obsessed with economic growth to the detriment of poverty eradication, environmental conservation and social justice. This GDP increase is overwhelmingly characterised by the accumulation and inflow of capital-intensive investment for the purpose of extracting and exporting natural resources. There is a visible absence of value addition on the African side, a situation similar to that described by Rodney in 1973. Giuliani makes reference to scholars such as Sen and Piketty who assert that economic growth does not necessarily lead to improved well-being but may rather lead to increased inequality.[7]

Africa's position in the global economy is, according to Taylor, still an unfavourable one characterised, to a large extent, by the export of agricultural products such as coffee, cocoa, peanuts, cotton, oil palm, fruits, oil and minerals such as diamonds, gold, copper and other rare metals, with China joining the list of the key consumers.[8] He asserts that commodities still make up 80 percent of Africa's exports, and despite the 'Africa rising trope' that is lauded around, its level of manufacturing and industrialisation is still very low. He affirms that the current growth and governance models that are pursued in sub-Saharan Africa are concerned with profit-making, capital accumulation and market-based actors, the key tenets of capitalism. As a result, despite seemingly rising GDP levels in sub-Saharan Africa, poverty, unemployment, hunger and starvation levels are still on the rise as well as unequal access to education, health, water and sanitation facilities, showing a decline in human development.

Stephens asserts that transnational corporations (TNCs) rely on and thrive on the divided nature of states, which is the case with Africa.[9] Corporations directly and indirectly influence negotiations on patent protection and trade agreements and subsequently, international economic policy. These enterprises use their advantages which include advanced bargaining skills, clearly formulated purposes and vast knowledge and experience to the detriment of inexperienced, divided governments. To their disadvantage, states tend to have varying interests that hinder the effective formulation of a common policy that could give them some form of advantage over multinational enterprises. The inherently international nature of international corporations ultimately presents a challenge in formulating regulations. Of course, there have been measures to alleviate this imbalance; although, it is not surprising that industrialised countries have been at the centre of discouraging the adoption of a binding treaty aimed at regulating activities of corporations.[10]

The *SERAC v Nigeria* case provides a vivid example of how a repressive African government colludes to protect the interest of an exploitative TNC against its own vulnerable population.[11] In the *SERAC* case, the military government of Nigeria together with Shell Oil, a subsidiary of Royal Dutch Shell was found to have contributed to environmental degradation resulting in health problems, including respiratory ailments, skin infections, risks of cancers, neurological and reproductive challenges among others, as a result of oil spills that contaminated the environment inhabited by the Ogoni people. Another example is the 1984 Bhopal incidence in India where the negligence of the US Company, Union Carbide led to the death and injury of thousands of people and animals in addition to causing environmental damage.[12] These are a few cases that have come to light on how the insidious ways in which corporate entities produce negative challenges on the people, countries and economies remain hidden.[13]

Africa as 'policy taker' instead of 'policy maker'

The African extractive sector is stuck in a cul-de-sac whereby attempts to emancipate itself from dependence and exploitation end up right where they started. Gumede emphasises that Africa has failed to understand the impact of ideological dominance as well as power relations between political economies of the colonially created micro-states with the West and other emerging economies.[14] Thus, there is need for an elaborate programme of action aimed at un-learning, un-thinking and re-thinking dominant thought patterns acquired over the last 500 years or so of interacting with imperial powers.

In order to accentuate Africa's role as a policy maker rather than a policy taker, efforts have been made to ensure extractive work sustainably develops the continent. An example of such an attempt was the elaboration of the African Mining Vision (AMV) in 2009 at the African Union Summit.[15] The AMV aims at the reformulation of energy and resource governance in Africa in order to ensure sustainable development in the industrialisation process. Additionally, in order to ensure that the AMV is practicalised, the African Minerals Development Centre (AMDC) was established in 2013.[16] The adoption of the AMV and the AMDC is a commendable step towards Africa owning its policy formulation processes. However, critics like Ambe-Uva have expressed wariness about the ability of the AMV in bringing about genuine transformation in the extractive sector.[17] They wonder whether it is not just 'new wine in old wine skin', especially owing to the fact that its implementation is based on the voluntary commitment of mining companies.

This criticism is not new as it also points to other policies that have been adopted at the African level but have largely not been successful due to the fact that they always envision their funding coming from without. Gumede critiques the African Union (AU) African Agenda 2063 for envisioning its assistance coming from without and forwards that it should have been steeped in a philosophical framework unique to Africa.[18] Indeed, while efforts have been undertaken to collect sources internally, the majority of such policies depend on external resources for their implementation. For instance, it was reported that only 33 percent of the AU's operational budget is funded by member states.[19] The bigger chunk is funded by donors; with the European Union (EU) marked as the biggest donor, a few EU member states and other emerging donors including China, South Korea, India and Turkey, among other sources.[20] The initial funding for the African Minerals Development Centre for funding the AMV came from Australia and Canada.

Another key factor in the AMV coming to fruition is for the implementing entities to ensure that whatever country's mining visions are adopted, such visions are extensively researched, non-elitist and localised (contextualised). Due to their failure to consider contextual challenging issues in areas where natural resources are found, the formalised externally formulated governance mechanisms have been noted to achieve little or no success.[21] For instance, the Kimberly Process—a certification scheme aimed at avoiding conflict and preventing 'blood diamonds' from getting into the global supply chain—has not achieved much in reversing the poor socio-economic conditions of the communities of Sierra Leone. Instead, the extraction of gold and diamond produced a meagre 0.1 percent of GDP from royalties in 2016, an unfortunate fact that has been attributed to tax evasion by the biggest diamond mining companies. The artisanal sector also has not fared much better in terms of achieving results aimed at improving human security.[22]

It has been forwarded that there is limited implementation of the AMV as a result of muted response from actors within and beyond Africa. One of the reasons advanced is the possibility of mining companies being discouraged from actively participating due to the requirement to invest large sums in projects aimed at preserving the environment and the dignity of the regions in which they operate.[23] This challenge points to the credence given, by mining companies, to capital accumulation over environmental and societal concerns.[24] Busia and Akong caution about the danger of over relying on Africa's geological endowment as the solution to Africa's challenges of underdevelopment, poverty alleviation and industrialisation.[25]

Relations of resource-rich countries in Africa with the developed world have not changed much from the colonial times. This is a relationship of mostly unequal playing fields always rife for exploitation and massive gain/profit for developed countries and little or worse-off situation for African resource-rich countries. A vivid example of this is the liberalisation programmes that took place in the 1980s and 1990s in the resource-rich countries aimed at attracting foreign direct investment.[26] The liberalisation mechanisms not only shifted power from the state to private actors but also redefined relations between elites and the local communities. Campbell and Hatcher argue that despite countless attempts to contest regulatory frameworks, 'neo-extractivist' mining methods have persisted, characterised by alignment to foreign market demands, limited domestic processing, enormous environmental and social costs whilst forwarding the discourse of poverty reduction and social justice.[27]

Guinea is a prime example, which boasts of enormous reserves of high-grade bauxite, large tonnes of high-grade iron ore, uranium, gold and diamond. In 1995, the World Bank strongly favoured a liberalised mining code that redefined the functions of the state and introduced new tax concessions aimed at attracting foreign investment. Not surprisingly, these non-transparent mining activities sparked popular discontent, which gradually increased over the years. These grievances were spurred on by unfavourable living conditions characterised by lack of employment and high prices for basic commodities.[28]

Adeleke forwards that lack of accountability in the extractive industry triggers illicit financial flows as estimated by the Africa Progress Panel, which reports that in 2013, Africa lost twice as much in illicit flows as it received in foreign aid.[29] Furthermore, tax rates have been noted to be on the lower side. In 2012, Zambia is reported to have collected only USD240 million in tax revenue from copper mining despite exporting USD10 billion. The tax revenue collected amounted to only 2.4 percent of the export rates. Similarly, Guinea collected a meagre USD48 million (0.4 percent of the export value) despite exporting USD1.4 billion. In 2012, the African Development Bank reported that while the average global tax was 45 to 65 percent, the average mining royalty in Africa, was an insignificant 3.5 percent. These figures are an indicator of revenue lost that could possibly have contributed to the development of other sectors.

Hilson and Maconachie further assert that the poor performance of Africa's resource-rich countries is more attributable to inequitable mining codes than poor governance.[30] They note that host countries derive nominal profits from their large-scale mining projects as a result of the favourable tax breaks and low royalty payments they were receiving from mining companies. The AU/Economic Commission for Africa (ECA) High Level Panel Report revealed excessive under-reporting of the quality and quantity of minerals extracted for export.[31] The report estimates that USD50 billion leave Africa annually, a sum that is higher than the overall external development assistance to the continent. This is as a result of tax avoidance, tax evasion, market power abuses and mis-invoicing.

While Africa participates in international fora, its influencing power is very minimal and thus, the adopted international tax rules rarely address Africa's key priorities. For example, Africa had minimal participation in the formulation of the Base Erosion and Profit Sharing Action Plan developed by the Organisation for Economic Cooperation and Development (OECD) and the G20 countries.[32] Ultimately, Africans have been unable to harness their natural resources to produce industrialised and diversified economies. In the period between 1980 and 2009, the percent of manufacturing value added to the African GDP fell from 16.6 percent to 12.6 percent, respectively, in sub-Saharan Africa. It is estimated that the value of a range of processed products was approximately 400 times higher than the equivalent unit value by weight of the raw material.[33]

Such unfavourable policies are usually covered up with misleading narratives such as the 'Africa rising' one. Ndlovu-Gatsheni decries the impulsive celebration of Africa's economic growth as development.[34] The 'Africa rising' trope conceals the new scramble for Africa's natural resources together with land grabbing perpetuated by Euro-north American powers and multinational corporations that have now been joined by emerging powers from the global south. He emphasises that in order for Africans to create their own futures, there is need to ceaselessly fight the all-encompassing matrix of global coloniality. Research has shown that there is no blueprint for policy making in countries with weak institutions.[35] Therefore, policies that have been successful in developed countries cannot blindly be transplanted to the African context. In such a context, the motivation to observe formal rules is often not present as policy choices are usually largely influenced by informal institutions.

Regrettably, advocacy efforts for local ownership mostly only focus on domestic ownership rather than local value addition and most have not considered the benefit of adding a regional dimension to their policies.[36] The process of local content development has also been captured by politically connected elites in countries like Mozambique and Ghana. These elites use their access to information to put in place small- and medium-sized enterprises just for the purposes of winning contracts for the importation of supplies to the mining industry, taking the opportunity away from those entrepreneurs that are genuinely keen to manufacture and add value to products locally.

Unmasking 'good governance' as a response to the natural 'resource curse' argument

The notion of Africa's inability to harness its natural resources has been referred to as the 'resource curse'. This rentier state theory forwards that resource-rich countries are usually characterised by corruption and authoritarianism and are subsequently, prone to conflict. These same governments also use strategies such as patronage and repression of the masses. This argument has been challenged by scholars such as Manzano and Rigobon who set out that the poor performance of resource-rich countries emanates from the fact that these countries acquired foreign debts in the 1970s while commodity prices were high, leading to a debt overhang when the prices fell in the 1980s. Subsequently, they attribute the poor performance of resource-rich countries to unfavourable macroeconomic policies rather than to natural resources.[37] Based on this, emphasis has increasingly been shifted to issues of leadership and good governance.

It is asserted that what ultimately brings forth the resource curse is the failure to govern natural resources effectively.[38] These governance issues include non-transparency of the signed agreements, divisive political interests, inadequate capacity to negotiate investment agreements, weak institutional capacity, inability to transform resource wealth into broader societal development and limited capacity to put in place and enforce a sound legal and regulatory framework.[39] In several cases, mining companies are more skilled and more resourced as they come to the negotiating table with larger terms as well as the advantage of having back-up support from well-equipped lead firms at their headquarters. This unequal footing is a breeding ground for exploitation and corruption.

Consequently, the 'good governance' concept has found its way to the centre of natural resource management, which in Africa faces the same dilemma outlined above, that is, the lack of local ownership. Ware alludes to the fact that in forwarding the good governance ideology, donor countries seem to be telling least developed countries that the way to achieve development is to magically turn into developed countries since the ideas of governance they outline are only feasible in high-income countries.[40] In some instances, the term is deliberately vague as it provides leeway for aid donors (for political reasons) to continue assisting governments with poor governance records. Ware quotes Issa Shivji who argues that governance is essentially an issue of power relations.[41]

Western nations underplay the role of power and power relations and instead present it as good governance.[42] Ultimately, good governance is presented as a moral paradigm sprinkled with achieving good outcomes and hence, the addition of the word 'good' in the governance paradigm. Kumawu-Abiwu also forwards that the good governance narrative is idealistically long with unrealistic demands but also has, for a long time, been based on Western-supported development agencies and then unsystematically embraced by developing nations.[43] In their misguided attempts to promote good governance as a solution for development and poverty alleviation, agencies like the World Bank give little direction on which measures are urgent and which ones are not, what can realistically be achieved in the short or long term and what may not be feasible. Grindle forwards that there is need for a more nuanced understanding of what governments are capable of and the process of evolution of institutions.

The good governance agenda is even more preposterous in the mining context Courmans illustrates that countries like Canada avail of public funds for programmes supposedly aimed at improving natural resource governance in countries considered to have 'weak governance'.[44] However, these programmes explicitly support extraction activities by mining companies that originate from their domain and also undertake to enhance their investment conditions in these countries even when their activities lead to adverse environmental and social consequences especially on the indigenous populations. Concerns raised on the negative impact of mining including militarisation of mine sites, environmental degradation, abuses by public and private mine security, and threats and attacks on environmental activists are thus usually ignored.

Courmans further contends that while the resource curse argument may have some merit to the extent that governments in resource-rich countries often lack the capacity to effectively manage and govern resources from mining to enable sustainable development through benefits sharing. This line of thought obscures the negative effects of the extractive industry itself which include diversion of wealth from extraction activities into global tax havens, evading taxes and royalties through confidential investor–state contracts, unnecessarily long tax holidays, maintaining taxes at minimum levels and undermining legislation aimed at protecting communities from the negative impacts of mining.

Even if they purport to advance good governance, these same companies undermine governance by interfering with national and local government's efforts to respond to communities' demands for protection of critical resources, such as water, from mining or enforcing regulations aimed at prohibiting mining in some water catchment areas. In their attempts to do so, the host countries find themselves in the midst of lawsuits by the well-resourced mining companies in what is referred to as investor–state dispute settlement tribunals. The threat and use of such tribunals as well as the exorbitant amounts requested as compensation ultimately undermine local governance efforts in the host states. When local communities attempt to uphold good governance ideals with consequences that don't favour proponents for mining, their attempts are met with violence from the police, the military and paramilitary forces. Not forgetting that the countries that purport to promote good governance in other developing states also have governance challenges accruing from extraction in their own countries.[45] For instance, Canada, as a result of its extensive extractive activities, faces challenges similar to those in the countries purported to have weak governance. These include conflict and protests by indigenous populations, tax evasion scandals and corporate capture of Canadian regulatory authorities.

The Extractive Industry Transparency Initiative (EITI) falls into this category as it is predicated on the idea of promoting good governance.[46] The EITI is based on the premise that development failures in resource-rich countries are a result of corruption and misappropriation of resources. The EITI was launched in 2002, as a tool meant to turn around the deteriorating economies of natural resource-rich continents like Africa, Latin America and Asia. This was to be done through publication and verification of company payments as well as government revenues of gas and mining, and oil as a means of fostering more effective management of such resources.

In their critique of the EITI, Hilson and Maconachie point out that the initiative was a welcome respite for the World Bank, mining companies and G8 governments as it diverted attention from the increasing condemnation that the World Bank was getting (through its financial branch, the International Finance Corporation) for its unscrupulous involvement in the extractive industry. The launch of the EITI drew attention away from the bank to the governments of the developing world. This trend has continued with donors shying away from holding Western countries and foreign companies that control operations accountable. The focus on promoting good governance in the developing world thus takes attention away from the insufficient incomes and revenues received from oil and mining activities to the mismanagement of the miniscule royalties they receive. Even more hypocritical is the fact that the calls for accountable governance do not detract developed nations from investing in countries that have been deemed to be highly corrupt and unaccountable. The mismanagement of resources and corruption in oil-rich countries such as Nigeria and Chad have not discouraged investment in these sectors. Political instability and autocracy in the Democratic Republic of Congo (DRC) and

Sierra Leone have also not reduced or even deterred Western interest in their diamonds and other natural resources.[47]

According to Poncian and Kigodi, the adoption of the EITI, while commendable, has not led to the lowering of corruption as well as improving equity as a result of the exploitation of natural resources.[48] Nigeria, one of the EITI member countries, has enacted various favourable legislation but is still considered one of the most corrupt countries in the world with worryingly high poverty and inequality levels. Research conducted in 2013 on monitoring the implementation of the EITI in Tanzania reveals that of the 59 mining companies operating there, only 16 had paid corporate tax. In the same financial year, it was reported that USD8 billion was estimated to have illegally been drained from the country, most of it as a result of tax evasion.

The inclusion of environmental clauses in country regulations such as section 24 of the South African Constitution and its various mineral and environmental management regulations have played a miniscule role in eradicating poverty, inequality and putting an end to untenable mining conditions.[49] These same environmental clauses have also not made a substantial contribution to transforming accountability and transparency initiatives into tangible results. Poncian and Kigodi note that the supposed benefits of transparency initiatives in the extractive industries are uncertain.[50] For instance, the process of making information available on finances received by the government from extractive activities does not necessarily result in taking the required action or an increase in demanding accountability for how the revenues are spent. The same government (in this case, the Tanzanian government) that purports to prioritise mechanisms aimed at improving transparency and accountability in the extractive sector has been noted to crack down or violently respond to such initiatives from the media, opposition and civil society.

Transformational leadership

The role of African leaders in undermining Africa's development cannot be underplayed. One of the major ways this happens is through lack of accountability, which leads to diversion of massive amounts of resources meant to benefit their national economies to their private coffers.[51] The OECD estimates that 25 percent of the GDP of African Countries, amounting to approximately USD148 billion is lost to corruption on an annual basis, exemplified by corrupt leaders sending money overseas equivalent to half of foreign debt and far above the amount of foreign aid to Africa.[52] The African ruling elites exploit the loopholes in the tax system, operating internationally, to illegally transfer diverted funds from the extractive sector to safe havens abroad. This is demonstrated by financial links in 2016, which revealed a list of 18 highly placed government officials in resource-rich countries with bank accounts stashed with multimillions held by major banks in Panama.[53]

This unfortunate set of circumstances was alluded to by the literary giant, Wole Soyinka. He stated that African dreams of peace and prosperity have been lost on account of corrupt and greedy strongmen and that African demands are basic in nature, including a clean environment, education, health care and poverty eradication. These have been continually ignored and sidelined by power-crazy leaders whose only aim is to achieve their selfish goals while oppressing the masses.[54] This often results in periodic spouts of political instability in the majority of resource-rich countries like the DRC and Sierra Leone that are characterised by opaque governance, corruption and gross mismanagement of revenue flows, with countries such as Nigeria and Chad ranking among the most corrupt in the world.[55] Cameroon and Gabon are characterised by the prolonged stay in power of their leaders while the governing elites are preoccupied with maintaining authority and control over valuable productive resources through patronage networks that guarantee their privileged positions.[56] In Cameroon for instance, President Paul Biya has been in office for a whopping 39 years, having assumed office in 1982. Not surprisingly, Cameroon is considered among the world's most corrupt countries.

With the exception of a few, most African countries inherited leadership systems that are for the most part flawed. Ndlovu-Gatsheni asserts that the several years of Africa's exploitation have only produced a style of leadership of the 'petit bourgeoisie', largely unable to transform political, economic and social institutions from colonialism into their own image.[57] Thus, African leaders' grip on power is enabled by their continuous straddling of external and internal powers, and often the external powers outweigh internal constituencies, expressed in the following wording:

> The power of coloniality as a structure of control, is that it speaks for us so forcefully that we see no recourse but to represent it, uphold its existence, to ratify its dispensing with ethics and with the sanctity of human life in everything we say and do as labour and resource.[58]

Gumede proposes a trio of instruments implemented together as vital for transforming African leadership and essentially moving Africa forwards; thought leadership, thought liberation and critical consciousness.[59] He defines thought leadership as the kind of leadership that is historically grounded, responsive to cultural diversity and therefore contextually situated based on progressive ideas with rational and impact appeal. Thought liberation is a call for the rediscovery of self (especially relevant in the context of Africa) as an individual capable of producing progressive thought, action and achievements. Critical consciousness is the ability to analyse and perceive economic, political and social contradictions and taking action against oppressive elements of reality. He further emphasises that raising Africans' critical consciousness requires deconstructing their sense of being away from political apathy, collusion with domestic and transnational elites, uncritically adopting external development strategies and leaders' refusal to take full accountability for their actions and consequences.

In forwarding the rare leadership theory as a response to Africa's leadership challenge, Ngambi emphasises the need for engaging with current realities and doing things differently and innovatively.[60] For him, Africa's leadership is characterised by incompetence, a strong sense of self-involvement, lack of concern for followers, is unethical, lacks vision and is immoral. He advocates for the kind of leadership that is responsible, accountable, relevant and ethical. This kind of leadership challenges the status quo, connects to people in a personal and spiritual way, and acknowledges that what may be relevant in one context may not be so in another context. It keeps in touch with the environment, embraces diversity and is adaptable and innovative.

De Sa highlights that transformational leadership is especially crucial because the extraction of mineral resources necessitates high initial investments coupled with undefined payback periods and the fact that mineral resources are not renewable.[61] Some of the initiatives that can be taken by governments to realise this include supporting local governments to better perform their roles (environmental and social management, monitoring mining operations), sharing infrastructure between sectors and supporting small and medium enterprises. Countries such as Botswana, Namibia and Ghana, despite the challenges they face, have prioritised such mechanisms to ensure that the gains from the sector are converted into other forms of capital and thus, enhance economic and social development.

Another vital aspect of transformational leadership is bringing communities back to the centre of the debate. Essentially, the ethos and values of indigenous African communities that reside in areas where resources are discovered should not be discounted or marginalised at the expense of profit maximisation. The process of ensuring that communities are the primary beneficiaries from the extractive industry is not an easy one. Sometimes, it involves protests against the governments in place and mining companies, which in some cases turn violent, leading to the massive loss of lives. One such violent protest was the wage-related protest between miners and the South African police at the Lonmin Marikana mine that led to the dead of 44 mine workers. The Commission of Inquiry that was set up to investigate the killings, found that the failure of the mining company (Lonmin) to comply with housing obligations created discontentment and unrest among the employees leading to the protests, which unfortunately became fatal.[62]

The importance of community inclusion in the extractive industry has been echoed over and over again, leading to the coining of the term 'community based natural resource management (CBNRM)'.[63] Harvey emphasises the need for the recognition of people's dignity through free and informed consent prior to commencing extractive work. There is need for comprehensive sharing of information with affected communities and equitable decision-making processes that build trust in communities about the system. One of the major drawbacks of the CBNRM has been the superficial implementation of downwards decentralisation, that is, the failure to effectively transfer decision-making power and control to the local level, which renders most decentralisation efforts superficial.[64] Usually, livelihood in these communities which are often already marginalised, deteriorates further upon the commencement of extracting operation.

Numerous injustices are meted out on them, ranging from eviction without adequate compensation, threats on life and environmental degradation perpetuated with adverse health consequences in the communities. The communitarian values of kinship, which could be utilised to enhance extraction, are usually disregarded. Obeng-Odoomuses Marx's capitalist critique to explain land grabs in Africa.[65] Large tracts of land continue to be appropriated from people for whom land is more than just a commodity for commercial farming, and developmental purposes. Odoom traces this to the transformation of societies from feudal to capitalist societies, entailing the displacement of peasants from their lands and transforming them into the industrial working class. A vivid example of this is the illegal land grabs and rampant evictions that have increased as a result of the discovery of oil in the Albertine region in Uganda.[66] The envisaged commencement in the extraction of the oil has triggered illegal and violent land evictions without compensation, threats on the lives of the occupants of the land and the use of police and the army to manhandle the occupants. In doing so, there has been blatant disregard of the 1995 Constitution, which recognises the right of occupancy of the bonafide occupants and the Land Amendment Act of 2010, which sets out that bonafide occupants can only be evicted upon failure to pay ground rent.

Conclusion

This chapter calls for the legitimate rather than superficial ownership of Africa's natural resources by Africans. A position where Africa is fully in charge of deciding the most appropriate strategies for its historical, environmental, cultural, political and socio-economic context. Legitimate ownership and management of its extractive sector is envisioned to translate into effective harnessing of the resources for the sustainable development of African communities and not just a few well-placed elites in collusion with the companies that operate within the extractive industry and their allies. It is suggested that there is a need for transformational leadership that is accountable to the local population, historically grounded, contextually relevant and capable of engaging deeply rather than superficially with the various facets of the extractive sector.

Notes

1 Pedro, Anthonio. 2016. 'The Africa Mining Vision as a Model for Natural Resource Governance in Africa'. In *Governance of Natural Resources for Africa's Development*, edited by Besada, Hany. New York: Routledge, 13–38.

2 Ingram, Joseph. 2016. 'Canada and China: Accepting Africa's Rise?' In *Governing Natural Resources for Africa's Development*, edited by Besada, Hany. New York: Routledge, 263–283.

3 Rodney, Walter. 1973. *How Europe Underdeveloped Africa*. Bogle-L'Ouverture Publications: London and Tanzanian Publishing House.

4 Ocheni, Stephen and Nwankwo, Basil. 2012. 'Analysis of Colonialism and Its Impact on Africa'. *Cross-Cultural Communication* 8(3): 46–54.

5 Hanson, Kebena. 2016. 'Capacity Development for Natural Resource Management: Findings from the 2013: Africa Capacity Indicators Report'. In *Governing Natural Resources for Africa's Development*, edited by Besada, Hany. New York: Routledge, 39–63.
6 Taylor, Ian. 2016. 'Dependency Redux: Why Africa Is Not Rising'. *Review of Black Political Economy* 43(147): 8–25.
7 Giuliani, Elisa. 2018. 'Regulating Global Capitalism amidst Rampant Corporate Wrong Doing—Reply to "Three Frames for Innovation Policy"'. *Research Policy* 47(9): 1577–1582.
8 Taylor, note 6.
9 Stephens, Beth. 2002. 'The Amorality of Profit: Transnational Corporations and Human Rights'. *Berkeley Journal of International Law* 20(1): 45–90.
10 Ruggie, John. 2007. 'Business and Human Rights: The Evolving International Agenda'. *The American Journal of International Law* 101(4): 819.
11 *Social and Economic Rights Action Centre and the Center for Economic and Social Rights v Nigeria (Communication No 155/96).*
12 Grear, Anna and Weston, Burns. 2015. 'The Betrayal of Human Rights and the Urgency of Universal Corporate: Reflections in a Post-Kiobel Landscape'. *Human Rights Law Review* 15(1): 21–44.
13 De Schutter, Olivier. 2015. 'Towards a New Treaty on Business and Human Rights'. *Business and Human Rights Journal* 1(1): 41–67.
14 Gumede, Vusi. 2015. 'Exploring Thought Leadership, Thought Liberation and Critical Consciousness for Africa's Development'. *Africa Development* 40(4): 91–111.
15 Pedro, note 1.
16 *Ibid*, 22.
17 Ambe-Uva, Terhemba. 2017. 'Whither the State? Mining Codes and Mineral Resource Governance in Africa'. *Canadian Journal of African Studies* 51(1): 92.
18 Gumede, note 14, 97.
19 Busia, Kojo and Akong, Charles. 2017. 'The African Mining Vision: Perspectives on Mineral Resource Development in Africa'. *Journal of Sustainable Development Law and Policy* 8(1):186–187.
20 Pharatlhatlhe, Kesa and Vanheukelom, Jan. 2019. 'Financing the African Union: On Mindsets and Money'. *Political Economy Dynamics of Regional Organisations in Africa—Discussion Paper 240*, 11.
21 Harvey, Ross. 2019. 'Governance with or Against the Grain? Challenges Confronting African Governments and Foreign Players in Africa's Extractives Sector'. *South African Journal of International of International Affairs* 26(2): 176–177.
22 Harvey, note 21, 176–177.
23 Shaw, Timothy. 2017. 'Transnational Initiatives towards Natural Resource Governance in Africa Post-2015'. *Journal of Sustainable Dev't Law and Policy* 8(1): 222–225.
24 Ambe-Uva, note 17, 82.
25 Busia and Akong, note 19, 157.
26 Campbell, Bonnie and Hatcher, Pascale. 2019. 'Neoliberal Reform, Contestation and Relations of Power in Mining: Observations from Guinea and Mongolia'. *The Extractive Industries and Society* 6(3): 642–653.
27 *Ibid.*
28 *Ibid.*
29 Adeleke, Fola. 2017. 'The Extractive Industries Transparency Initiative and Africa's Mineral Governance Regime'. *South African Institute of International Affairs—Occasional Paper 266*, 5–6.
30 Hilson, Gavin and Maconacie, Roy. 2008. 'Good Governance and the Extractive Industries in Sub-Saharan Africa'. *Mineral Processing and Extractive Metallurgy Review* 30(1): 86–87.
31 Busia and Akong, note 19, 173.
32 *Ibid*, 183.
33 *Ibid*, 184.

34 Ndlovu-Gatsheni, Sabelo. 2014. 'Global Coloniality and the Challenges of Creating African Futures'. *Strategic Review for Southern Africa* 36(2): 181–202.
35 Harvey, Ross. 2018. 'Looking Back to the Future: Mining for Development in Africa'. *South African Institute of International Affairs—Occasional Paper 283*, 8–9.
36 Busia and Akong, note 19, 188–190.
37 Manzano, Osmel and Rigobon, Roberto. 2001. 'Resource Curse or Debt Overhang?' *National Bureau of Economic Research—Working Paper No 8390.*
38 Pedro, note 1.
39 De Sa, Paulo. 2016. 'Sustainable Development of Mineral Resources in Africa: Challenges and Policies'. In *Governance of Natural Resources for Africa's Development*, edited by Besada, Hany. New York: Routledge, 204.
40 Ware, Helen. 2018. 'Africa and the Rhetoric of Good Governance'. *Australasian Review of African Studies* 39(2): 198–221.
41 *Ibid*, 204–206.
42 Helen, Ware. 2018. 'Africa and the Rhetoric of Good Governance'. *Australasian Review of African Studies* 39(2): 198–221.
43 Kumawu-Abiwu, Felix. 2017. 'Democratic Institutions, Natural Resource Governance, and Ghana's Oil Wealth'. *Social Sciences* 6(1): 7.
44 Couman, Catherine. 2019. 'Minding the "Governance Gaps": Re-thinking Conceptualizations of Host State "Weak Governance" and Re-Focusing on Home State Governance to Prevent and Remedy Harm by Multinational Mining Companies and Their Subsidiaries'. *The Extractive Industries and Society* 6(3): 675–687.
45 *Ibid*.
46 Hilson and Maconachie, note 30, 53–54.
47 *Ibid*.
48 Poncian, Japhace and Kigodi, Henry. 2018. 'Transparency Initiatives and Tanzania's Extractive Industry Governance'. *Development Studies Research* 5(1): 108–113.
49 Murombo, Tumai. 2016. 'The Effectiveness of Initiatives to Promote Good Governance, Accountability and Transparency in the Extractives Sector in Zimbabwe'. *Journal of African Law* 60(2): 237–238.
50 Poncian and Kigodi, note 46, 108–113.
51 Ware, note 40.
52 *Ibid*, 200.
53 Busia and Akong, note 19, 184.
54 Ware, note 40.
55 Hilson and Maconachie, note 30.
56 Nelson, Fred and Agrawal, Arun. 2008. 'Patronage or Participation? Community-based Natural Resource Management Reform in Sub-Saharan Africa'. *Development and Change* 39(4): 562.
57 Ndlovu-Gatsheni, note 34, 181–202.
58 *Ibid*.
59 Gumede, note 14.
60 Ngambi, Hellicy. 2011. 'RARE Leadership: An Alternative Leadership Approach for Africa'. In *Perspectives on Thought Leadership for Africa's Renewal*, edited by Kwandiwe, Kondlo. London: African Books Collective, 110–129.
61 De Sa, note 39 (2016).
62 Adeleke, note 29, 7.
63 Nelson and Agrawal, note 54.
64 *Ibid*, 562.
65 Obeng-Odom, Franklin. 2015. 'Understanding Land Grabs in Africa: Insights from Marxist and Georgist Political Economics'. *Review of Black Political Economy* 42(3): 337–354.
66 Mbabazi, Pamela *et al*. 2016. 'Winners and Losers: Contestations over Land in the Advent of Oil Exploration in Uganda'. In *Governance of Natural Resources for Africa's Development*, edited by Besada, Hany. New York: Routledge, 243–159.

References

Adeleke, Fola. 2017. 'The Extractive Industries Transparency Initiative and Africa's Mineral Governance Regime'. South African Institute of International Affairs—*Occasional Paper 266.*

Ambe-Uva, Terhemba. 2017. 'Whither the State? Mining Codes and Mineral Resource Governance in Africa'. *Canadian Journal of African Studies* 51(1): 81–101.

Busia, Kojo and Akong, Charles. 2017. 'The African Mining Vision: Perspectives on Mineral Resource Development in Africa'. *Journal of Sustainable Development Law and Policy* 8(1): 186–187.

Campbell, Bonnie and Hatcher, Pascale. 2019. 'Neoliberal Reform, Contestation and Relations of Power in Mining: Observations from Guinea and Mongolia'. *The Extractive Industries and Society* 6(3): 642–653.

Couman, Catherine. 2019. 'Minding the "Governance Gaps": Re-thinking Conceptualizations of Host State "Weak Governance" and Re-Focusing on Home State Governance to Prevent and Remedy Harm by Multinational Mining Companies and their Subsidiaries'. *The Extractive Industries and Society* 6(3): 675–687.

De Sa, Paulo. 2016. 'Sustainable Development of Mineral Resources in Africa: Challenges and Policies'. In *Governance of Natural Resources for Africa's Development*, edited by Besada, Hany. New York: Routledge, 197–219.

De Schutter, Olivier. 2015. 'Towards a New Treaty on Business and Human Rights'. *Business and Human Rights Journal* 1(1): 41–67.

Giuliani, Elisa. 2018. 'Regulating Global Capitalism amidst Rampant Corporate Wrong Doing—Reply to "Three Frames for Innovation Policy"'. *Research Policy* 47(9): 1577–1582.

Grear, Anna and Weston, Burns. 2015. 'The Betrayal of Human Rights and the Urgency of Universal Corporate: Reflections in a Post-Kiobel Landscape'. *Human Rights Law Review*, 15(1): 21–44.

Gumede, Vusi. 2015. 'Exploring Thought Leadership, Thought Liberation and Critical Consciousness for Africa's Development'. *Africa Development* 40(4): 91–111.

Hanson, Kebena. 2016. 'Capacity Development for Natural Resource Management: Findings from the 2013: Africa Capacity Indicators Report'. In *Governing Natural Resources for Africa's Development*, edited by Besada, Hany. New York: Routledge, 39–63.

Harvey, Ross. 2018. 'Looking Back to the Future: Mining for Development in Africa'. *South African Institute of International Affairs—Occasional Paper 283.*

Harvey, Ross. 2019. 'Governance with or Against the Grain? Challenges Confronting African Governments and Foreign Players in Africa's Extractives Sector'. *South African Journal of International of International Affairs* 26(2): 176–177.

Helen, Ware. 2018. 'Africa and the Rhetoric of Good Governance'. *Australasian Review of African Studies* 39(2): 198–221.

Hilson, Gavin and Maconacie, Roy. 2008. 'Good Governance and the Extractive Industries in Sub-Saharan Africa'. *Mineral Processing and Extractive Metallurgy Review* 30(1): 86–87.

Ingram, Joseph. 2016. 'Canada and China: Accepting Africa's Rise?' In *Governing Natural Resources for Africa's Development*, edited by Besada, Hany. New York: Routledge, 263–283.

Kumawu-Abiwu, Felix. 2017. 'Democratic Institutions, Natural Resource Governance, and Ghana's Oil Wealth'. *Social Sciences* 6(1): 7.

Manzano, Osmel and Rigobon, Roberto. 2001. 'Resource Curse or Debt Overhang?' National Bureau of Economic Research—*Working Paper No 8390.*

Mbabazi, Pamela *et al.* 2016. 'Winners and Losers: Contestations over Land in the Advent of Oil Exploration in Uganda'. In *Governance of Natural Resources for Africa's Development*, edited by Besada, Hany. New York: Routledge, 243–259.

Murombo, Tumai. 2016. 'The Effectiveness of Initiatives to Promote Good Governance, Accountability and Transparency in the Extractive Sector in Zimbabwe'. *Journal of African Law* 60(2): 237–238.

Ndlovu-Gatsheni, Sabelo. 2014. 'Global Coloniality and the Challenges of Creating African Futures'. *Strategic Review for Southern Africa* 36(2): 181–202.

Nelson, Fred and Agrawal, Arun. 2008. 'Patronage or Participation? Community-based Natural Resource Management Reform in Sub-Saharan Africa'. *Development and Change* 39(4): 562.

Ngambi, Hellicy. 2011. 'RARE Leadership: An Alternative Leadership Approach for Africa'. In *Perspectives on Thought Leadership for Africa's Renewal*, edited by Kwandiwe, Kondlo. London: African Books Collective, 110–129.

Obeng-Odom, Franklin. 2015. 'Understanding Land Grabs in Africa: Insights from Marxist and Georgist Political Economics'. *Review of Black Political Economy* 42(3): 337–354.

Ocheni, Stephen and Nwankwo, Basil. 2012. 'Analysis of Colonialism and its Impact on Africa'. *Cross-Cultural Communication* 8(3): 46–54.

Pedro, Anthonio. 2016. 'The Africa Mining Vision as a Model for Natural Resource Governance in Africa'. In *Governance of Natural Resources for Africa's Development*, edited by Besada, Hany. New York: Routledge, 13–38.

Pharatlhatlhe, Kesa and Vanheukelom, Jan. 2019. 'Financing the African Union: On Mindsets and Money'. *Political Economy Dynamics of Regional Organisations in Africa*—Discussion Paper 240.

Poncian, Japhace and Kigodi, Henry. 2018. 'Transparency Initiatives and Tanzania's Extractive Industry Governance'. *Development Studies Research* 5(1): 108–113.

Rodney, Walter. 1973. *How Europe Underdeveloped Africa*. Bogle-L'Ouverture Publications: London and Tanzanian Publishing House.

Ruggie, John. 2007. 'Business and Human Rights: The Evolving International Agenda'. *The American Journal of International Law* 101(4): 819–840.

Shaw, Timothy. 2017. 'Transnational Initiatives towards Natural Resource Governance in Africa Post-2015'. *Journal of Sustainable Development Law and Policy* 8(1): 222–225.

Stephens, Beth. 2002. 'The Amorality of Profit: Transnational Corporations and Human Rights'. *Berkeley Journal of International Law* 20(1): 45–90.

Taylor, Ian. 2016. 'Dependency Redux: Why Africa is not Rising'. *Review of Black Political Economy* 43(147): 8–25.

Ware, Helen. 2018. 'Africa and the Rhetoric of Good Governance'. *Australasian Review of African Studies*, 39(2): 198–221.

5 Resource constitutionalism for socio-economic development in Africa's petrostates

Re-imagining prior-informed consent

Aaron Olaniyi Salau

Introduction

The upstream petroleum industry is pivotal to Africa's socio-economic development as existing production capacity, reserves and revenues in countries like Nigeria, Angola and Algeria portend. New oil discoveries in the Gulf of Guinea and Eastern Africa attest thereto.[1] Expectedly, the ultimate beneficiaries of Africa's huge resource endowments ought to be its otherwise economically impoverished and politically marginalised peoples. This not withstanding, African petrostates remain underdeveloped while most of their populations live in abject poverty because the revenues accruing from oil and gas extraction are either mismanaged or outright stolen by the ruling elites.[2]

Interestingly, the United Nations Declaration on the Right to Development (UNDRD), which recognises Permanent Sovereignty over Natural Resources (PSNR) as a component of the right to development,[3] is corroborated by the African Charter on Human and Peoples' Rights[4] and the African Union's Agenda 2063.[5] Consequently, most African countries vest the exclusive legal ownership and control of land, solid minerals and petroleum in the state for overall national development with respect to the UN principle of PSNR. Rather than dislodge foreign interests and promote economic growth, state ownership has engendered power accumulation by rent-seeking elites and petroleum operations in conjunction with transnational corporations (TNCs) without any benefit to host communities.[6]

In order to wrest de facto control of their oil wealth from TNCs, African petrostates have since the 1990s embarked on 'resource nationalism' through the legal-contractual instrumentality of Production Sharing Contracts (PSCs) or Production Sharing Agreements (PSAs)[7] and local content initiatives.[8] These efforts have so far produced mixed results because the deeper challenge is how to prize-open government ownership[9] and institutionalise community participation and 'free, prior and informed consent' (FPIC) for equitable benefit and sustainable development of petroleum resources.[10] Addressing this problem through a comparative policy analysis, this chapter argues that emergent African petrostates must embrace 'resource constitutionalism' for efficient management of their vast petroleum resources to escape the 'resource curse' syndrome and actualise Africa's developmental imperatives.

The chapter proceeds in the section 'Sovereignty over natural resources: an appraisal' to reassess the currency of the United Nations (UN) principle of PSNR from an African perspective, including the question of FPIC to resource extraction. The section 'Resource constitutionalism and governance in Africa' undertakes an analysis of the idea of resource constitutionalism and how it can be engineered through constitution-building to protect oil communities' rights. The section 'Resource constitutionalism and oil production—an analysis of model PSCs in Nigeria, Angola and South Sudan' engages with resource constitutionalism vis-à-vis PSCs' utility to petroleum extraction in Nigeria, Angola and South Sudan. The section 'Conclusion' concludes that the use of PSCs in emerging African petrostates is not totally compliant with the principle of PSNR. It therefore, recommends public participation and consultation in oil production through resource constitutionalism for overall sustainable development.

Sovereignty over natural resources: an appraisal

This section re-appraises the normative content, meaning, scope and modern trends of the right to PSNR, including FPIC as its ingredient, and examines how these have been contextualised, constitutionalised and institutionalised in African law as regards state participation in petroleum exploitation.[11]

Sovereignty over natural resources as an international norm

Starting with Resolution 626(VII) of 21 December 1952,[12] on people's 'right to exploit freely natural wealth and resources' and the historic Resolution 1803(XVII) of 14 December 1962,[13] which later dovetailed into allied Resolutions,[14] Declarations,[15] Charters,[16] and states' practice,[17] the UN principle of peoples' and nations' right to PSNR is now firmly established in international law. The principle of PSNR emerged as a precondition to realising both the right to self-determination[18] and the right to development.[19] At its emergence, PSNR applied to interstate relationships,[20] and was intended to protect newly independent states' territorial rights and the quest for development through the exploitation of natural resources.[21] Several writers have attempted to draw out the essential requirements of the PSNR. Ng'ambi says it is 'contingent upon a state utilizing the resources for national development' and entails rights to freely dispose natural resources, explore and exploit natural resources based on national law.[22] Atsegbua argues similarly that Resolution 1803 enunciates at least three basic legal requirements,[23] one of which is that an investment 'must be in the interest of national development and well-being of people of the state concerned'.[24] Furthermore, Banerjee,[25] Rajan,[26] and Elian[27] all set out the position of developing states, which is that territorial sovereignty encompasses an absolute right to dispose of natural resources within the territorial limits of a state, devoid of any external control.

However, Duruigbo contends, from a rights-based point of view that 'if the phrase "rights of peoples" has any independent meaning, it must confer rights on peoples against their own governments' to restrain mismanagement and ensure use of resources for peoples' benefit.[28] In a similar perspective, Onifade argues that international law on PSNR posits a government's 'sovereign rights to manage natural resources on behalf of citizens' which may be translated into 'peoples-based permanent sovereignty over natural resources (PPSNR)'.[29] For Hossain, 'the inherent and overriding right of a state to control the exploitation and the use of its natural resources' must be exercised for the benefit of its citizens.[30] Accordingly, as Onifade extrapolates, citizens' marginalisation can be obviated indirectly through the deployment of 'common ownership and environmental justice claims' to 'advance distributive justice rights of citizens to returns from natural resources exploitation'.[31]

Conversely, the PSNR is part of international law rules governing international economic co-operation.[32] Hence, Armstrong says natural resources 'disposal' is at the instance of nation-states within the limits of international economic co-operation.[33] Interestingly, some African states have now executed Bilateral Investment Treaties (BITs) or contracts which contain international rules that guarantee protection to foreign property through conditions imposed for the legality of takings of property.[34] The principle of PSNR has undergone several other re-interpretations. Okeke posits that the people's right to exercise permanent sovereignty over their wealth and natural resources has attained the level of international customary law or the status of *jus cogens* based on common article 1(2) of the International Covenant on Civil and Political Rights (ICCPR) and the International Covenant on Economic, Social and Cultural Rights (ICESCR) which guarantee that 'All peoples may, for their own ends, freely dispose of their natural wealth and resources without prejudice to any obligations arising out of international economic co-operation, based upon the principle of mutual benefit and international law'.[35]

Thus, '[a]ll peoples' means 'the collection of individuals within state territories' having the right to self-determination which encompasses the sovereign right to dispose of natural resources.[36] Corroborating Okeke, Duruigbo submits that the principle of PSNR has evolved 'from a political claim to an accepted principle of customary international law'[37] and now has jurisprudential foundations, having been accepted by the International Court of Justice.[38] Clearly, the roots of PSNR burrow into customary international law, international development and economic law, and lately, into international human rights law.[39] The principle of PSNR can thus be described as a shield and a sword; serving not only as a minelayer, but also a mine sweeper against foreign domination of natural resources.[40] The legal and institutional devises of sovereignty employed by African states to govern and manage petroleum resources for overall national well-beingare examined in what proceeds.

Sovereignty over petroleum resources—a recurring African concern

State participation in the oil sector resulted from UN resolutions declaring the right to PSNR.[41] Chekera and Nmehielle regard people-oriented development as underlying the exercise of PSNR in the oil sector,[42] an argument which a combined reading of the identical articles 1(2) of ICCPR and ICESCR and article 2(2) of ICCPR[43] strengthens. The latter provisions further align with articles 21(1) and 21(5) of the African Charter. Carleton argues that the intendment of article 21(5) of the African Charter is to ensure that 'the primary beneficiaries of the wealth lying in the land are those living on the land', which comes with a freedom to dispose or retain, but in Africa 'shall be exercised by the states parties individually and collectively'.[44] Accordingly, efforts made in the 1970s and around the early to mid-2000s involving the transfer of oil assets held by TNCs to the state apparently resulted in assertion of 'greater national control over natural resource development'.[45] For instance, Nigeria nationalised British Petroleum in 1979, albeit with political undertones,[46] as the trend has not always been sustained. It is now characterised by legislative reforms, trade liberalisation and lately, by new foreign hegemony.[47]

However, the reality is that state ownership is becoming a scam whereby oil revenues are siphoned off into Swiss banks by conniving politicians of African states,[48] and as Shaxson observes, sub-Saharan Africa's established oil producers have not performed economically better than their less-endowed counterparts.[49] Moreover, the communities most affected by the pernicious effects of oil extraction are scarcely afforded any say there indespite the increasing international concern for the people's right to FPIC to resource extraction.

FPIC—model diffusion of an emerging norm

The right of communities including minority, ethnic or other groups to be affected by a development project to give FPIC based on antecedent disclosure is an emerging concern of international economic co-operation.[50] The right to FPIC is multidimensional with roots deeply embedded in international environmental law, the right to development and international human rights law.[51] For instance, the preamble to the UNDRD conceives of development as a comprehensive socio-economic, cultural and political process which aims at the constant improvement of general and individual well-being as achievable only through active, free and meaningful participation. This makes public participation crucial to sustainable development.

Accordingly, Agenda 21, the UN Programme of Action for Sustainable Development adopted at the Rio Earth Summit in 1992,[52] declares that '[o]ne of the fundamental prerequisites for the achievement of sustainable development is broad public participation in decision-making. [...] and to know about and participate in decisions, particularly those that potentially affect the communities in which they live'.[53] Public participation itself is, however, meaningless if people lack adequate information about the inner workings of public institutions or private interests performing public functions. Thus, the diffusion and global acceptance of the people's right-to-know pursuant to international and comparative law leverages the principle of FPIC.[54]

Furthermore, the norm of improved resource revenue governance through the twin pillars of transparency and accountability, which has developed global reach, has steered crucial developments on the continent. These include the 'Publish What You Pay' initiative in 2002 and the Extractive Industry Transparency Initiative (EITI) created in 2003 to foster strict transparency in revenue reporting in the extractive industries.[55] In sub-Saharan Africa, there are 24 countries engaging with the EITI process, including 18 compliant, 6 candidates and 1 suspended country.[56]

Within the African juridical context, the African Commission on Human and Peoples' Rights (African Commission) has in the *Endorois* case, specified FPIC, although not expressly mentioned in the African Charter, as a requirement of the right to development.[57] The African Commission held in particular that the right to development 'requires fulfilling five main criteria: it must be equitable, non-discriminatory, participatory, accountable, and transparent, with equity and choice as important, over-arching themes'.[58] The African Commission adjudged the alleged removal of the Endorois community from their ancestral lands and resources in 1973 by the Kenyan government to have violated their right to development, including the right to freely dispose of their natural resources. According to Tadeg, the reiteration of the right to be consulted and of participation in the development process as keys to the right to development under article 22 of the African Charter gives impetus to the justiciability of the right in Africa.[59]

Not withstanding the policy shift from statism to 'resource nationalism' that has occurred in recent times in the African oil and minerals sectors, the essence of PSNR remains unfulfilled. Consequently, it is imperative for African petrostates to embrace resource constitutionalism.

Resource constitutionalism and governance in Africa

This section advances the argument that resource constitutionalism presupposesa constitution-building process popularly designed to uphold the dynamics of democratic accommodation of institutional norms and values that underlie decentralised resource ownership and governance for socio-economic development.

Resource constitutionalism—foundation, design and challenges

An institutionalised bottom-up approach to constitution-making can motivate long-lasting resolution of natural resource ownership, control and revenue-sharing problems.[60] The value of constitutionalism, denoting a dynamic composition of democratisation, good governance and constitution-building, is undeniable.[61] The good governance-sustainable development nexus is also indispensable.[62] Good governance enables a democratic government to meet societal needs and effectively allocate national resources through broad consensus and participatory decision-making.[63] Thus, the mere writing of a constitution cannot guarantee constitutionalism.[64] Constitutionalism envisions a more functional, inclusive and evolutionary process of institution-building with firm societal foundations.[65]

Constitution-building for resource governance in Africa

Entrenching constitutional principles on the plural rights of ownership, management and distribution of revenue from mineral resources is an increasingly important feature of constitutional design in Africa.[66] Africa's third and fourth waves of democratisation seem to have ushered in a new age of constitutionalism through a process of constitution-building informed by the hard lessons of the past from corruption and conflicts associated with state opacity in natural resource management.[67] The challenge of how to evolve a genuine people-owned and people-centred constitution cognisant of expectations of economic emancipation through participatory resource governance however remains.[68] Unfortunately, very few constitutions in Africa conform to this view, but largely uphold the colonial legacy of vesting land and natural resource ownership in the state. This is being unmindful of Mangu's admonition that 'the consolidation of constitutionalism and democracy in African countries should also aim at improving the socio-economic situation of the African people'.[69] Yet, state consolidation in Africa brought about the economic marginalisation of oil-bearing communities. Resource constitutionalism should therefore address the perennial problems of legitimacy, legal validity, efficacy in constitution-making, institution-building and constitutional design or conditions that can underwrite the establishment of an acceptable system of resource governance for economic development in Africa.

Resource constitutionalism depends on the re-imagining of the concepts of PSNR and FPIC based on the tenets of modern constitutionalism which incorporate mechanisms that placet limitations on state power. Resource constitutionalism emphasises a shift from classical notions of sovereignty over resources which, in Africa, elevates the interest of ruling classes as 'national interest' towards decentralisation of resource control powers, transparency, financial accountability and participatory decision-making. Moreover, resource constitutionalism as proposed here draws from the modern concept of constitutionalism and Africa's fourth wave of constitutionalism.[70] Fombad, for instance, summarises its essentials viz fundamental rights protection; separation of powers, judicial independence; constitutionality of laws review; control of powers to amend the constitution; and democracy-supporting institutions.[71]

Nwabueze argues that constitutionalism has shifted emphasis from liberalism and the concept of limited government[72] to one that obligates government to subject itself to constitutional limitations.[73] Resource constitutionalism then implies a resource governance regime based on democratic principles of accountability, transparency and peoples' claims to natural resources.

Resource constitutionalism mustenable an institutionalised framework that ensures petroleum contracts are negotiated with full disclosure of their terms (duration, operationalisation and financial), address concerns on risks, provide for equitable participation by host communities and specify dispute resolution mechanisms. Consequently, resource constitutionalism is posited to be a constitutionally entrenched state recognition of host communities' sovereign participatory rights and a standard for testing the merits of legal-contractual mechanisms for oil production in African petrostates.

Resource constitutionalism and oil production—an analysis of model PSCs in Nigeria, Angola and South Sudan

This section engages with a comparative analysis of PSCs as the main instrumentality of PSNR in the post-concession era for petroleum production in Nigeria, Angola and South Sudan. It considers the PSCs'compliance with the idea of resource constitutionalism and Africa's goal of economic development. The selected countries share many commonalities; apart from being among Africa's biggest oil producers,[74] they represent a sort of incremental progression of the resource curse syndrome (South Sudan, Angola and Nigeria).[75] They are also marked by OPEC membership (except South Sudan), civil wars/militancy over control of oil[76] and oil dependency for economic development. Moreover, these countries typify Africa's main legal systems—Angola (civil law) and Nigeria and South Sudan (common law) while their investment policies equally emphasise national control of oil wealth.[77]

Understanding the PSC

Several African countries capitalised on the 1990s oil boom to renegotiate traditional concessions and sign PSCs[78] for the control of petroleum development, equity participation in TNCs and increased government revenue, among others.[79] Under the PSCs, the host state owns the oil while TNCs act as contractors or operators and financiers who bear all the risks attendant on oil production. The host state allows the contractor to recoup all running costs and expenses through a share of the oil production which is allocated in agreed proportions as cost oil, tax oil, royalty oil and revenue oil and satisfied in that order. The PSC model has been implemented in Nigeria, Angola and South Sudan.

Nigeria—ownership, control of oil and model PSCs

Nigeria possesses abundant hydrocarbons—Africa's second largest proven oil and highest gas reserves of 37.5 billion barrels and 188.8 trillion cubic feet (tcf), respectively—and is Africa's largest, but the world's number 11th oil producer considering its average daily production of 2 million barrels (bpd) from 2008 to 2018.[80] Despite previous oil windfalls and later volatility of oil prices, successive governments continued to mismanage oil revenues and perpetrate the criminal neglect of host communities.[81] So, instead of translating to economic success, Nigeria became afflicted by the 'Dutch disease' while oil dependency also resulted in other 'economic woes betiding the country'.[82]

Meanwhile, a state ownership structure is firmly entrenched in section 44(3) of the Constitution of the Federal Republic of Nigeria, 1999 (CFRN 1999) (as amended), which provides that

> [t]he entire property in and control of all minerals, mineral oils and natural gas in under or upon any land in Nigeria or in, under or upon the territorial waters and the Exclusive Economic Zone of Nigeria shall vest in the Government of the Federation.

Section 1(1) of the Petroleum Act of 1969 provides that '[t]he entire ownership and control of all petroleum in, under or upon any lands to which this section applies shall vest in the State'. These provisions underlie the pernicious gradual reduction of the earnings by the oil regions from proceeds of their resource beginning from 50 percent in 1969 under the Constitution of Nigeria 1963, to 1.5 percent under the military regime in 1992.[83] This was later reversed in 2002 with the implementation of the 13 percent derivation prescribed by the CFRN 1999.[84] The vesting of ownership of all lands and right to compulsory acquisition thereof by the state for 'overriding public interest', including for 'mining purposes or oil pipelines and for purposes connected therewith'[85] further yielded into total disregard for host communities' rights.[86]

Since 1971, oil exploitation was undertaken by the government through the Nigerian National Petroleum Corporation (NNPC) in conjunction with TNCs through NNPC's acquisition of a minimum of 55 to 60 percent equity stakes in their operations and under joint venture contracts (JVCs). According to Atsegbua, this was PSNR at play.[87] In any event, the grant of new oil blocks covered by Oil Exploration Licences (OELs), Oil Prospecting Licences (OPLs) or Oil Mining Licences (OMLs) subsumed under joint operating agreements (JOAs) or PSCs signed in 1993 have been criticised since TNCs take 'all profit after payment of taxes', hence resulting in huge economic losses for Nigeria.[88] Unfortunately, since the PSCs that have significant effects on Nigeria's public revenue are kept secret, such 'can facilitate abuse in the negotiation process, prevent public scrutiny, andrepresent fiscal risk'.[89] Consequently, the Nigeria's National Assembly initiateda Petroleum Industry Governance Bill (PIGB) in 2016, aside other bills, aimed at promoting transparency and accountability in the administration of petroleum resources.[90]

Angola—ownership, control of oil and model PSCs

With crude oil and natural gas reserves of 8.160 billion barrels and 383 billion cubic metres (cu.m.), respectively, and crude oil production of 1.473 million b/d as at the end of 2018—nearly all of which it exported[91]—Angola became

Africa's second biggest oil producer behind Nigeria.[92] Angola's economic dependency on mineral exports, particularly oil, resembles that of Nigeria.[93] Oil contributes about 45 percent of Angola's gross domestic product (GDP), 90 percent of exports and over 75 percent of government revenues.[94]

Oil extraction in Angola is governed by a diverse set of laws and regulations in addition to PSAs.[95] Moreover, reforms are being implemented since 2018 to increase transparency, modernise old laws and rejuvenate the sector. These include the main instruments for petroleum operations.[96] Also, Presidential Decree No 86/18 of 2 April 2018, simplifies the procurement processes for concessions and licences and eliminates the requirement of 'pre-qualification' for bidders. Both Presidential Decrees No 6/18 of 18 May 2018, and No 7/18 of 18 May 2018, provide legal frameworks for tax incentives to promote natural gas exploration, development and production. The National Agency for Oil, Gas and Biofuels (ANPG), created through President João Lourenço's Presidential Decree No 49/19 of 6 February 2019, now replaces Sonangol as national concessionaire and overall oil sector coordinator under the amended General Petroleum Activities Law.[97] The government's regulatory and oversight body responsible for regulating the oil and gas sector remains the Ministry of Mineral Resources and Petroleum.

However, Angola's model of state ownership of oil is deeply flawed. All petroleum onshore and offshore Angolan territory, exclusive economic zone and continental shelf belong to the state and is assigned to, controlled and managed by the ANPG.[98] Theoretically, oil extraction is carried out only by Sonangol, although in reality, in partnership with Exxon Mobil, BP, Shell and Total, under the PSCs and JVCs,[99] especially in the highly lucrative deep water and ultra-deep water–oil blocks awarded since 1999.[100] Confidentiality clauses in the PSAs encourage secretive oil deals and non-disclosure of revenues earned thereby,[101] which are eventually siphoned by the ruling elites.[102] The Cabinda enclave, with offshore oil deposits accounting for half of Angola's hydrocarbon output, is one of Angola's poorest provinces and hence, the region's neglect has bred separatist agitations.[103] Nonetheless, expectations are rife, both locally and internationally, that ongoing reforms in the oil sector would invigorate the economy.

South Sudan

Being the world's most oil-dependent nation, revenue and taxes from oil account for over 98 percent of government revenues for South Sudan.[104] While South Sudan's oil infrastructure is controlled by Sudan, the entire industry is completely dominated by three major Asian TNCs: Chinese National Petroleum Company (CNPC), Petronas of Malaysia and Oil and National Gas Company (India).[105] This exclusive foreign control of oil activities is deeply concerning because South Sudan 'could become an oil importer in 6–8 years' if no new discoveries are made.[106] Nevertheless, South Sudan's legal and institutional frameworks for governance and management of the oil sector are promising.

Article 172(1) of the Transitional Constitution of the Republic of South Sudan, 2011 (TCSS 2011) (as amended),[107] which provides for the

> Guiding Principles for Petroleum and Gas Development and Management' vests '[o]wnership of petroleum and gas [...] in the people of South Sudan [...] [to] be developed and managed by the National Government on behalf of and for the benefit of the people.[108]

Natural resources, including petroleum and gas resources on any land in South Sudan belong to the National Government.[109] The state is, however, enjoined to 'involve communities in decisions relating to the exploitation of natural resources in their areas and promote a safe and healthy environment'[110] while ensuring the entitlement of 'all states, localities and communities [...] to equitable development'.[111]

Moreover, oil and gas development and management are guided by the principles of national interest,[112] sustainable resource management,[113] oil revenue-based economic diversification,[114] transparency and accountability.[115] The TCSS 2011 assures that 3 percent out of the 5 percent of national oil revenue payable to oil-producing states is allocated 'to the [host] communities' while the remaining 2 percent is allocated to the oil-producing states as regulated by law.[116] The Petroleum Act 2011 contains important clauses for transparency of oil contracts. Section 18(2) provides for 'an open, transparent, non-discriminatory and competitive tender process' prior to the signing of petroleum exploration contracts. Under section 18(4), '[p]etroleum agreements may only be entered into with company having the requisite technical competence and able to fulfil all obligations of the petroleum agreements'. Moreover, there are detailed provisions on governmental obligations to publicly disclose pertinent information including oil sector revenue, petroleum agreements and proof of the requisite technical competence.[117] Oil blocks are worked based on exploration and production-sharing agreements (EPSAs) whereby Asian transnationals hold majority shares alongside South Sudan's National Petroleum and Gas Corporation (NPGC) as a minority shareholder. Unfortunately, as of 2014, South Sudan had the lowest per capita electricity consumption in Africa while over 96 percent of the population depends on firewood for household use.[118]

Conclusion

The cases of Nigeria and Angola as sub-Saharan Africa's dominant producers of oil and gas, and significant new discoveries and production in South Sudan, are used as exemplars of the vital importance of oil revenues to Africa's socio-economic development. However, expanding resource extraction comes with numerous costs such as there source curse highlighted in literature as the supposed incompatibility of resource boom and sustainable economic growth, which must be offset. Such concern is substantiated by petroleum or mineral legislations in African petrostates which entrench confidentiality clauses that provide legal cover for the shielding of oil proceeds from the people.

Scholarship on the right to development and prevailing trends on PSNR stand in opposition to the exclusive control of natural resources by political elites under the shadow of sovereignty.[119] From the countries under consideration, only South Sudan evinces a move towards resource constitutionalism as the TCSS 2011 and its Petroleum Act 2011 portend, though it is still beset by institutional incapacity and non-implementation of transparency provisions.[120] Ongoing reforms in Nigeria and Angola may produce palliative effects which may not be long-lasting. Hence, constitutional reform must be at the heart of consensual state regulation of the African oil sector.[121] Resource constitutionalism, with its potential to guarantee equitable distribution of oil revenue and host community participation in contractual arrangements for oil extraction and the joint management and ownership of oil, is hereby recommended.

Beyond legislative frameworks, robust and constitutionalised participatory decision-making in the drafting of oil contracts is imperative to accommodate marginalised communities.[122] The public scrutiny of government activities involved in the popular participation in the processes of negotiation, execution and payment for oil will promote democracy and good governance. African petrostates must therefore learn from countries that have increased local resource control and decision-making like multi-ethnic Ethiopia, India and Pakistan.[123] Article 161 of the Pakistan Constitution of 1973 as amended by Constitution (Eighteenth Amendment) Act of 2010 assigns the net proceeds of federal excise duty levied on natural gas and oil together with royalties to the province of derivation. Article 172(3) thereof, vests equal and joint ownership of mineral oil and natural gas within a province or the territorial waters adjacent thereto in the province and federal government.[124]

In India, articles 294–295 of the Constitution of 1949 vest ownership of land and onshore natural resources, including petroleum in each federating territory subject to their control and development by the central government while article 297 thereof, vests offshore natural resources exclusively in the central government.[125] For Ethiopia, the 1994 Constitution of the Federal Democratic Republic empowers the federating states to administer and utilise land and natural resources subject to federal laws.[126] Indeed, to realise sustainable development based on resource extraction, African countries must overcome their inability to capitalise on their natural resource endowments.[127]

Notes

1 Katz-Lavigne, Sarah. 2017. 'The Renegotiation Window: Resource Contract Renegotiations in the Mining Industry in Africa from 2000 to 2013'. *Resources Policy* 51(1): 23.

2 Omeje, Kenneth. 2016. 'Natural Resource Rent and Stakeholder Politics in Africa: Towards a New Conceptualisation'. *Commonwealth and Comparative Politics* 54(1): 101–102.

3 UN General Assembly Official Records (GAOR) Doc Resolution A/RES/41/128 adopted 4 December 1986.
4 African Charter on Human and Peoples' Rights 21 ILM 58 (1982) adopted June 1981, arts 20 and 21(1).
5 African Union (AU) Commission. 2015. *Agenda 2063: The Africa We Want*. Addis Ababa: AU Commission.
6 Hanson, Kobena. 2017. 'Managing Africa's Natural Resource Endowments: New Dispensations and Good-Fit Approaches'. *Afe Babalola University Journal of Sustainable Development Lawand Policy* 8(1): 121.
7 Okeke, Nwachukwu. 2008. 'The Second Scramble for Africa's Oil and Mineral Resources: Blessing or Curse?' *The International Lawyer* 42(1): 193. PSC and PSA are used interchangeably.
8 Ovadia, Jesse. 2014. 'Local Content and Natural Resource Governance: The Cases of Angola and Nigeria'. *The Extractive Industries and Society* 1(2): 137–146.
9 Johnston, David. 2007. 'How to Evaluate the Fiscal Terms of Oil Contracts'. *Escaping the Resource Curse*, edited by Macartan, Humphreys; Sachs, Jeffrey and Stiglitz, Joseph. New York: Columbia University Press, 8.
10 Duruigbo, Emeka. 2013. 'Community Equity Participation in African Petroleum Ventures: Path to Economic Growth?' *North Carolina Central Law Review* 35(2): 114.
11 Ejims, Okechukwu. 2013. 'The Impact of Nigerian International Petroleum Contracts on Environmental and Human Rights of Indigenous Communities'. *African Journal of International and Comparative Law* 21(3): 355.
12 GA Resolution 626(VII) UN Doc A/PV.411.
13 GA Resolution 1803(XVII) UN Doc A/5217 'Declaration on Permanent Sovereignty over Natural Resources'.
14 GA Resplution 2158(XXI) UN Doc A/6518 28 November 1966; GA Resolution 2692(XXV) (1970); GA Resolution 3171(XXVIII) (1973); GA Resolution 3082 (XXVIII), art 2; Organisation of Petroleum Exporting Countries (OPEC) 'Declaratory Statement of Petroleum Policy on States' Invocation of the Principle of Permanent Sovereignty over Natural Resources' OPEC Resolutions and Official Press Releases 1960–1990.
15 United Nations General Assembly (UNGA) Declaration on the Establishment of a New International Economic Order Resolution A/RES/S-6/3201 adopted 1 May 1974 (NIEO Declaration).
16 GA Resolution 3281(XXIX) Charter of Economic Rights and Duties of States (CERD); UNGA Resolution on a World Charter for Nature adopted 28 October 1982.
17 Art 18 of Energy Charter Treaty (ECT) signed in 1994 between states of the former Soviet Union and Japan, Russia and Turkey.
18 UN Charter 1945, art 2; International Covenant on Civil and Political Rights adopted 16 December 1966, art 1; International Covenant on Economic, Social and Cultural Rights adopted 16 December 1966, art 1; African Charter, art 20.
19 Chekera, Yolanda and Nmehielle, Vincent. 2013. 'The International Law Principle of Permanent Sovereignty over Natural Resources as an Instrument for Development: The Case of Zimbabwean Diamonds'. *African Journalof Legal Studies* 6(1): 69–101.
20 Aponte, Miranda. 2012. 'The Role of International Law in Intrastate Natural Resource Allocation: Sovereignty, Human Rights, and Peoples-Based Development'. *Vanderbilt Journalof Transnational Law* 45(3): 789.
21 Onifade, Temotope. 2015. 'Peoples-Based Permanent Sovereignty over Natural Resources: Toward Functional Distributive Justice?' *Human Rights Review* 16(4): 350.
22 Ng'ambi, Sangwani. 2015. 'Permanent Sovereignty over Natural Resources and the Sanctity of Contracts, from the Angle of Lucrum Cessans'. *Loyola University Chicago International Law Review* 12(2): 154.

23 Atsegbua, Lawrence. 1993. 'Principle of Permanent Sovereignty over Natural Resources and its Contribution to Modern Petroleum Development Agreements'. *Journal of the Indian Law Institute* 35(1): 116.
24 *Ibid*, 123.
25 Banerjee, Sanjay. 1968. 'The Concept of Permanent Sovereignty over Natural Resources'. *Indian Journal of International Law* 8(1): 517.
26 Rajan, Mannaraswamighala. 1978. *Sovereignty over Natural Resources*. New Delhi: Radiant Publishers, 11.
27 Elian, George. 1979. *The Principle of Sovereignty over Natural Resources*. The Hague: Martinus Nijhoff Publishers, 12.
28 Duruigbo, Emeka. 2006.'Permanent Sovereignty and Peoples' Ownership of Natural Resources in International Law'. *George Washington International Law Review* 38(1): 67.
29 Onifade, note 21, 343.
30 Chowdhury, Subrata. 1984. 'Permanent Sovereignty over Natural Resources'. In *Permanent Sovereignty over Natural Resources in International Law*, edited by Hossain, Kamal and Chowdhury, Subrata. London: Pinter, 66–67.
31 Onifade, note 21.
32 Mughraby, Muhamad. 1966. *Permanent Sovereignty over Oil Resources*. Beirut: The Middle East Research and Publishing Centre, 161.
33 Armstrong, Chris. 2015. 'Against "Permanent Sovereignty" over Natural Resources'. *Politics Philosophy and Economics* 14(2): 130.
34 UK/Kenya: Agreement for the Promotion and Protection of Investments [TS No.2/2018] art 5; Mozambique-Netherlands BIT. 2001, signed 18 December 2001, art 6 <https://investmentpolicy.unctad.org/international-investment-agreements/countries/143/mozambique>.
35 Okeke, note 7, 193, 202.
36 Onifade, note 21, 335.
37 Duruigbo, Emeka. 2011. 'Realizing the People's Right to Natural Resources'. *Whitehead Journal of Diplomacy and International Relations* 12(1): 111.
38 *East Timor (Portugal v Australia)* [1995] ICJ 90; *Armed Activities on the Territory of the Congo (Democratic Republic of the Congo v Uganda)* ICJ Rep [2005] 168 [244].
39 Duruigbo, note 37.
40 Chowdhury, Subrata. 1988. 'Permanent Sovereignty over Natural Resources: Substratum of the Seoul Declaration'. In *International Law and Development*, edited by de Waart, Pim; Peters, Paul and Denters, Erik.Dordrecht: Martinus Nijhoff Publishers, 62.
41 Atsegbua, note 23.
42 Chekera and Nmehielle, note 19.
43 The provision obligates each state party to adopt such laws or other measures as may be necessary to give effect to the rights recognised in the Covenant.
44 African Charter, note 4, art 21(4); Carleton, Alexandra. 2014.'Defining Peoples under the ICCPR and African Charter: Identification of Collective Claims to Natural Mineral Wealth'. *International Journal for Minority Group Rights* 21(2): 133, 134, 137.
45 Stephens, Paul. 2008.'National Oil Companies and International Oil Companies in the Middle East: Under the Shadow of Government and the Resource Nationalism Cycle'. *The Journal of World Energy Law and Business* 1(1): 5, 10.
46 Duruigbo, note 37, 112.
47 Okeke, note 7, 195.
48 McFerson, Hazel. 2009. 'Governance and Hyper-Corruption in Resource-Rich African Countries'. *Third World Quarterly* 30(8): 1542.
49 Shaxson, Nicholas. 2005. 'New Approaches to Volatility: Dealing with the "Resource Curse" in Sub-Saharan Africa'. *International Affairs* 81(2): 311.
50 International Financial Corporation. 2006. 'Performance Standards on Social and Environmental Sustainability'. <www.ifc.org/ifcext/enviro.nsf/AttachmentsByTitle/pol PerformanceStandards2006fulUSFILE/IFC+Performance+Standards.pdf>.

51 ICESCR, art 1(2); Declaration on the Rights of Indigenous Peoples, UN Doc A/RES/61/295 13 September 2007.
52 UN GAOR 46th Sess., Agenda Item 21, UN Doc A/Conf.151/26 (1992).
53 *Ibid.*
54 Sand, Philip. 2011. 'The Right to Know: Freedom of Environmental Information in Comparative and International Law'. *Tulane Journal of Internationall and Comparative Law* 20(1): 203.
55 Besada, Hany and Philip, Martin. 2015. 'Mining Codes in Africa: Emergence of a Fourth Generation?' *Cambridge Review of International Affairs* 28(2): 267–268.
56 EITI. 2019. 'How We Work'. *Extractive Industries Transparency Initiative* <https://eiti.org/about/how-we-work#upholding-the-standard-internationally-validation>.
57 *Centre for Minority Rights Development (Kenya) and Minority Rights Group International on Behalf of Endorois Welfare Council v Kenya* (*Endorois*) Comm 276/2003; *African Commission on Human and Peoples' Rights v Republic of Kenya* (May 2017) (*Ogiek Community*) Comm 006/2012 ACtHPR (2017).
58 *Ibid*, paras 277–282.
59 Tadeg, Mesenbet. 2010. 'Reflections on the Right to Development: Challenges and Prospects'. *African Human Rights Law Journal* 10(2): 332.
60 Haysom, Nicholas and Sean, Kane. 2009. *Negotiating Natural Resources for Peace: Ownership, Control and Wealth-Sharing*. Geneva: Centre for Humanitarian Dialogue.
61 Mangu, Andre. 2005. 'Challenges to Constitutionalism and Democratic Consolidation in Africa'. *Politeia* 24(3): 316.
62 Hatchard, John; Muna, Ndulo and Slinn, Peter. 2004. *Comparative Constitutionalism and Good Governancein the Commonwealth: An Eastern and Southern African Perspective*. Cambridge: Cambridge University Press, 2.
63 Ogboye, Lateefand Abubakri, Yekini. 2014. 'Constitutionalism and Good Governance in Nigeria(1999–2014)'. *Nnamdi Azikiwe University Journal of International Law and Jurisprudence* 5(1): 125.
64 Fombad, Charles. 2008. 'Post-1990 Constitutional Reforms in Africa A Preliminary Assessment of the Prospects for Constitutional Governance and Constitutionalism'. In *The Resolution of African Conflicts: The Management of Conflict Resolution and Post-Conflict Reconstruction*, edited by Nhema, Alfredand Zeleza, Paul. Athens: Ohio University Press, 179, 180.
65 *Rates Action Group v City of Cape Town* 2004 (12) BCLR 13286 (Budlender AJ) (arguing that the South African Constitution of 1996 provides 'a […] framework and […] blueprint for transformation of our society from its […] unequal past').
66 Choudhry, Sujitand Richard, Stacey. 2015. *Oil and Natural Gas: Constitutional Frameworks for the Middle East and North Africa*. Stockholm: Centre for Constitutional Transitions, International IDEA and UNDP, 15.
67 Egede, Edwin. 2005. 'Who Owns the Nigerian Offshore Seabed: Federal or States: An Examination of *Attorney-General of the Federation v Attorney-General of Aba State and 35 Others case*'. *JournalofAfrican Law* 49(1): 91.
68 Shivji, Issa. 2013. 'Paradoxes of Constitution-Making in Tanzania'. Paper presented at the East African Law Society Conference, Mombasa, November 15–16.
69 Mangu, note 61, 327.
70 Hessebon, Gedion. 2014. 'The Fourth Constitution-Making Wave of Africa: Constitutions 4.0?' *Temple Internationaland Comparative Law Journal* 28(2): 186.
71 Fombad, Charles. 2007. 'Challenges to Constitutionalism and Constitutional Rights in Africa and the Enabling Role of Political Parties: Lessons and Perspectives from Southern Africa'. *American Journalof Comparative Law* 55(1): 2–3, 7–9, 11–24.
72 Nwabueze, Ben. 1973. *Constitutionalism in the Emergent States*. London: Hurst and Co, 1–2.
73 Fombad, note 71.
74 Nigeria and Angola accounted for 75 percent of sub-Saharan Africa (SSA)'s oil production in 2012 (Energy and Information Administration, 2013).

75 Patey, Luke. 2010. 'Crude Days Ahead? Oil and the Resource Curse in Sudan'. *African Affairs* 109(435):618.
76 Obi, Cyril. 2010. 'Oil as the "Curse" of Conflict in Africa: Peering through the Smoke and Mirrors'. *Review of African Political Economy* 37(127): 490.
77 Arbatli, Ekim. 2018. 'Resource Nationalism Revisited: A New Conceptualization in Light of Changing Actors and Strategies in the Oil Industry'. *Energy Research and Social Science* 40(1): 102.
78 Katz-Lavigne, note 1, 22–30.
79 Alexander, Karin and Stefan, Gilbert. 2008. *Oil and Governance Report: A Case Study of Chad, Angola, Gabon, and Sao Tome and Principe*. Cape Town: IDASA.
80 British Petroleum. 2019. *Statistical Review of World Energy 2019*. (68th edn), London: British Petroleum, 30; Encyclopaedia Britannica. 2019. 'Major Oil-Producing Countries'. *Britannica* <www.britannica.com/science/petroleum/Major-oil-producin g-countries>; OPEC. 2019. 'Nigeria Facts and Figures'. *Organisation of Petroleum Exporting Countries* <www.opec.org/opec_web/en/about_us/167.htm>.
81 Ikelegbe, Augustine. 2001. 'Civil Society, Oil and Conflict in the Niger Delta Region of Nigeria: Ramifications of Civil Society for a Regional Resource Struggle'. *The Journal of Modern African Studies* 39(3): 437–469.
82 Adegbite, Idowu. 2015. 'Law, Development and the Nigerian Oil and Gas Refineries'. *Nigerian Law Journal* 18(1): 75.
83 Omorogbe, Yinka. 2002. 'The Legal Framework for Public Participation in Decision-Making on Mining and Energy Development in Nigeria: Giving Voices to the Voiceless'. In *Human Rights in Natural Resource Development: Public Participation in the Sustainable Development of Mining and Energy Resources*, edited by Zillman, Donald; Alastair, Lucas and Pring, George. Oxford: Oxford University Press, 549, 560.
84 Sect 162(2) and (3).
85 Land Use Act 1978, Cap 202 LFN 1990, sect 1.
86 Ejobowah, John. 2000. 'Who Owns the Oil? The Politics of Ethnicity in the Niger Delta of Nigeria'. *Africa Today* 47(1): 28.
87 Atsegbua, note 23, 122.
88 NEITI. 2019. *Policy Brief*. Abuja: NEITI.
89 International Monetary Fund. 2019. *Country Report No 19/93—Nigeria Selected Issues*. Washington: IMF.
90 *Ibid*, 13.
91 Organisation of the Petroleum Exporting Countries. 2019. *Annual Statistical Bulletin of the Organisation of the Petroleum Exporting Countries*. Vienna: OPEC.
92 Azinge, Epiphany. (Ed). 2010. *A Book of Communiques, 2010 Roundtables of Nigerian Institute of Advanced Legal Studies*. Lagos: NIALS, 130.
93 Kiala, Carine. 2010. 'China-Angola Aid Relations: Strategic Cooperation for Development?' *South African Journalof International Affairs* 17(3): 324.
94 Hammond, John. 2011. 'The Resource Curse and Oil Revenues in Angola and Venezuela'. *Science and Society* 75(3): 349, 354.
95 Ovadia, Jesse. 2012. 'The Dual Nature of Local Content in Angola's Oil and Gas Industry: Development Vs Elite Accumulation'. *Journal of Contemporary African Studies* 30(3): 399–402.
96 Petroleum Activities Law No 10/04 of 12 November 2004, amended by Law 5/19 and Presidential Decree 5/2018 of 18 May; Angolan Petroleum Operations Decree 1/09 of 27 January (empowers the issuance of oil prospecting licences only by authority of the Ministry for Mineral Resources and Petroleum (MIMREPET) or a mining lease by ANPG.
97 No 13/78 of 26 August 1978, see Hirst, Caty. 2020. 'In Summary: Angola's Petroleum Reforms'. <https://africaoilandpower.com/2018/12/21/in-summary-angolas-petro leum-reforms/>.
98 ICLG.com. 2020. 'Angola: Oil and Gas Regulations 2020'. <https://iclg.com>.

99 Hammond, note 96, 355.

100 Amundsen, Inge. 2014. 'Drowning in Oil: Angola's Institutions and the "Resource Curse"'. *Comparative Politics* 46(2): 177.

101 Frynas, Jedrzej and Wood, Geoffrey. 2001. 'Oil and War in Angola'. *Review of African Political Economy* 28(90): 600.

102 Amundsen, note 102, 175.

103 Cropley, Ed. 2016. 'Rebels Alive and Kicking in Angolan Petro-Province, Oil Workers Say'. *Reuters*, June 14 <www.reuters.com/article/angola-oil-security/rebels-alive-and-kicking-in-angolan-petro-province-oil-workers-say-idUSL8N1952C9>.

104 Belloni, Roberto. 2011. 'The Birth of South Sudan and the Challenges of Statebuilding'. *Ethnopolitics* 10(3): 411–429.

105 Patey, note 77, 623.

106 *Ibid.*

107 Constitution of the Republic of South Sudan 2012 (as amended), sect 173(1).

108 But TCSS 2011, sect 170(4) protects state ownership of 'rights over all subterranean and other natural resources throughout South Sudan, including petroleum and gas resources and solid minerals' while sects 170(10) and (11) assures of Communities' rights to be consulted in decisions that may affect their rights in lands and resources and to 'prompt and equitable compensation' for government acquisition of land.

109 TCSS 2011, sect 170(4).

110 *Ibid*, sect 165(6)(j) now sect 166(6)(j).

111 *Ibid*, sect 168(5) now sect 169(5).

112 *Ibid*, sect 172(2)(a) now sect 173(2)(a).

113 *Ibid*, sect 172(2)(c) now sect 173(2)(c).

114 *Ibid*, sect 172(2)(d) now sect 173(2)(d).

115 Sect 172(2)(e) now sect 173(2)(e).

116 Sect 178(1)(a)(b).

117 Sect 79 clause 1(a)–(c).

118 Patey, note 77, 757.

119 UNOHCHR. 2013. *Realizing the Right to Development: Essays in Commemoration of 25 Years of the United Nations Declaration on the Right to Development*. Geneva: UN Publication, 76, 101–102, 363.

120 Savage, Emily. 2013. *South Sudan's Petroleum Revenue Management Act*. Juba: The SUDD Institute, 5–6.

121 Wennmann, Achim. 2011. 'Breaking the Conflict Trap? Addressing the Resource Curse in Peace Processes'. *Global Governance* 17(2): 267.

122 Omotola, Shola. 2007. 'From the OMPADEC to the NDDC: An Assessment of State Responses to Environmental Insecurity in Niger Delta, Nigeria'. *Africa Today* 51(1): 79.

123 Adangor, Zacchaeus. 2015. 'The Petroleum Industry Bill 2012 and the Niger Delta Region of Nigeria: Panacea or Placebo'. *Journal of Law, Policy and Globalization* 38(1): 55–56.

124 *Ibid.*

125 *Ibid.*

126 Arts 51(5) and 52(2)(d).

127 Hanson, note 6.

References

Adangor, Zacchaeus. 2015. 'The Petroleum Industry Bill 2012 and the Niger Delta Region of Nigeria: Panacea or Placebo'. *Journal of Law, Policy and Globalization* 38(1): 47–62.

Adegbite, Idowu. 2015. 'Law, Development and the Nigerian Oil and Gas Refineries'. *Nigerian Law Journal* 18(1): 72–106.

Alexander, Karin and Stefan, Gilbert. 2008. *Oil and Governance Report: A Case Study of Chad, Angola, Gabon, and Sao Tome and Principe*. Cape Town: IDASA.

Amundsen, Inge. (2014). 'Drowning in Oil: Angola's Institutions and the "Resource Curse"'. *Comparative Politics* 46(2): 169–189.

Aponte, Miranda. 2012. 'The Role of International Law in Intrastate Natural Resource Allocation: Sovereignty, Human Rights, and Peoples-Based Development'. *Vanderbilt Journal of Transnational Law* 45(3): 785–840.

Arbatli, Ekim. 2018. 'Resource Nationalism Revisited: A New Conceptualization in Light of Changing Actors and Strategies in the Oil Industry'. *Energy Research and Social Science* 40(1): 101–108.

Armstrong, Chris. 2015. 'Against "Permanent Sovereignty" over Natural Resources'. *Politics Philosophy and Economics* 14(2): 129–151.

Atsegbua, Lawrence. 1993. 'Principle of Permanent Sovereignty over Natural Resources and its Contribution to Modern Petroleum Development Agreements'. *Journal of the Indian Law Institute* 35(1): 115–126.

Azinge, Epiphany. (Ed). 2010. *A Book of Communiques, 2010* Roundtables of Nigerian Institute of Advanced Legal Studies. Lagos: NIALS.

Banerjee, Sanjay. 1968. 'The Concept of Permanent Sovereignty over Natural Resources'. *Indian Journal of International Law* 8(1): 517.

Belloni, Roberto. 2011. 'The Birth of South Sudan and the Challenges of Statebuilding'. *Ethnopolitics* 10(3): 411–429.

Besada, Hany and Philip, Martin. 2015. 'Mining Codes in Africa: Emergence of a Fourth Generation?' *Cambridge Review of International Affairs* 28(2): 263–282.

Carleton, Alexandra. 2014. 'Defining Peoples under the ICCPR and African Charter: Identification of Collective Claims to Natural Mineral Wealth'. *International Journal for Minority Group Rights* 21(2): 131–177.

Chekera, Yolanda and Nmehielle, Vincent. 2013. 'The International Law Principle of Permanent Sovereignty over Natural Resources as an Instrument for Development: The Case of Zimbabwean Diamonds'. *African Journal of Legal Studies* 6(1): 69–101.

Choudhry, Sujit and Richard, Stacey. 2015. *Oil and Natural Gas: Constitutional Frameworks for the Middle East and North Africa*. Stockholm: Centre for Constitutional Transitions, International IDEA and UNDP.

Chowdhury, Subrata. 1984. 'Permanent Sovereignty over Natural Resources'. In *Permanent Sovereignty over Natural Resources in International Law*, edited by Hossain, Kamal and Chowdhury, Subrata. London: Pinter.

Chowdhury, Subrata. 1988. 'Permanent Sovereignty over Natural Resources: Substratum of the Seoul Declaration'. In *International Law and Development*, edited by de Waart, Pim; Peters, Paul and Denters, Erik. Dordrecht: Martinus Nijhoff Publishers.

Duruigbo, Emeka. 2006. 'Permanent Sovereignty and Peoples' Ownership of Natural Resources in International Law'. *George Washington International Law Review* 38(1): 33–100.

Duruigbo, Emeka. 2011. 'Realizing the People's Right to Natural Resources'. *Whitehead Journal of Diplomacy and International Relations* 12(1): 111.

Duruigbo, Emeka. 2013. 'Community Equity Participation in African Petroleum Ventures: Path to Economic Growth?' *North Carolina Central Law Review* 35(2): 111–159.

Egede, Edwin. 2005. 'Who Owns the Nigerian Offshore Seabed: Federal or States: An Examination of *Attorney-General of the Federationv Attorney-General of Aba State and 35 Others Case*'. *Journal of African Law* 49(1): 73–93.

Ejims, Okechukwu. 2013. 'The Impact of Nigerian International Petroleum Contracts on Environmental and Human Rights of Indigenous Communities'. *African Journal of International and Comparative Law* 21(3): 355.

Ejobowah, John. 2000. 'Who Owns the Oil? The Politics of Ethnicity in the Niger Delta of Nigeria' *Africa Today* 47(1): 28–47.

Elian, George. 1979. *The Principle of Sovereignty over Natural Resources*. The Hague: Martinus Nijhoff Publishers.

Fombad, Charles. 2007. 'Challenges to Constitutionalism and Constitutional Rights in Africa and the Enabling Role of Political Parties: Lessons and Perspectives from Southern Africa'. *American Journal of Comparative Law* 55(1): 1–24.

Fombad, Charles. 2008. 'Post-1990 Constitutional Reforms in Africa a Preliminary Assessment of the Prospects for Constitutional Governance and Constitutionalism'. In *The Resolution of African Conflicts: The Management of Conflict Resolution and Post-Conflict Reconstruction*, edited by Nhema, Alfred and Zeleza, Paul. Athens: Ohio University Press.

Frynas, Jedrzej and Wood, Geoffrey. 2001. 'Oil and War in Angola'. *Review of African Political Economy* 28(90): 587–606.

Hammond, John. 2011. 'The Resource Curse and Oil Revenues in Angola and Venezuela'. *Science and Society* 75(3): 348–378.

Hanson, Kobena. 2017. 'Managing Africa's Natural Resource Endowments: New Dispensations and Good-Fit Approaches'. *Afe Babalola University Journal of Sustainable Development Law and Policy* 8(1): 121–144.

Hatchard, John; Muna, Ndulo and Slinn, Peter. 2004. *Comparative Constitutionalism and Good Governance in the Commonwealth: An Eastern and Southern African Perspective*. Cambridge: Cambridge University Press.

Haysom, Nicholas and Sean, Kane. 2009. *Negotiating Natural Resources for Peace: Ownership, Control and Wealth-Sharing*. Geneva: Centre for Humanitarian Dialogue.

Hessebon, Gedion. 2014. 'The Fourth Constitution-Making Wave of Africa: Constitutions 4.0?' *Temple International and Comparative Law Journal* 28(2): 185–431.

Ikelegbe, Augustine. 2001. 'Civil Society, Oil and Conflict in the Niger Delta Region of Nigeria: Ramifications of Civil Society for a Regional Resource Struggle'. *The Journal of Modern African Studies* 39(3): 437–469.

Johnston, David. 2007. 'How to Evaluate the Fiscal Terms of Oil Contracts'. *Escaping the Resource Curse*, edited by Macartan, Humphreys; Sachs, Jeffrey and Stiglitz, Joseph. New York: Columbia University Press.

Katz-Lavigne, Sarah. 2017. 'The Renegotiation Window: Resource Contract Renegotiations in the Mining Industry in Africa from 2000 to 2013'. *Resources Policy* 51(1): 22–30.

Kiala, Carine. 2010. 'China-Angola Aid Relations: Strategic Cooperation for Development?' *South African Journal of International Affairs* 17(3): 313–331.

Mangu, Andre. 2005. 'Challenges to Constitutionalism and Democratic Consolidation in Africa'. *Politeia* 24(3): 315–334.

McFerson, Hazel. 2009. 'Governance and Hyper-Corruption in Resource-Rich African Countries'. *Third World Quarterly* 30(8): 1529–1547.

Mughraby, Muhamad. 1966. *Permanent Sovereignty over Oil Resources*. Beirut: The Middle East Research and Publishing Centre.

Ng'ambi, Sangwani. 2015. 'Permanent Sovereignty over Natural Resources and the Sanctity of Contracts, from the Angle of Lucrum Cessans'. *Loyola University Chicago International Law Review* 12(2): 153–172.

Nwabueze, Ben. 1973. *Constitutionalism in the Emergent States*. London: Hurst and Co.

Obi, Cyril. 2010. 'Oil as the "Curse" of Conflict in Africa: Peering through the Smoke and Mirrors'. *Review of African Political Economy* 37(127): 483–495.

Ogboye, Lateef and Abubakri, Yekini. 2014. 'Constitutionalism and Good Governance in Nigeria (1999–2014)'. *Nnamdi Azikiwe University Journal of International Law and Jurisprudence* 5(1): 118–130.

Okeke, Nwachukwu. 2008. 'The Second Scramble for Africa's Oil and Mineral Resources: Blessing or Curse?' *The International Lawyer* 42(1): 193–209.

Omeje, Kenneth. 2016. 'Natural Resource Rent and Stakeholder Politics in Africa: Towards a New Conceptualisation'. *Commonwealth and Comparative Politics* 54(1): 92–114.

Omorogbe, Yinka. 2002. 'The Legal Framework for Public Participation in Decision-Making on Mining and Energy Development in Nigeria: Giving Voices to the Voiceless'. In *Human Rights in Natural Resource Development: Public Participation in the Sustainable Development of Mining and Energy Resources*, edited by Zillman, Donald; Alastair, Lucas and Pring, George. Oxford: Oxford University Press, 548–587.

Omotola, Shola. 2007. 'From the OMPADEC to the NDDC: An Assessment of State Responses to Environmental Insecurity in Niger Delta, Nigeria'. *Africa Today* 51(1): 73–89.

Onifade, Temotope. 2015. 'Peoples-Based Permanent Sovereignty over Natural Resources: Toward Functional Distributive Justice?' *Human Rights Review* 16(4): 343–368.

Ovadia, Jesse. 2012. 'The Dual Nature of Local Content in Angola's Oil and Gas Industry: Development Vs Elite Accumulation'. *Journal of Contemporary African Studies* 30(3): 395–417.

Ovadia, Jesse. 2014. 'Local Content and Natural Resource Governance: The Cases of Angola and Nigeria'. *The Extractive Industries and Society* 1(2): 137–146.

Patey, Luke. 2010. 'Crude Days Ahead? Oil and the Resource Curse in Sudan'. *African Affairs* 109(435): 617–636.

Rajan, Mannaraswamighala. 1978. *Sovereignty over Natural Resources*. New Delhi: Radiant Publishers.

Sand, Philip. 2011. 'The Right to Know: Freedom of Environmental Information in Comparative and International Law'. *Tulane Journal of Internationall and Comparative Law* 20(1): 203–232.

Savage, Emily. 2013. *South Sudan's Petroleum Revenue Management Act*. Juba: The SUDD Institute.

Shaxson, Nicholas. 2005. 'New Approaches to Volatility: Dealing with the "Resource Curse" in Sub-Saharan Africa'. *International Affairs* 81(2): 311–324.

Shivji, Issa. 2013. 'Paradoxes of Constitution-Making in Tanzania'. Paper presented at the East African Law Society Conference, Mombasa.

Stephens, Paul. 2008. 'National Oil Companies and International Oil Companies in the Middle East: Under the Shadow of Government and the Resource Nationalism Cycle'. *The Journal of World Energy Law and Business* 1(1): 3–23.

Tadeg, Mesenbet. 2010. 'Reflections on the Right to Development: Challenges and Prospects'. *African Human Rights Law Journal* 10(2): 325–344.

UNOHCHR. 2013. *Realizing the Right to Development: Essays in Commemoration of 25 Years of the United Nations Declaration on the Right to Development*. Geneva: UN Publication.

Wennmann, Achim. 2011. 'Breaking the Conflict Trap? Addressing the Resource Curse in Peace Processes'. *Global Governance* 17(2): 265–279.

6 Peoples' right to natural resources in Africa and state sovereignty

A quest for equilibrium

Osuntogun Abiodun Jacob

Introduction

There are protracted conflicts relating to the ownership of natural resources in Africa. A careful inquiry into some of these conflicts, from the past to the present, shows that the conflicts have been fought by applying two strategies. One of these strategies is violence. Those who adopt this method take the path of war and intransigence in fighting their cause. Others are more civil in their approach as they make the courts their battlefields. They seek justice from national, regional and international courts to truncate the status quo or the regulatory frameworks in their countries that put the use and control of natural resources and the appropriation of its benefits beyond their reach. However, the African Commission on Human and Peoples' Rights (Africa Commission) and the African Court on Human and Peoples' Rights (African Court) have developed an impressive jurisprudence to protect the rights of indigenous peoples on the continent. The jurisprudence conveys the rule that indigenous peoples can claim ownershipover the natural resources within their territories. In addition, where they are dispossessed of their land, restitution can be ordered.

This chapter interrogates the jurisprudence and its compatibility with domestic, regional and international law, particularly, the right to development (RTD). It argues that although the jurisprudence satisfies the yearnings of indigenous peoples and human rights enthusiasts, it falls short of the expectation of other sections of the population and may affect the actualisation of the RTD negatively if not well implemented. It revisits and critiques the opposition of the African Group to the United Nations Declaration on the Rights of Indigenous Peoples. It further explains the intersection between state sovereignty and ownership of natural resources and argues that the jurisprudence conflicts not only with state sovereignty and the spirit of communalism in Africa, but also with the fundamental law of the land.

Natural resources in Africa

Africa is naturally endowed with an abundance of natural wealth and mineral resources. It shares 30 percent of the mineral reserves in the world.[1] Specifically, for mineral reserves in the production of diamonds, its share of the world stock is 40 percent;[2] for oil, it is 10 percent;[3] for natural gas, it is 8 percent[4] and for uranium, it is 19 percent.[5] Therefore, Africa's extractive industry for both renewable and non-renewable minerals is second to none in the world.

Natural resources 'play a dominant economic, social and political role in the lives of 3.5 billion people living on'[6] the African continent. Natural resources are a major source of engaging in global trade as it is the purveyor of foreign investment inflows to Africa. As raw materials, even without major refinement, they are exported to the world as commodities and thereby contribute immensely to grow the economy and balance of payments in the international market. These are the reasons why states in Africa participate in the world trade, the consequence of which is that most of the exporters take advantage of special preferential treatments under multilateral or bilateral agreements. In addition, most African states do not only rely on natural resources for their public spending; their budgets are prepared based on the forecast of impending profits that might accrue to them by circulating their prices on the world market. It has been reported that the total average of export in Africa is 70 percent and this has contributed to about 28 percent of gross domestic product.[7]

The right to development—controversy at its inception

At the level of the international community, the RTD is still a controversial right despite its endorsement as an inalienable human right in 1986 by the United Nations (UN).[8] Debates on the RTD show that its endorsement has not been without fierce resistance.[9] Resistance to the emergence of the RTD as a human right dates back to the period of its conception right through to the moment for its endorsement at international fora. If not for the resistance, the idea of development as a human right could have been embraced by the UN as far back as 1966. In that year, the General Assembly went into retrospection on the inability of the UN to achieve most of its objectives. To be specific, on 23 September 1966, the Senegalese Foreign Minister, Doudou Thiam, remonstrated on the failure of the UN to achieve its third objective, which is to promote economic and social development for mankind.[10] Deep in that retrospection, he fathomed a way out by conceiving the idea of the right to development.[11] He noted 'we reaffirm our right to development but we must also take the steps which will enable this right to become a reality'.[12]

Almost six years thereafter, nothing significant had happened. It was as if the idea of development as a right had not been suggested. Consequently, in 1972, the then-Chief Justice of the Supreme Court of Senegal, Kéba M'baye took the gauntlet to brush the idea afresh.[13] It is important to note that the resistance against the RTD was not orchestrated because those who suggested it were only from Africa. In the course of the drafting of the Universal Declaration of Human Rights (UDHR), Eleanor Roosevelt argued that it is an essential right which could attract negative consequences if jettisoned in the process of drafting a bill of universal rights for the world.[14] In spite of her persistence in ensuring that equal attention was given to economic, social and cultural rights in the UDHR, this was not so as civil and political rights carried the field day with minimal reference to the latter. As Constantinides argues, the UDHR 'devotes 22 articles to civil and political rights and only five articles to economic, social and cultural rights. As a result, civil liberties are heavily emphasized and privileged, while economic, social and cultural rights, albeit included, are given less attention'.[15]

Even when the adoption of the International Covenant on Economic, Social and Cultural Rights (ICESCR), on 16 December 1966, could have prepared the Western world for easy reception of the RTD in the 1970s and 1980s since both are essential rights that could enhance the attainment of 'freedom from fear and want' (the ideals of the UDHR), it failed to do so. Of course, the opposing view against its recognition as a right was not without its strength of argument as they argued forcefully in contrast to supplant its conception and put an abrupt end to its metamorphosis into a right. M'baye's view, even though endorsed by the General Assembly, was recognised as a moral adventure without legal force and the purported link between it and the UDHR was regarded as a ruse.[16]

The reason behind this resistance seems to be tripartite; it is theoretical, economical and political. With regard to the theoretical aspect, the debate between neo-liberal theorists[17] and the cosmopolitan school of thought[18] influenced the controversy greatly. The neo-liberal theory supports free trade and market-oriented policies without state intervention. It questions the rationale behind the argument that favourable opportunities should be provided for certain peoples and communities to develop socially, culturally and economically through a regulatory reform of the global trade architecture rather than by open competition.[19] To the neo-liberals, the market is an ideal place 'for the production and distribution of wealth' and 'the optimal vehicle for social mobility'[20] and consequently, any super-imposed concessions facilitated under the guise of human rights to redistribute wealth are absurd and contrary to classical international law.

In contrast, cosmopolitanism adopts a practical approach in its response and argument in support of the RTD. It argues on the basis of deductive reasoning using syllogism tinctured with the cause and effects approach. If the neo-liberal school supports and advocates economic globalisation and economic globalisation gives birth to inequalities and divides the world into rich and poor nations with its attendant negative consequences, then, it is only logical and equitable that the neo-liberal school supports the corollary that rights be extended to those who suffered poverty as a result of economic globalisation. This, according to cosmopolitanism, will ensure that those who are negatively affected by the system have a claim not only against the system that produces it but also against those who benefit from it and who are responsible for its perpetuation.[21]

The theoretical discourse is similar to what operates in the political and economic spheres of the debate. With regard to adopting the Declaration on the Right to Development (DRTD), four groups emerged in the UN fora along ideological leanings with each group protecting its common interest. The first group, known as the Non-Aligned Movement (NAM), supported the RTD with the aim to use it as a mechanism to correct the global injustices and other defects associated with the international trade regime.[22] The second group, consisting of some developing countries, fell into the same category as the first group in the sense that it also supported the RTD but distinguished itself from the first group, which appeared too revolutionary. The second group sought, in contrast, to maintain the status quo, as a means to access funding from 'the donor community, the international development agencies, and financial institutions'.[23] In addition, this group yearned to transplant the RTD into the domestic legal systems.[24] The third group consisted of the European Union (EU) and a number of other developed countries which support the RTD with the aim to foster amiable dialogue between the developed and developing countries. [25] For this purpose, the third group desired some level of advancement in its results.

The EU 'sometimes expresses skepticism and occasionally sees its role in the Commission as damage-limitation'.[26] Of course, the third group intended to go along with the draft if nothing particularly objectionable was inserted but more likely than not, abstained from voting in favour of the RTD.[27] The group sought to prevent the opposing group led by the United States of America. It seeks to prevent the successful outcome of efforts towards birthing the RTD. Consequently, the group objected to and opposed everything that could advance the move towards recognition of the RTD. Most of the countries in this group abstained from voting for the adoption of the DRTD, with the United States casting the only dissenting vote.[28]

It is crucial to note that the bedrock of controversy on the RTD is the unwillingness of developed countries to take responsibility for its implementation. Consequently, the controversy splitting the nations of the world into four groups can actually be subsumed into a two-sided argument between two ideological leanings of the North and South divide. The North concedes to recognition of the RTD as a right that is entitled only to individuals as members of a grough and thus, refuses to acknowledge that the state is an equivalent beneficiary of the right.[29] Their argument is also that the RTD is a domestic concern and, therefore, only the national government has the responsibility to implement it and not at the international level and that each state must develop its own mechanism to implement it.[30] States cannot abandon their obligation to enforce the RTD to the international community, although it might provide the necessary development assistance.[31] In a nutshell, developed countries reject a global responsibility to implement the RTD, insisting that no state can compel another state to provide a helping hand, even if the outcome of that gesture would accelerate global development.

Conversely, the countries of the South distinguish between the RTD and collective rights like minority rights and argue that states are beneficiaries of the

RTD.[32] They reject the argument advanced by countries of the North that the jurisdictional scope of the RTD is territorial and not global. In the alternative, it posits that the obligation of the international community to co-operate has gone beyond mere charity endeavours or occasional global development aids, even if they are aimed at eradicating poverty, to a compelling duty demanding commitment to implement it.[33] Countries of the South note that in the midst of globalisation, the outcome of the cumulative effects of the multilateral trade architecture, intellectual property challenges, foreign indebtedness by states and 'foreign domestic economic policies' render them too weak to discharge their responsibilities, and, hence, the need for the international community to assist.[34]

Natural resource curse

That natural resources have become a curse in Africa is not something new to report about. In 1999, Human Rights Watch published a report on Nigeria where it narrated the orgies of violence, miseries, loss of lives, agonies, evils, wanton destruction of properties, and ecological and environmental degradation that the peoples of the Niger-Delta are subjected to just because they are endowed with oil on the irland.[35] It is not an experience to be envied. The Niger-Delta experience is not unlike what other African countries that are endowed with natural resources go through, although, the causes of such nasty experiences differ from one country to another. The civil war in Sierra Leone lasted almost eleven years. The duration was longer in Liberia, where it took twenty-five years and in Angola, twenty-seven years. Similarly, genocides occurred in Rwanda and Sudan to mention but a few. It is worth noting that in all of these conflicts where an uncountable number of peoples lost their lives, precious properties were destroyed and many people were maimed for life, the exploitation of natural resources was a major catalyst.

Thus, the United Nations Environment Programme (UNEP) in its report notes that for the past sixty years, exploitation of natural resources has been the significant factor in at least 40 percent of all civil wars or internal conflicts that occurred in the world.[36] Another report from OXFAM illustrates that 'the recent discovery of gold in the Jebel Amir area in North Darfur has resulted in tribal militias fighting over control of mines'.[37] Of course, the realisation of the role that resource control plays in fuelling civil wars in Africa informed the coinage of words like 'conflict minerals' and 'blood diamonds' and the 'Kimberly Process'[38] mechanism that is deplored to discourage the sale of diamond tainted with conflicts and civil wars.

This negative experience, particularly in Africa, has motivated some scholars[39] to propound a thesis that an adverse effect like a curse indeed runs with natural resource endowments in the same way covenants run with land. This thesis is not without its defects. It is too sweeping and ties the effect of natural resources to not only negativity but to fate, which is unacceptable. What is intended to be a blessing is not meant to be a curse. It depends on the way it is handled. Consequently, some scholars are of the contrary view that '[n]atural resources are a potential asset rather than a pre-ordained curse'.[40]

African jurisprudence in resolving conflicts over resource ownership

It is crucial to note that the preamble to the African Charter gives insight into the path it intends to chart on how to apply natural resources to implementing the RTD.[41] In doing that, it departs from the Western approach of categorisation of human rights according to priority to adopt the concept of 'indivisibility, interdependency, and interrelatedness' of human rights, a concept that was later approved by the United Nations.[42] Consequently, it provides access for all peoples to enjoy 'the common heritage of mankind' in the same dimension.[43] It also guarantees to them the right 'to their economic, social and cultural development' and the freedom to 'dispose of their wealth and natural resources' without hindrance.[44] The DRTD[45] follows the same approach as it accords full sovereignty to peoples 'over all their natural wealth and resources' as an 'inalienable right of the peoples' to development.[46] The DRTD further notes that it includes the right of a person and of peoples to participate in and to contribute to the modalities that can be adopted to 'enjoy economic, social, cultural and political development' for the purpose of attaining the full realisation of 'all human rights and fundamental freedoms'.[47]

The RTD has been the subject of legal analysis in cases decided by the African Commission and the African Court where it has been deplored in support of minority and indigenous peoples, who are marginalised and dispossessed by their governments.[48] In *Centre for Minority Rights Development and Others v Kenya* (Endorois case),[49] the government of Kenya, in order to expropriate the land of the Endorois' community for a game reserve, evicted them from their ancestral lands situated in the 'Lake Bogoria area of the Baringo and Koibatek Administrative Districts, as well as in the Nakuru and Laikipia Administrative Districts within the Rift Valley Province in Kenya'.[50] Besides the eviction that was executed without due consultation, the government failed to compensate them adequately for the land. Disenchanted by the government's failure to compensate them, the Endorois people filed an action against the government at a national court.

In spite of their argument that former President Arap Moi had issued a directive ordering for the said compensations to be paid with a 25-percent share of annual income on the land, the high court dismissed their case and held that the duty owed by the government to the Endorois community for the expropriation of their ancestral land had been discharged with the payment of some money 'in 1986 to 170 families for the cost of relocating' them.[51] In contrast, the African Commission in its judgment found that the government had violated most of the rights of the Endorois community, which are recognised and protected in the African Charter. These are the rights to practice their religion, to culture, to property, to the free disposal of their natural resources and to socio-economic and cultural development.

Consequently, the Commission ordered the government of Kenya to restitute the land to the Endorois and compensate them adequately for the damages they incurred in the process of eviction.

In a similar case, also from Kenya, the *African Commission on Human and Peoples' Rights v Republic of Kenya* (*Ogiek case*),[52] which dealt with the forceful eviction of the Ogiek community from the Mau Forest (their ancestral land), the African Court held that the right of ownership of the Ogiek community to their ancestral land had also been violated. In the Court's finding, the manner in which the forceful evictions were carried out had a negative impact on the livelihood of the Ogiek community and consequently, violated their right to development.

Between peoples' rights and state sovereignty

In all the cases dealing with natural resources and the RTD, the debate or litigation has been between two opposing groups; the community or peoples on the one hand and the state on the other hand. Apart from cases or petitions that are brought on grounds of public interest, it is ironical that petitioners or plaintiffs in these cases are predominantly the peoples whose traditional lands are endowed with natural resources that generate wealth for the entire country. To mention a few, the Mau Forest inhabited by the Ogiek has, with the aim to generate wealth for the country, been allowed to be exploited through logging concessions granted to corporations. The ancestral land of the Endorois community was also expropriated for a game reserve and ruby mining by the government for income purposes. In Nigeria, the oil in the Ogoni land was the mainstream of the country's wealth. In such a situation, how do we resolve the ownership tussle between the peoples and the state over natural resources?

This undoubtedly calls for caution, particularly, when the African Union has consistently pursued the vision that because natural resources constitute the main source of wealth on the continent, they should be explored and utilised 'for the development of our economies for the benefit of our peoples'[53] to ensure that their aspirations for development are not defeated.[54] It seems the arguments of the Kenyan government in the *Endorois and Ogiek* cases tilted towards this view.[55] In the *Endorois* case, the government argued that it had been using the 'proceeds from the game reserve […] to finance a number of projects in the area, such as schools, health facilities, wells and roads'.[56] Of course, the African Commission attests to the fact, which the Endorois community had no objection to, that the 'proceeds from the game reserve have been used to finance a lot of useful projects'.[57] In spite of these developmental initiatives, the African Commission held that the RTD of the Endorois community had been violated by the state.

In the *Ogiek* case, the Kenyan government argued that it had not failed in its obligation to initiate developmental projects to the benefit of the Ogiek community and threw a challenge to the complainant to disprove the claim.[58] The government further argued that it had embarked on a reform of its legislation in order to ensure development for the Ogiek community. For this purpose, the government referred to its 2010 Constitution, its Vision 2030 and 'its budget statements as proof of development for the Ogieks'.[59] The Court, however, held that the RTD of the Ogieks had also been violated. In the *Endorois* case, the African Commission, relied majorly on the regional human rights law in arriving at its conclusion, while in the *Ogiek* case, the African Court relied majorly on international law. In *Endorois*, the Commission considered two sets of precedents, one from its jurisprudence and the other from the Inter-American Court of Human Rights (IACHR). These are the *Saramaka* case[60] decided by the IACHR and the *Ogoni* case, which the African Commission adjudicated on.

Based on the *Saramaka* case, the African Commission noted that the right to property in article 21 of the American Convention is similar to the right to natural resources in article 21 of the African Charter and consequently, the jurisprudence adopted to interpret the provision in the *Saramaka* case could equally be adopted for interpretation in the *Endorois* case.[61] On that assumption, the African Commission held that 'the Endorois peoples have the right to freely dispose of their wealth and natural resources in consultation with the Respondent State'.[62] That decision is important as it settled the issue of ownership of natural resources in favour of the Endorois community, a decision that was reached by relying on the *Ogoni* case. The Commission made reference to the *Ogoni* case, which establishes the principle that 'the right to natural resources contained within their traditional lands is also vested in the indigenous people'.[63] The Commission further noted that such a right, which is conferred on the indigenous peoples or a section of the population by virtue of 'inhabiting a specific region' is a legal right within the territory of a domestic state that entitles them 'to claim the protection of article 21 of the African Charter'.[64]

With regard to the *Ogiek* case, the African Court stated that article 22 of the African Charter must be considered in light of article 23 of the United Nations Declaration on the Rights of Indigenous Peoples (UNDRIP) which provides that indigenous peoples shall have the right to determine the path in which the implementation of their 'development priorities and strategies' will take and that they must 'have the right to be actively involved in developing and determining health, housing and other economic and social programmes affecting them and, as far as possible, to administer such programmes through their own institutions'.[65] Consequently, the Court held that the Ogiek peoples' right to development was violated because they were not informed before they were evicted, nor were they carried along as active participants in developmental projects that would affect them.[66]

At this juncture, it is important to determine the extent of the peoples' right to ownership of natural resources on their traditional land; this is essential in order to avoid confrontation with the state. It is obvious that there is no clear-cut answer to this as the outcome differs from one law to another depending on whether it is domestic, regional or international law. Under domestic law, there is no doubt that minority or indigenous peoples' right to ownership is limited by statutory law that vests the right of ownership to states in most countries of the world including African states.[67] In contrast, proprietary rights are generally vested on the community or the individual by domestic law,[68] which differs from the rights protected under article 21 but falls into the same category as the right to property under article 14 of the African Charter.[69] Even so, the right to property under article 21 is not absolute in the sense that it is susceptible to alienation by the state on the grounds of public interest, although some caveats are built into the exercise of such powers in order to limit power excesses by the state.[70]

On that basis, the argument of the Kenyan government in the *Endorois* case was similar to that of the state of Suriname in the *Saramaka* case in the sense of the non-permissible approach of domestic law which does not recognise the right of minority members or indigenous peoples 'to the use and enjoyment of property in accordance with their system of communal property, but rather a privilege to use land'.[71] In contrast, the Kenyan government departed from this approach in the *Ogiek* case by expounding how the 2010 Constitution accords special protection to the indigenous populations as part of the 'marginalised community', an argument that instigated the Court to holding that in such a change of legal regime, 'the Ogieks could theoretically fit into that category' of those who will benefit from 'such constitutional safeguards'.[72]

The African Court, however, noted that although the so-called constitutional safeguards for the protection of indigenous peoples in Kenya are commendable, they could only prevent the government from liability that might occur in the future and were not able to atone for its past misdeeds 'between the time the Respondent became a Party to the Charter and when the Respondent's new Constitution was enacted'.[73] Consequently, the government was found liable. Of course, the new existing framework in Kenya also exposes a very serious concern in the sense that it is not effective in implementing the rights of indigenous peoples as protected under international law with respect to national resources. This perhaps should be the yardstick adopted by the African Court in evaluating the new regulatory framework when it opined that the failure of the government to put a stop to the eviction of the Ogieks and the unwillingness of the government 'to comply with the decisions of the national courts demonstrate that the new Constitution and the institutions which the Respondent has set up to remedy past or on-going injustices are not fully effective'.[74]

Be that as it may, international law, as noted earlier provides a paradigm shift in this regard. According to the African Commission, 'the jurisprudence under international law bestows the right of ownership' as enunciated by the DRTD 'rather than mere access'.[75] Consequently, the Commission reasoned that if international law were to adopt the domestic approach and grant mere access, 'indigenous peoples would remain vulnerable to further violations/dispossession by the State or third parties'.[76] In order to forestall such unreasonable consequences and to empower the indigenous peoples to 'engage with the state and third parties as active stakeholders rather than as passive beneficiaries', international law departed from such an approach.[77] The African Commission could not but adopt the same approach in its jurisprudence on resource control and ownership. Similarly, the African Court could also not depart from the same approach laid down by the African Commission, since to do otherwise would spell doom to the vulnerable groups. Because of that uniformity, it could be said, without controversy that international human rights law for the protection of indigenous peoples and regional law contained in the African Charter adopts the same approach.

Are they sitting on gunpowder?

During the negotiations for the adoption of the DRTD at the UN, the issue of natural resources and development played a formidable role in determining the position taken by most states, particularly those in Africa, which took the path of no compromise. Most countries remained unperturbed by the efforts of the lobbyists to seek their voting for the DRTD. For this purpose, they are called 'negative countries'.[78] During the debate at the UN, this group maintained a hostile approach against the adoption of the DRTD and never made an attempt to support it. The reason for their indifference was not unconnected with the issue of development and natural resources. There was a deep fear ricocheting on their minds on what would be the fate of their development as a nation if ownership of natural resources in their countries was vested in indigenous peoples. The most vocal members of this group in Africa were Kenya and Nigeria.[79] As noted earlier, Nigeria and Kenya 'depend heavily for a major source of foreign exchange' on petroleum and wildlife-based tourism, which are found in the territories or domains of indigenous peoples. This negative group was supported by another group known as the CANZUS group.[80]

The opposing views against the adoption of the DRTD associated with Africa were championed by Kenya and Nigeria. One issue was paramount in their opposition. It was the issue of conflict between the DRTD and the constitutional provisions of the domestic states on the right to ownership of natural resources by indigenous peoples. This issue was supported with grounds of argument. It may be important to mention five. First, the word 'indigenous' is familiar with Africa but not in the context to which it was used in the DRTD.[81] Therefore, failure to define what 'indigenous' means was considered to be dangerous to Africa.[82]

Second, recognition of the rights of indigenous peoples to self-determination without qualification could be misconstrued and then used as an avenue for secession and disintegration of the country.[83] Third, they contended that the requirement for consultation and to seek 'free, prior and informed consent' of indigenous peoples before 'legislative and administrative' decisions that affect them are taken was also considered to be counter-productive to development as they could veto crucial decisions the governments might want to take.[84] Fourth, they argued that to bestow on indigenous peoples the right to belong to an indigenous community might be misconstrued to mean that they can belong to one country and 'inhabit another' at the same time.[85] Lastly, they claimed that to return to indigenous peoples the lands, territories and resources they had once occupied or acquired would be contrary to the spirit of the Constitution, which prescribes that the power to control natural resources belongs to the states and not indigenous peoples.[86]

Quest for equilibrium

This chapter demonstrates that indigenous peoples and the state are the major contending parties for the soul of natural resources in Africa. With the emergence of the DRTD and liberal jurisprudence of the African Commission and the African Court, there is no doubt that indigenous peoples have found an effective legal instrument to confront the state. The consequence is that the sovereignty of the state is further whittled down. However, it is crucial to note that with respect to control and management of natural resources all over the world, states have a responsibility that cannot be substituted. It is not only because national constitutions confer the power on states to manage resources in the country for the minimum welfare attainable for all the peoples within their territories[87] but also because regional and international human rights instruments depend on them for implementation. The DRTD for example, provides that it is the responsibility of states to implement the guaranteed rights and must take 'the appropriate measures, including legislative measures, to achieve the ends of this Declaration' with consultation and the consent of indigenous peoples.[88]

In such a situation, implementation of the RTD with respect to natural resources calls for a balanced scale that will not undermine the interests protected. In that respect, the DRTD has done justice to the demands of African states by making provision for safeguard mechanisms in article 46(1) as follows:

> Nothing in this Declaration may be interpreted as implying for any State, people, group or person any right to engage in any activity or to perform any act contrary to the Charter of the United Nations or construed as authorizing or encouraging any action which would dismember or impair, totally or in part, the territorial integrity or political unity of sovereign and independent States.

This safeguard mechanism is directed mostly to the judicial officers whose duties are to interpret the law. It is a hint on the limit of the court's powers. The African Court took cognizance of this hint when it opined that there is no justifiable reason why indigenous peoples could not be protected under the African Charter provided that 'such groups or communities do not call into question the sovereignty and territorial integrity of the State without the latter's consent'.[89] To think otherwise, according to the Court, is to assume the unthinkable that

> the drafters of the Charter intended, for example, to automatically recognise for the ethnic groups and communities that constitute their population, the right to self-determination and independence guaranteed under article 20(1) of the Charter, which in this case would amount to a veritable right to secession.[90]

According to the Court, article 20(1) of the Charter sets the boundary line between the quantum of rights that could be protected for the entire peoples of the state and that of the constituent units. It provides that all people shall have 'the unquestionable and inalienable right to self-determination' and 'they shall freely determine their political status and shall pursue their economic and social development according to the policy they have freely chosen'.[91] In contrast, it is arguable that even a glance through articles 4, 5 and 6, (not to talk of a critical look) of the DRTD shows that the same right has been given to them. That is the reason behind the apprehension of African states as discussed earlier. However, it is important to note that article 46 is added to allay the fears of African states and that the right to self-determination in article 20(1) of the African Charter is said to be 'unquestionable'. This is because it relates to the sovereignty of the state. In view of the foregoing, there is a need for appropriate balancing in navigating through these rights on a case-by-case basis. Otherwise, more likely than not, the end result will not be achieved.

Conclusion

In arriving at an equilibrium for the resolution of conflict on natural resources in Africa, the position of the government also deserves special attention. This is what the government of Kenya referred to when it argued that 'while the right of ownership and control of natural resources belongs to the people, states are the entities that would ultimately exercise the enjoyment of the right in the interest of the people'.[92] The people here go beyond indigenous peoples and extend to the entire population of the country.

The Kenyan government's argument is undoubtedly sound, but for the problem of insincerity in implementation.[93] In spite of this, the ball is still in the court of the states. The protection of indigenous peoples is essential; more often, they deserve special attention but at the same time, the overall development of the continent must also not be sacrificed. Navigating these seemingly conflicting interests through equilibrium is possible if state governments in Africa are sincere. However, if states indulge in hypocritical stances, it would be difficult to discharge their constitutional responsibility of allocating resources among their constituent units equitably. Therefore, it is recommended that international and regional administrative and judicial bodies must thoroughly consider the safeguard mechanism before taking decisions that relate to natural resources.

Notes

1 African Natural Resource Centre. No date. 'Catalyzing Growth and Development through Effective Natural Resources Management'.<www.afdb.org/fileadmin/uploads/afdb/Documents/Publications/anrc/AfDB_ANRC_BROCHURE_en.pdf>.
2 This can be found in South Africa, Botswana, Namibia, Angola and Zaire.
3 World Bank Extractive Industries. 2020. 'Overview'. *The World Bank*, 16 April<www.worldbank.org/en/topic/extractiveindustries/overview>; found in Nigeria, Cameroon, Gabon, Congo and Angola.
4 *Ibid.*
5 Dittmar, Michael. 2013. 'The End of Cheap Uranium'. *Sci Total Environ* 461(462): 792.
6 World Bank, note 3.
7 *Ibid.*
8 United Nations General Assembly Declaration on the Right to Development (A/RES/41/128) 4 December 1986.
9 Carty, Anthony. 1984. 'From the Right to Economic Self-Determination to the Right to Development: A Crisis in Legal Theory'. *Third World Legal Studies* 3(5): 75.
10 Whelan, Daniel. 2015. '"Under the Aegis of Man": The Right to Development and the Origins of the New International Economic Order'. *Humanity an International Journal of Human Rights, Humanitarianism and Development* 6(1): 93.
11 United Nations General Assembly, Official Records, Twenty-first session, 1414 Plenary Meeting, September 23, 1966 (228).
12 *Ibid.*
13 M'baye, Kéba. 1983. 'Le Droit au Développment en Droit International'. In *Études de Droit International en L'Honneur du Juge Manfred Lachs*, edited by Makarczyk, Jerzy. The Hague : Martinus Nijhoff, 165.
14 Johnson, Glen. 1987. 'The Contributions of Eleanor and Franklin Roosevelt to the Development of International Protection for Human Rights'. *Human Rights Quarterly* 9(1): 36; Sengupta, Arjun. 2000. 'Realizing the Right to Development'. *Development and Change* 31(3): 555.
15 Constantinides, Aristotelles. 2008. 'Questioning the Universal Relevance of the Universal Declaration of Human Rights'.*Cuadernos Constitucionales de la Cátedra Fadrique Furió Ceriol* 62(63): 60.
16 Donnelly, Jack. 1985. 'In Search of the Unicorn: The Jurisprudence and Politics of the Right to Development'. *California Western International Law Journal* 15: 475.
17 Moyn, Samuel. 2014. 'A Powerless Companion: Human Rights in the Age of Neoliberalism'. *Law and Contemporary Problems* 77(4): 147.
18 Salomon, Margot. 2008. 'Legal Cosmopolitanism and the Normative Contribution of the Right to Development'. In *Implementing the Right to Development: The Role of International Law*, edited by Marks, Stephen. Geneva: Friedrich-Ebert-Stiftung, 17.
19 Haque, Shamshul. 1999. 'The Fate of Sustainable Development under Neo-liberal Regimes in Developing Countries'. *International Political Science Review* 20(2): 218.
20 Espinal, Rosario. 1992. 'Development, Neo-liberalism and Electoral Politics in Latin America'. *Development and Change* 23(4): 27 and 32.
21 Margot, note 18, 22.
22 Marks, Stephen. 2004. 'The Human Right to Development: Between Rhetoric and Reality'. *Harvard Human Rights Journal* 17(1): 141–143.
23 *Ibid.*
24 *Ibid.*
25 *Ibid.*
26 *Ibid.*
27 *Ibid.*
28 DRTD, note 8.

29 Piron, Laure-Hélène. 2002. 'The Right to Development a Review of the Current State of the Debate for the Department for International Development'. *Overseas Development Institute* <www.odi.org/sites/odi.org.uk/files/odi-assets/publications-opinion-files/2317.pdf>.
30 *Ibid.*
31 *Ibid.*
32 *Ibid*, 18–19.
33 *Ibid.*
34 *Ibid.*
35 Human Rights Watch. 1999. 'The Price of Oil: Corporate Responsibility and Human Rights Violations in Nigeria's Oil Producing Communities'. *Human Rights Watch* <www.hrw.org/reports/1999/nigeria/>.
36 XinhuaNet. 2017. 'UNEP: 40 Percent of Internal Conflicts Linked to Exploitation of Natural Resources'. 6 November <www.xinhuanet.com/english/2017-11/07/c_136732699.htm>.
37 OXFAM. 2014. 'We No Longer Share the Land: Agricultural Change, Land and Violence in Darfur'. 29 April.
38 Howard, Audrie. 2016. 'Blood Diamonds: The Successes and Failures of the Kimberley Process Certification Scheme in Angola, Sierra Leone and Zimbabwe'. *Washington University Global Studies Law Review* 15(1): 137.
39 Sachs, Jeffrey and Warner, Andrew. 2001. 'The Curse of Natural Resources'. *European Economic Review* 45(4–6): 828.
40 Gelb, Alan. 2014. 'Should Canada Worry about a Resource Curse?' *School of Public Policy Research Papers* 7(2): 1–26.
41 African Charter on Human and Peoples' Rights adopted in Nairobi, Kenya on 27 June 1981 (1982) 21 ILM 58.
42 Vienna Declaration and Programme of Action adopted at the World Conference on Human Rights in Vienna, 25 June 1993, para 5.
43 African Charter, note 41, art 22.
44 *Ibid*, art 21.
45 DRTD, note 8.
46 *Ibid*, art 1(2).
47 *Ibid*, art 1(1).
48 *Centre for Minority Rights Development (Kenya) and Minority Rights Group (on Behalf of Endorois Welfare Council v Kenya* Comm 276/03; *African Commission on Human and Peoples' Rights v Republic of Kenya* ACtHPR Appl No 006/2012 (2017) (*Ogiek*).
49 *Endorois*, note 48.
50 *Ibid.*
51 *William Yatich Sitetalia, William Arap Ngasia and Others v Baringo Country Council*, High Court Judgment of 19th April 2002, Civil Case No 183 of 2000, para 6.
52 *Ogiek*, note 48.
53 OAU Lagos Plan of Action for the Economic Development of Africa 1980–2000 adopted in Lagos, April 1980, art 3(f).
54 AU Commission. 2015. 'Agenda 2063: The Africa We Want'. African Union, paras 8 and 10.
55 *Endorois*, note 48.
56 *Ibid*, para 253(a).
57 *Ibid*, para 256.
58 *Ogiek*, note 48, para 205.
59 *Ibid*, para 203.
60 *Saramaka People v Suriname*, 2007 Inter-American Court of Human Rights (ser. C) No 172, 28 November 2007, para 95.
61 *Endorois*, note 48, para 256.

62 *Ibid*, para 268.
63 *Ibid*, para 255, 267.
64 *Ibid*.
65 *Ogiek*, note 48, para 209–210.
66 *Ibid*, para 211.
67 Sect 1 of the Land Use Act, chap L5, Laws of the Federation of Nigeria 2004 and on Kenya before the 2010 Constitution, *Endorois*, note 48, paras 174–177.
68 Protected by sect 43 of the 1999 Nigerian Constitution on the right to acquire and own immovable property.
69 African Charter, note 41.
70 *Ibid*, art 14.
71 *Endorois*, note 48, paras 174–177, 194.
72 *Ogiek*, note 48, para 143.
73 *Ibid*, para 143.
74 *Ibid*, para 144.
75 *Endorois*, note 48, para 204.
76 *Ibid*.
77 *Ibid*.
78 Kipuri, Naomi. 2009. 'The UN Declaration on the Rights of Indigenous Peoples in the African Context'. In *Making the Declaration Work: The United Nations Declaration on the Rights of Indigenous Peoples*, edited by Charters, Claire and Stavenhagen, Rodolfo. Copenhagen: International Work Group for Indigenous Affairs.
79 *Ibid*.
80 The CANZUS Group comprises the list of developed countries like Australia, Canada, New Zealand and the USA.
81 To Africa, 'indigenize the economy means to free the economy from imperialism'. UN Department of Public Information press release: 'General Assembly adopts Declaration on Rights of Indigenous Peoples; "major step forward" towards Human Rights for All, says President'. UN Doc GA/10612, 13 September 2007.
82 Africa Group. 2006. 'Draft Aide Memoire on the United Nations Declaration on the Rights of Indigenous Peoples'. 9 November, 2.0.
83 *Ibid*, 3.0.
84 *Ibid*, 6.0.
85 *Ibid*, 5.0.
86 *Ibid*, 7.0.
87 For Nigeria, sect 16(1a and b) of the 1999 Constitution.
88 DRTD, note 11, art 8.
89 *Ogiek*, note 48, para 199.
90 *Ibid*.
91 African Charter, note 41.
92 *Ogiek*, note 48, para 194.
93 *Ibid*, 130.

References

Carty, Anthony. 1984. 'From the Right to Economic Self-Determination to the Right to Development: A Crisis in Legal Theory'. *Third World Legal Studies* 3(5): 73–86.

Constantinides, Aristotelles. 2008. 'Questioning the Universal Relevance of the Universal Declaration of Human Rights'. *Cuadernos Constitucionales de la Cátedra Fadrique Furió Ceriol* 62(63): 49–63.

Dittmar, Michael. 2013. 'The End of Cheap Uranium'. *Science of the Total Environment* 461(462): 792–798.

Donnelly, Jack. 1985. 'In Search of the Unicorn: The Jurisprudence and Politics of the Right to Development'. *California Western International Law Journal* 15(3): 473–509.

Espinal, Rosario. 1992. 'Development, Neo-liberalism and Electoral Politics in Latin America'. *Development and Change* 23(4): 27–48.

Gelb, Alan. 2014. 'Should Canada Worry about a Resource Curse?' *School of Public Policy Research Papers* 7(2): 1–28.

Haque, Shamshul. 1999. 'The Fate of Sustainable Development under Neo-liberal Regimes in Developing Countries'. *International Political Science Review* 20(2): 197–218.

Howard, Audrie. 2016. 'Blood Diamonds: The Successes and Failures of the Kimberley Process Certification Scheme in Angola, Sierra Leone and Zimbabwe'. *Washington University Global Studies Law Review* 15(1): 137–159.

Johnson, Glen. 1987. 'The Contributions of Eleanor and Franklin Roosevelt to the Development of International Protection for Human Rights'. *Human Rights Quarterly* 9(1): 19–48.

Kipuri, Naomi. 2009. 'The UN Declaration on the Rights of Indigenous Peoples in the African Context'. In *Making the Declaration Work: The United Nations Declaration on the Rights of Indigenous Peoples*, edited by Charters, Claire and Stavenhagen, Rodolfo. Copenhagen: International Work Group for Indigenous Affairs.

M'baye, Kéba. 1983. 'Le Droit au Développment en Droit International'. In *Études de Droit International en L'Honneur du Juge Manfred Lachs*, edited by Makarczyk, Jerzy. The Hague: Martinus Nijhoff.

Marks, Stephen. 2004. 'The Human Right to Development: Between Rhetoric and Reality'. *Harvard Human Rights Journal* 17(1): 137–168.

Moyn, Samuel. 2014. 'A Powerless Companion: Human Rights in the Age of Neoliberalism'. *Law and Contemporary Problems* 77(4): 147–170.

Sachs, Jeffrey and Warner, Andrew. 2001. 'The Curse of Natural Resources'. *European Economic Review* 45(4–6): 827–838.

Salomon, Margot. 2008. 'Legal Cosmopolitanism and the Normative Contribution of the Right to Development'. In *Implementing the Right to Development: The Role of International Law*, edited by Marks, Stephen. Geneva: Friedrich-Ebert-Stiftung, 17–26.

Sengupta, Arjun. 2000. 'Realizing the Right to Development'. *Development and Change* 31(3): 553–578.

Whelan, Daniel. 2015. '"Under the Aegis of Man": The Right to Development and the Origins of the New International Economic Order'. *Humanity an International Journal of Human Rights, Humanitarianism and Development* 6(1): 93–108.

7 Colonial extraction of natural resources and the impact on the right to development in Africa

Annelie de Man and Carol Chi Ngang

Introduction

This chapter explores the extent to which the extraction of natural resources during colonialism has impacted on the right to development in Africa. The world is geopolitically structured into First, Second and Third World countries, categorised as such in terms of levels of economic growth and industrialisation. Those that fall into the 'First World' category have had and continue to experience high levels of economic and human development. While some of the countries in the Second World category like Russia, China and India have managed to rise to the status of global prominence in terms of economic strength and technological and infrastructural advancement, many Third World countries continue to experience major rdevelopment challenges. This necessitates a look back into history to determine the root causes of underdevelopment in developing countries, particularly Africa, which Havinden and Meredith think, is often wrongly blamed on colonisation.[1] For them, blame should equally be apportioned to the 'failure of developing countries to adjust adequately to the responsibilities of independence'.[2]

In other words, Havinden and Meredith wonder whether it is reasonable to continue attributing Africa's underdevelopment to colonialism when independence provided the occasion to right the wrongs committed by the colonial administration. They argue that the peaceful and stable environment that existed during colonialism was more conducive to development as compared to the instability and uncertainties that followed decolonisation.[3] Their seemingly good-intentioned positive account of British colonialism unfortunately obliterates the darker side of colonialism, which we revisit and highlight in this chapter.

Without discounting the impact of lopsided economic practices that contribute to development setbacks in Africa, we articulate the argument that the foot prints of colonisation on the prevailing development challenges that Africa continues to experience cannot be denied, particularly with regard to the extensive extraction of natural resources that took place on the continent during the colonial period. Egregious human rights violations committed under the guise of aiming to 'civilise the natives' extensively impoverished indigenous populations and rendered them susceptible to underdevelopment, particularly in terms of human and social progress. The extraction of and trade in natural resources by colonising nations played a significant role in this regard. The aim of this chapter is to examine how colonial practices, characterised by the abusive exploitation of natural resources contributed to and currently limit prospects to achieve the right to development in Africa.

The chapter commences with an overview of colonialism and the contradictions of the European civilising mission on the basis of which the colonisers justified their invasion of the African continent. The discussion then proceeds into examining what post-independence Africa envisaged achieving and how the right to development aims to contribute to that purpose, which as we point out, is hampered by unprecedented tensions over ownership rights and control over natural resources. The chapter concludes with a discussion of the importance of asserting sovereign ownership rights by the peoples of Africa over their natural resources as a means of fulfilling the right to development.

Colonialism and the contradictions of European civilisation

The ideology of social Darwinism that emerged in the late 1800s legitimised the idea that western and northern nations were superior to other countries and their indigenous populations.[4] This led to a Eurocentric construction of the belief that as 'superiors', they had the divine responsibility to 'civilise' the populations of the 'inferior' parts of the world, condescendingly referred to as 'natives' or 'indigenes'.[5] Europeans perceived colonisation as a means by which to enable Africans whom they considered 'uncivilised' to achieve 'progress, enlightenment and modernisation'.[6] The European civilisation theory created the false believe that prior to their encroachment into Africa, no form of enlightenment or progress existed on the continent, which we contend was an unfortunate distortion of historical facts. Africa is known to be the cradle of great ancient civilisations that were constructed exclusively by Africans and with the use of resources extracted exclusively from the African continent. The same is not true for European civilisation, which wholly depended on natural resource extraction from Africa and other parts of the world.

Interestingly, international law (which of course, was nothing other than a transposition of European customary law) and the mechanisms thereof became the vehicle for propagating colonialism. With the establishment of the League of Nations in 1920 and the creation of the mandates system, a 'universalizing language of "trusteeship" of developed and modern nations towards their colonies' was entrenched.[7] Under article 22 of the Covenant of the League of Nations, it was agreed that the development of colonised territories and their peoples who were 'not yet able to stand by themselves' should be placed in 'a sacred trust of civilisation' to be 'entrusted to advanced nations'.[8] This provided a 'humanitarian-religious' justification for intervention based on the idea of *supposedly* universal values of 'civilization', 'material and moral well-being' and 'social progress'.[9]

If the colonisers genuinely intended for Africa to achieve progress, enlightenment and modernisation as was the claim in justification of the colonial project, we supposed that their actions and conduct should have demonstrated evidence of the intention to see those objectives achieved. The realities, however, proved far from the point. The actual motives of colonialism manifested in the manner in which natural resources were abusively extracted from the colonies throughout the entire period of over seven decades of effective colonial occupation[10] and there after. Contrary to the argument that colonisation brought technical benefits and political advancement to the colonised territories, Albert Memmi depicts colonisation as nothing more than an 'arrest' of the human and social development of the colonised through a 'succession of unbearable constraints' and grave human rights violations.[11] As Joe Oloka-Onyango rightly observes, the Europeans only set development goals for the continent, such as the construction of roads, railways and seaports that would result in the realisation of their economic interests.[12]

To understand the point Memmi makes, it is important to have a comprehensive view of development; defined as a process that results in 'a rise in average living standards which involves not merely an increase in material goods, but an enlargement of social and cultural opportunities and widespread access to education, health and recreational facilities', denoting economic, human and social development.[13] According to Lawrence Bamikole, the processes of development must be conceived in the context of an entire social system, which entails moving away from a dire situation largely perceived to be restrictive and exploitative towards a situation that may be considered progressive in every dimension.[14] This process was truncated for the peoples of Africa with the advent of colonialism, which despite justification by its perpetrators with the backing of international law, severely undermined development prospects for the peoples of Africa.

Contrary tò Western ideologies in justification of the colonial enterprise, Emmanuel Eze portrays colonialism as a 'crisis disproportionately suffered by African people in their tragic encounter with the European world over two centuries'.[15] Not only Africa but also countries in Asia, the Middle East and Latin America were affected by the European expansionist policy of colonialism, which Robert Eno describes as

> a system of conquest by force, followed by direct or indirect rule. It entailed an administration or a process of exploitation, and a production system often geared towards the creation of capitalist relations, and the economic and socio-cultural aggrandizement of the coloniser. This may be done by covert or overt, psychological, legal, and military mechanisms.[16]

Colonisation was 'above all economic and political exploitation',[17] based on the practice of taking advantage of labour and land and 'the deliberate undermining of local culture and institutions'.[18] As Memmi rightly points out, '[c]olonization involved theft on all levels',[19] characterised by a complete denial of the human rights and inherent dignity of indigenous populations.[20] The Africans were viewed as sub-human and therefore, not entitled to any sort of entitlement or benefit of progress.[21] As Jean-Paul Sartre rightly states, 'Colonialism denies human rights to human beings whom it has subdued by violence, and keeps them by force in a state of misery and ignorance that Marx would rightly call a subhuman condition'.[22] The 'native' was regarded as a faceless being and thus valueless with no political importance; someone against whom violence was not worthy of justification.[23] In order to suppress uprisingand ensure continued forced labour, police brutality became a general practice.[24] This included killings, torture, unjust imprisonment and degradation of the colonised by those acting on behalf of the colonisers.[25]

The colonised were also denied citizenship rights and not allowed to participate in the administration of their own societies.[26] They were denied the 'most precious rights granted to most men', that is, the liberty and the choice to decide one's own destiny, and more importantly, the freedom to decide one's own path towards progress and what that progress should look like.[27] In effect, the colonised peoples of Africa were, to say the least, deprived of every opportunity for development and hence, the right to develop, which fundamentally contrasted with the foundational ideology of the civilising mission to bring enlightenment, progress and modernity to the supposedly uncivilised native Africans.

Not only civil and political rights, but also economic, social and cultural rights of the colonised were disregarded, including the rights to healthcare, education and living standards.[28] This is evident in the increase in poverty and famine in various countries during the colonial era.[29] Moreover, the reinforcement of entrenched inequalities by the colonisers impeded any type of sustainable economic or social development in the respective colonial territories.[30] According to Richard Gbedoah, 'discriminatory education policies' severely influenced the ability of people to develop themselves, even to this present day.[31]

It was only after the end of the Second World War and with the drawing up of the Universal Declaration of Human Rights in 1948[32] that inherent human dignity and the rights of colonised peoples gradually gained recognition.[33] This recognition was, however, only in principle in the sense that despite the range of rights enshrined in the Declaration, cruel, inhumane and dehumanising treatment, gross violations and abuse of inherent rights persisted even beyond decolonisation. It eventually became evident that colonialism was an intolerable and illegitimate system to be abolished. The post-colonial era then progressively evolved into ensuring that development is guaranteed to all peoples as a human right as reflected in the Declaration on the Right to Development (DRTD), which stipulates that:

> Considering that the elimination of the massive and flagrant violations of the human rights of the peoples and individuals affected by situations such as those resulting from colonialism, neocolonialism, apartheid, all forms

of racism and racial discrimination, foreign domination and occupation, aggression and threats against national sovereignty, national unity and territorial integrity and threats of war would contribute to the establishment of circumstances propitious to the development of a great part of mankind.[34]

Colonial exploitation of natural resources in Africa

Colonial extraction and exploitation as they are used in this chapter, describe not only the incidence where natural resources were looted and taken out of Africa by non-settler colonisers but also the instance where settler colonisers equally dispossessed the African peoples of the resources on which they depended as a source of livelihood and sustenance. Africa has since pre-historic times been richly blessed with an enormous endowment of terrestrial and aquatic resources, including wildlife, gas and fossil fuels, precious minerals, cash crops and foodstuffs, among others. Not withstanding the extent to which these resources are wantonly extracted from Africa, the continent continues to supply the rest of the world in a seemly inexhaustible manner. Although some of these natural resources are highly sought-after export products that fetch huge incomes for state governments, many communities in Africa have from time immemorial relied on them for sustenance and livelihood as well as for traditional, cultural, religious and spiritual purposes.[35]

On account of the fact that natural resources constitute a major contributing factor to development, relating to the 'civilising mission' that brought the Europeans to Africa, one would imagine the colonisers to have shown the colonised peoples (if at all they did not know) how to use the natural resources at their disposal to create or at least improve on their own civilisation, which of course existed prior to colonialism. The colonised peoples of Africa were generally only perceived as an economic utility to advance the colonial enterprise, which thrived principally on mining, commercial farming and logging. The people were, through the enforcement of colonial laws, which sanctioned systemic deprivation, subjugated and used as a source of cheap labourto facilitate extraction of the natural resources, which traditionally and by right of first priority belonged to the local communities.

Veiled in the justification of colonisation as a civilising mission, natural resources, varying from minerals to agricultural products were extracted on a broadscale and exported to Western markets to expand the economic horizons of the colonising nations.[36] Mahmood Mamdani points out that colonial rule happened in two forms; namely settler and extractive colonies.[37] In the extractive colonies, especially those along west and central Africa, because of harsh climatic conditions and the proliferation of tropical diseases, the colonisers mostly embarked on the extraction of natural resources primarily for export to Europe. In the settler colonies, most of them in eastern and southern African, the colonisers established a permanent base and effective control, which allowed them to forcibly expropriate huge portions of land and natural resources from the local and indigenous populations. The European settler communities in these parts of Africa crafted prescriptive laws and authoritarian political institutions that enabled them take 'ownership' of indigenous lands and natural resources,[38] which they exploited solely for their economic interests and benefits.

In South Africa for instance, various Land Acts adopted between 1913 and 1955 legally authorised the appropriation and setting aside of about 80 percent of the country's land exclusively to white ownership and by so doing, stripped the indigenous populations of entitlement or access to the lands, which naturally belonged to them prior to the arrival of the Europeans.[39] In Zimbabwe also, the Land Apportionment Act of 1930 gave ownership of almost half of the agricultural land and its natural reserves to a relatively small number of large-scale white commercial farmers.[40] Nigeria saw the reallocation of ownership over natural resources from the 'primordial public' (local communities) to the state, which included European settlers.[41] As explained earlier, indigenous populations were suppressed and hindered in creating opportunities for their own development.[42]

Local communities were, in most instances, coerced with the use of the gun and forcefully displaced, which as Joshua Settles observes, left the African continent with significant development challenges.[43] When independence was eventually achieved, most of the decolonised peoples of Africa were left with little of their own natural resources or at best significant damage to the natural environment as a result of abusive exploitation.[44] Robert Eno writes that the continent was left 'swindled, pillaged, plundered, exploited, and balkanized'.[45] This is evident in countries with large European settler populations like South Africa and Zimbabwe where several years after independence, the dispossessed peoples have remained impoverished and denied the legitimate claim to land repossession or redistribution. Empirical findings by Heldring and Robinson indicate that 'in most cases post-independence economic decline in Africa can be explicitly attributed to colonialism because the types of mechanisms that led to this decline were creations of colonial society'.[46]

Advancement for local and indigenous communities was severely hindered during the colonial era as a result of loose colonial laws that authorised arbitrary invasion of the commons, which subsequently impacted negatively on prospects for development on the continent. The 1950s saw the birth of anti-colonial struggles for self-determination and claims for sovereign ownership over their natural resources.[47] This eventually culminated in the adoption of various instruments affirming the rights of colonised peoples to permanent sovereignty over their natural resources as a means to achieve socio-economic development and to secure improved living standards for the peoples.[48]

Postcolonial anticipations of permanent sovereignty

Right to socio-economic and cultural development

In asserting the right to self-determination, the peoples of Africa envisaged rising beyond colonial subjugation to exercise permanent territorial sovereignty, including over their natural resources. For the colonised, the struggles for decolonisation meant acknowledgement of the capabilities to make their own development choices against those imposed by colonial administrators. The quest for independence defined by the entitlement to self-determination, provided assurance that all colonised peoples and territories would be able to autonomously set their own development priorities. The quest for independence provided assurance that colonised peoples and territories would be able to autonomously set their own development priorities. As opposed to colonialism, independence, which for Fanon, meant 'the birth of a new humanity'[49] envisaged creating an enabling environment for greater freedoms and expanded opportunities, initially articulated through the campaign for a new international economic order and eventually through asserting the right to development.[50] The United Nations (UN) Secretary General at the time, Kurt Waldheim notes that:

> [T]he central purpose of development is the realization of the potentialities of the human person in harmony with the community; the human person is the subject not the object of development; both material and non-material needs must be satisfied; respect for human rights is fundamental; the opportunity for full participation must be accorded; the principles of equality and non-discrimination must be respected; and a degree of individual and collective self-reliance must be achieved.[51]

In 1986, the DRTD was adopted and served as formal recognition of the relationship between human rights and development at the international level. Prior to the DRTD, the right to development had initially been given formal legal recognition and protection in the African Charter adopted earlier in 1981.[52] The DRTD is, however, more comprehensive in its conceptualisation of the right to development as a fundamental human right and an embodiment of the right to self-determination and entitlement to natural resource ownership for its realisation. It stipulates in article 1 that:

1. The right to development is an inalienable human right by virtue of which every human person and all peoples are entitled to participate in, contribute to, and enjoy economic, social, cultural and political development, in which all human rights and fundamental freedoms can be fully realized.
2. The human right to development also implies the full realization of the right of peoples to self-determination, which includes, subject to the relevant provisions of both International Covenants on Human Rights, the exercise of their inalienable right to full sovereignty over all their natural wealth and resources.

The DRTD not only defines development from a human rights viewpoint, but it also emphasises the importance of the process, which Arjun Sengupta says, must be rights-based.[53] The rights-based conception of development is intended to fulfil the ultimate purpose of ensuring improved human well-being and better living standards. The DRTD additionally embodies a purposive dimension to development, which includes entitlement to full sovereignty over natural wealth and resources, entailing equitable access to and distribution of the benefits derived therefrom. African state governments are accordingly imposed with the duty to ensure that the natural wealth and resources are utilised in a manner that progressively contributes to the realisation of the right to development for the populations across the continent. It imposes an obligation to formulate appropriate policies that are informed by effective participationof the peoples of Africa in the decision-making and governance of natural resources as well as in the equitable distribution of the available resources.[54]

It is noted that the right to development, especially as it is conceived within the African context is not about charity or a solicitation for development assistance, but rather an entitlement to socio-economic and cultural self-determination.[55] Contrary to the understanding of the DRTD as essentially non-binding,[56] it has progressively attained the status of customary international law,[57] meaning that the natural resource requirement for its realisation imposes a binding international law obligation on all states to respect and comply in ensuring equitable redistribution.[58] However, while the right to development is envisioned to provide the mechanism for redress of the development imbalances resulting from the colonial extraction of natural resources in Africa, it is important to highlight the ambiguity in the DRTD, which fails to define actual right holders and the duty bearers to whom responsibility may be imputed to remedy colonial wrongs.

Looking at the international dimension of the right to development and the question of duty bearer at that level, it is not inappropriate to question why the obligation is not apportioned to the European countries that perpetrated colonialism rather than loosely anticipate developing countries to achieve comprehensive development through the undefined mechanism of development cooperation. Even though the DRTD highlights the requirement of sovereign ownership of natural wealth and resources, many of its provisions rather project development cooperation as the principal mechanism for its realisation, thereby masking the quintessential relevance of natural resources for development and the need to assert ownership over it. While it took the natural resources from Africa to drive and sustain European industrialisation and economic advancement, presently, African countries are mostly only required to look up to these developed countries for assistance in their efforts to achieve comprehensive development.

The African Charter enshrines the right to development as a legally binding entitlement, which guarantees in article 22 that '[a]ll peoples shall have the right to their economic, social and cultural development with due regard to their freedom and identity and in the equal enjoyment of the common heritage of mankind'. The natural resource requirement for the realisation of the right to development in Africa is embodied in the concept of the 'common heritage',[59] which read in conjunction with article 21 of the African Charter, guarantees exclusive natural resource ownership rights to the peoples of Africa. On the basis of this understanding, the African Commission affirmed in the *Endorois* case that by dispossessing the Endorois peoples of their ancestral land, the Kenyan government violated their right to development.[60]

In addition, the Commission identified two associated requirements for the realisation of the right to development. The procedural element requires the participation and 'free, prior, and informed consent' of all affected in the development processes while the substantive element requires communities to benefit from the development processes that impact on their livelihood. By this two-pronged test, the precedent is set that no state party to the African Charter can justifiably dispose of natural resources without the active and meaningful involvement of the peoples in the decision-making to that effect and also without adequate guarantee that they would substantively benefit from the development gains resulting from the exploitation of their natural resources. Accordingly, the African Commission held that the affected Endorois community must be able to 'reasonably share in the benefits made as a result of a restriction or deprivation of their right to the use and enjoyment of their traditional lands and of those natural resources necessary for their survival'.[61] The importance of considering ownership over natural resources as integral to the realisation of the right to development is further explained in the next section.

Natural resource ownership

Natural resource ownership as a core component of the right to development equips the peoples of Africa with the right to self-determination in shaping their own development choices. It entails determining how the natural resource endowments on the African continent could productively be utilised for the purpose of socio-economic and cultural development. Common article 1 of the International Covenant on Civil and Political Rights and the International Covenant on Economic, Social and Cultural Rights provide that:

> 1. All peoples have the right of self-determination [...].
> 2. All peoples may, for their own ends, freely dispose of their natural wealth and resources without prejudice to any obligations arising out of international economic co-operation, based upon the principle of mutual benefit, and international law. In no case may a people be deprived of its own means of subsistence.[62]

These provisions suggest that the right to ownership and entitlement to freely dispose of natural resources may not negatively affect a state's international cooperation obligations. By emphasising that '[i]n no case may a people be deprived of its own means of subsistence', assurance is given that in entering into international cooperation agreements, the state must ensure that the terms of cooperation arrangements do not amount to depriving the peoples of their natural resources, which is recognised as avital component for their survival.

The UN Charter makes several references to the right to self-determination;[63] which represents the right to choose how to dispose of natural wealth and resources, and has been accepted by the UN General Assembly as constituting a basic component of the right to self-determination.[64] As stated earlier, the right to self-determination forms the foundation of the right to development. In addition to article 22 of the African Charter, articles 20 and 21 also affirm the right to self-determination guaranteed to the peoples of Africa as also constituting the right to freely dispose of their natural resources. Priscilla Claeys explains that the right to dispose of natural resources, conceived as a collective right, includes the right of peoples to participate in and to benefit from the exploitation of the natural resources over which they have sovereignty.[65] The rights of a community or peoples to 'full and complete' sovereignty over their natural wealth and resources are stated both in the preamble and in the text of the DRTD.

In line with the right to self-determination is the principle of permanent sovereignty over natural resources that emerged within the context of decolonisation as a principle of international law within the decades after the adoption of the Universal Declaration of Human Rights. After much contestation at the level of the UN General Assembly, especially by the industrialised countries of the north, it was eventually agreed that recognition of the right to self-determination and the right to sovereignty over natural resources were essential for the process of decolonisation and an imperative element for development.[66] The principle of permanent sovereignty guarantees to all people the freedom to determine whether the natural resources that traditionally belong to them may be disposed of.[67]

Nicolaas Schrijver explains that the principle of permanent sovereignty was conceived with the purpose to restrict the unlawful extraction of natural resources by colonial powers.[68] The idea was to provide a 'legal shield against infringements of [indigenous communities] economic sovereignty as a result of property rights or contractual rights claimed by other States (often the former colonial Powers)'.[69] Eventually, the Declaration on Permanent Sovereignty over Natural Resources[70] was adopted on 14 December 1962. It is interesting to note that South Africa was one of only two countries that voted against the adoption of the Declaration, which of course, reflected the apartheid policy that overtly dispossessed the black population of their lands and the natural resources thereon. Voting in favour of the Declaration on Permanent Sovereignty over Natural Resources would have meant for the apartheid government, an acknowledgement to return the lands and natural resources that were forcefully taken from the indigenes.

While articles 3, 10, 25 and 26 of the Declaration on the Rights of Indigenous Peoples provide for the right to self-determination over their traditional lands and resources for indigenous peoples, Schrijver writes:

> it should be noted that none of these provisions vests indigenous peoples expressis verbis with permanent sovereignty over their natural wealth and resources or entails exclusive rights for indigenous peoples over the natural resources within their territories. Rather, they vest indigenous peoples with clear-cut rights to consultation in decision-making and to benefit-sharing.[71]

Several other UN Declarations that have been adopted also give expression to rights associated with natural resources.[72] Together with the instruments discussed above, Schrijver identifies six criteria for determining sovereignty over natural wealth and resources:

> (a) natural resources should be employed for national development and the well-being of the people; (b) the rights of indigenous peoples to their habitat and its natural resources should be protected; (c) natural resources should be properly and prudently managed, based upon the principle of sustainable use; (d) nationalization and marine resource-related policies should be implemented 'in accordance with international law'; (e) due care should be paid to the environment without compromising the rights of future generations; (f) states should cooperate for worldwide sustainable development.[73]

These criteria provide justification for the peoples across Africa to legitimately assert the right to sovereignty over the natural resources which they were deprived of during colonialism and continue to be deprived of even to this present day. As stated earlier, the right to natural resource ownership forms a central component of the right to self-determination, which also is an essential integral component of the right to development. By implication, to achieve the right to development necessitates claiming not only the right to self-determination but also the right to permanent sovereignty over natural resources, which constitutes a means of subsistence essential for creating better living standards. By dispossessing the peoples of Africa through the wanton extraction of their natural resources during the colonial period, both their right to self-determination and the right to development were fundamentally also denied. As a consequence, that colonial legacy has continued to impact on development in Africa to this present day.

Returning to the *Endorois* case it is worth noting that the right to permanent sovereignty over natural resources is not only essential to facilitate or accelerate development, it is also intricately linked to the rights to freedom of conscience, beliefs and religious practices and to freely take part in cultural community life.[74] In that case, it was argued that the indigenous Endorois peoples regarded their traditional land as sacred and thus, was used for a variety of traditional and religious practices associated with their customary way of life. Accordingly, the African Commission found that by prohibiting access of the Endorois peoples to their ancestral lands, the Kenyan government violated not only their right to development but also all the other associated rights.[75]

Conclusion

In this chapter, we have shown how the extraction of natural resources during the colonial period impacted on the African peoples' right to utilise those resources for their own development and thus contributed to the development challenges that many African countries have continued to experience to this day. The discussion commenced with an overview of Havinden and Meredith's argument in justification of colonisation.[76] It is over half a century since colonialism was abolished; giving the peoples of Africa the opportunity to shape their own development future. To do so necessitates revisiting the question of natural resource ownership, without which socio-economic and cultural advancement on the continent continues to be hampered. While it is appropriate to apportion part of the blame for Africa's current state of retarded progress to the peoples of Africa, the reality cannot be ignored that the extraction of natural resources during the colonial era significantly limited prospects for development on the continent.

A discussion on the right to development as an integral part of customary international law is then undertaken to illustrate its correlation with natural resource ownership as a fundamental prerequisite for socio-economic and cultural development. The right to development conveys the understanding that the pursuit of development goals and the processes thereof must comply with human rights standards, which includes the duty on states to guarantee equitable access to and enjoyment of the natural resources needed for development, for which recognised as a means of subsistence, contemporary international law guarantees that under no circumstance may a people be deprived of.

Notes

1 Havinden, Michael and Meredith, David. 1993. *Colonialism and Development: Britain and its Tropical Colonies, 1850–1960*. New York: Routledge, xiii.

2 *Ibid.*

3 *Ibid.*

4 Hewitt, Vermon. 2013. 'Empire, International Development and the Concept of Good Government'. In *Empire Development and Colonialism: The Past in the Present*, edited by Duffield, Mark and Hewitt, Vermon. Pretoria: HSRC Press, 32.

5 Rist, Gilbert. 2008. *The History of Development: from Western Origins to Global Faith* (3rd edn, Camiller, P. trl). London: Zed Books, 47. On the side of morality, two main arguments were put forth supporting colonisation: First, that colonisation was the fulfilment of a moral obligation that rests on all great nations 'to employ the strengths and advantages that they have received from Providence for the general good of humanity' and second, on fait accompli (moral responsibility), entailed that where a country already has colonies under its control it is the duty of that country to 'preserve' them (Rist, 55). Fieldhouse, David. 1983. *Colonialism, 1870–1945: An Introduction*. London: Palgrave Macmillan; Elkins, Caroline and Pedersen, Susan. (Eds). 2005. *Settler Colonialism in the Twentieth Century: Projects, Practices, Legacies*. London/New York: Routledge; McMichael,

Philip. 2012. *Development and Social Change: A Global Perspective* (5th edn). London: Sage Publication, 3 and 26–38; Gbedoah, Richard. 2018. 'On Colonialism and Development—Why the Underdevelopment of the South Cannot Be Delinked from the Experience of the Past'. *African Journal of Governance and Development* 7(1): 7; Maseland, Robert. 2018. 'Is Colonialism History? The Declining Impact of Colonial Legacies on African Institutional and Economic Development'. *Journal of Institutional Economics* 14(2): 259, 261.

6 Reddy, Thiven. 2016. *South Africa, Settler Colonialism and the Failures of Liberal Democracy*. London: Zed Books, 54, 88; Richard, note 5, 7.

7 Hewitt, note 4, 32; Emerson, Rupert. 1965. 'Colonialism, Political Development and the UN'. *International Organization* 19(3): 484, 485.

8 Covenant of the League of Nations adopted on 29 June 1919.

9 Rist, note 5, 62.

10 Eno, Robert. 2008. 'Human Rights, Human Development, and Peace: Inseparable Ingredients in Africa's Quest for Prosperity'. PhD Thesis, University of Witwatersrand, 69; Caouette, Dominique and Kapoor, Dipp (Eds). 2016. *Beyond Colonialism, Development and Globalization: Social Movements and Critical Perspectives*. London: Zed Books, 2.

11 Memmi, Albert. 2006. *Decolonization and the Decolonized* (Bononno, R. trl). Minneapolis: University of Minnesota Press, 21, 49.

12 Oloka-Onyango, Joe. 1999. 'Heretical Reflections on the Right to Self-determination: Prospects and Problems for a Democratic Global Future in the New Millennium'. *American University International Law Review* 15(1): 151, 171.

13 Havinden and Meredith, note 1, 5.

14 Bamikole, Lawrence. 2012. 'Nkrumah and the Triple Heritage Thesis and Development in Africana Societies'. *International Journal of Business, Humanities and Technology* 2(2): 68, 70.

15 Eze as quoted in Reddy, note 6, 54; Richard, note 5, 7.

16 Eno, note 10, 66.

17 Memmi, Albert. 1965. *The Colonizer and the Colonized* (Greenfeld, H. trl). London: Orion Press, 149.

18 Reddy, note 6, 53; Lange, Matthew; Mahoney, James and vom Hau, Matthias. 2006. 'Colonialism and Development: A Comparative Analysis of Spanish and British Colonies'. *American Journal of Sociology* 111(5): 1412, 1439.

19 Césaire, Aimé. 2000. *Discourse on Colonialism* (Pinkham, J.trl). New York: Monthly Review Press, 9, 36–37; Memmi 2006, note 11, 20; Memmi 1965, note 17, 74; Lange, Mahoney and vom Hau, note 18, 1446; Reddy, note 6, 56.

20 Eno, note 10, 69, 71.

21 Sartre, Jean-Paul. 1965. 'Introduction'. In *The Colonizer and the Colonized*, edited by Memmi, Albert and Greenfeld, Howard. (trl), London: Orion Press, xxvi–xxv.

22 *Ibid*, xxiv.

23 Reddy, note 6, 88.

24 *Ibid*, 90; Lange, Mahoney and vom Hau, note 18, 1439; Eno, note 10, 69, 73; Richard, note 5, 7.

25 Césaire, note 19, 32, 35.

26 Memmi 1965, note 17, 96.

27 *Ibid*, 85, 91.

28 Havinden and Meredith, note 1, 196–198; Césaire, note 19, 46; Lange, Mahoney and vom Hau, note 18, 1419–1420; Eno, note 9, 71.

29 Sheldon, Richard. 2013. 'Development, Poverty and Famines: The Case of the British Empire'. In *Empire Development and Colonialism: The Past in the Present*, edited by Duffield, Mark and Hewitt, Vermon. Pretoria: HSRC Press, 38.

30 Lange, Mahoney and vom Hau, note 18, 1446; Aspengren, Henrik. 2013. 'Empire: A Question of Hearts? The Social Turn in Colonial Government, Bombay c. 1905–1925'. In *Empire Development and Colonialism: The Past in the Present*, edited by Duffield, Mark and Hewitt, Vermon. Pretoria: HSRC Press, 48.

31 Richard, note 5, 12.
32 Universal Declaration of Human Rights adopted on 10 December 1948.
33 Eno, note 10, 71.
34 Declaration on the Right to Development Resolution A/RES/41/128 adopted by the United Nations General Assembly (UNGA) on 4 December 1986.
35 Mkumbukwa, Abdallah. 2008. 'The Evolution of Wildlife Conservation Policies in Tanzania during the Colonial and Post-independence Periods'. *Development Southern Africa* 25(5): 590–591.
36 Eno, note 10, 72.
37 Mamdani, Mahmoud. 1996. *Citizen and Subject: Contemporary Africa and the Legacy of Late Colonialism*. Princeton: Princeton University Press, 17.
38 Easterly, William and Levine, Ross. 2012. 'The European Origins of Economic Development'. NBER Working Paper Series Working Paper 18162, 1; Lange, Mahoney and vom Hau, note 17, 1452.
39 Encyclopaedia Britannica. 'Apartheid Social Policy'. *Britannica* <www.britannica.com/topic/apartheid#ref210036>.
40 Chigonda, Tanyaradzwa. 2018. 'More than Just Story Telling: A Review of Biodiversity Conservation and Utilisation from Precolonial to Postcolonial Zimbabwe'. *Scientifica* 8: 4.
41 Umejesi, Ikechukwu. 2015. 'Collective Memory, Coloniality and Resource Ownership Questions: The Conflict of Identities in Postcolonial Nigeria'. *Africa Review* 7(1): 42, 43; Göcke, Katja. 2013. 'Protection and Realization of Indigenous Peoples' Land Rights at the National and International Level'. *Goettingen Journal of International* 5(1): 87, 91–96.
42 Easterly and Levine, note 38, 1.
43 Settles, Dwayne. 1996. 'The Impact of Colonialism on African Economic Development'. University of Tennessee Honors Thesis Projects, 12.
44 Eno, note 10, 4.
45 *Ibid*, 297.
46 Heldring, Leander and Robinson, James. 2012. 'Colonialism and Economic Development in Africa'. *National Bureau of Economic Research—Working Paper 18566*, 10–12.
47 Schrijver, Nicolaas. 2013. 'Self-determination of Peoples and Sovereignty over Natural Wealth and Resources'. In *Realizing the Right to Development: Essays in Commemoration of 25 Years of the United Nations Declaration of the Right to Development*, edited by UN Human Rights. Geneva: UN Publication, 97.
48 UNGA Resolution 626(VII) on the Right to Exploit Freely Natural Wealth and Resources adopted on 21 December 1952; UNGA Resolution 1803(XVII) of 14 December 1962 on Permanent Sovereignty over Natural Resources, para 1; International Covenant on Economic, Social and Cultural Rights (ICESCR) adopted by GA Resolution 2200A(XXI) 1966, art 1(2) and the International Covenant on Civil and Political Rights (ICCPR) adopted by GA Resolution 2200A(XXI) 1966, art 1(2); Resolutions 2158(XXI), 2386(XXIII) and 2692(XXV) adopted in 1966, 1968 and 1970, respectively; Declaration on the Right to Development Resolution A/RES/41/128 adopted by the UNGA on 4 December 1986, art 1(2).
49 Fanon, Frantz. 1968. *The Wretched of the Earth* (Farington, C. trl). New York: Grove Press, 35.
50 Cornwall, Andrea and Nyamu-Musembi, Celestine. 2004. 'Putting the "Rights-based Approach" to Development into Perspective'. *Third World Quarterly* 25(8): 1422; Gready, Paul and Ensor, Jonathan. 2005. *Reinventing Development: Translating Rights-Based Approaches from Theory into Practice*. London: Zed Books, 18.
51 Quoted in Gready and Ensor, note 50, 18.
52 African Charter on Human and Peoples' Rights, adopted in 1981 and came into effect on 21 October 1986.
53 Sengupta, Arjun. 2002. 'On the Theory and Practice of the Right to Development'. *Human Rights Quarterly* 24(4), 846–848; Sengupta, Arjun. 2004. 'The Human Right to Development'. *Oxford Development Studies* 32(2): 180–181.
54 DRTD, note 34, preamble and art 10.

55 Ngang, Carol. 2018. 'Towards a Right-to-Development Governance in Africa'. *Journal of Human Rights* 17(1): 112.
56 Uvin, Peter. 2007. 'From the Right to Development to the Rights-based Approach: How "Human Rights" Entered Development'. *Development in Practice* 17(4/5): 598.
57 Mokhiber, Graig. 2001. 'Toward a Measure of Dignity: Indicators for Rights-based Development'. *Statistical Journal of the United Nations Economic Commission for Europe* 18(2–3): 156; Davis, Thomas. 2009. 'The Politics of Human Rights and Development: The Challenge for Official Donors'. *Australian Journal of Political Science* 44(1): 173, 176.
58 Vienna Declaration and Programme of Action adopted on 25 June 1993, para 8; Uvin, note 56, 598.
59 Ngang, Carol. 2020. 'Right to Development in Africa and the Common Heritage Entitlement'. *Journal for Juridical Science* 45(1): 41–42.
60 *Centre for Minority Rights Development (Kenya) and Minority Rights Group International on Behalf of Endorois Welfare Council v Kenya* Comm 276/2003; African Commission on Human and Peoples Rights, 27th Activity Report (2009).
61 *Endorois*, note 60, para 295.
62 International Covenant on Economic, Social and Cultural Rights adopted on 16 December 1966; International Covenant on Civil and Political Rights adopted on 16 December 1966.
63 Charter of the United Nations adopted on 26 June 1945, arts 1(2) and 55.
64 Chekera, Yolanda and Nmehielle, Vincent. 2013. 'The International Law Principle of Permanent Sovereignty over Natural Resources as an Instrument for Development: The Case of Zimbabwean Diamonds'. *African Journal of Legal Studies* 6(1): 69, 74; Claeys, Priscilla. 2015. 'Rights to Sovereignty over Natural Resources, Development and Food Sovereignty'. *FIAN International Briefing* 1, 3; UN General Assembly Resolution 626 (VII) of 21 December 1952; Resolution 1314 (XIII) of 12 December 1958.
65 Claeys, note 64, 3.
66 Schrijver, note 47, 100; Chekera and Nmehielle, note 64, 74.
67 Schrijver, note 47, 96.
68 *Ibid*, 95.
69 *Ibid*.
70 Declaration on Permanent Sovereignty over Natural Resources adopted on 14 December 1962.
71 Schrijver, note 47, 99; Declaration on the Rights of Indigenous Population adopted on 13 December 2007, art 32.
72 Declaration of the United Nations Conference on the Human Environment adopted on 16 June 1972; Convention on the Law of the Sea adopted on 10 December 1982; Rio Declaration on Environment and Development adopted at the 1992 UN Conference on Environment and Development; Johannesburg Declaration on Sustainable Development adopted at the 2002 World Summit on Sustainable Development; Charter of Economic Rights and Duties of States adopted by UN General Assembly Resolution 3281 (XXIX) on 12 December 1974.
73 Schrijver, note 47, 101.
74 African Charter, note 52, arts 8 and 17.
75 Ashamu, Elizabeth. 2011. 'Centre for Minority Rights Development (Kenya) and Minority Rights Group International on Behalf of Endorois Welfare Council v Kenya: A Landmark Decision from the African Commission'. *Journal of African Law* 55(2): 305–309.
76 Havinden and Meredith, note 1, xiii.

References

Ashamu, Elizabeth. 2011. 'Centre for Minority Rights Development (Kenya) and Minority Rights Group International on Behalf of Endorois Welfare Council v Kenya: A Landmark Decision from the African Commission'. *Journal of African Law* 55(2): 300–313.

Aspengren, Henrik. 2013. 'Empire: A Question of Hearts? The Social Turn in Colonial Government, Bombay c. 1905–1925'. In *Empire Development and Colonialism: The Past in the Present*, edited by Duffield, Mark and Hewitt, Vernon. Pretoria: HSRC Press, 45–58.

Bamikole, Lawrence. 2012. 'Nkrumah and the Triple Heritage Thesis and Development in Africana Societies'. *International Journal of Business, Humanities and Technology* 2(2): 68–76.

Caouette, Dominique and Kapoor, Dipp (Eds). 2016. *Beyond Colonialism, Development and Globalization: Social Movements and Critical Perspectives*. London: Zed Books.

Césaire, Aimé. 2000. *Discourse on Colonialism* (Pinkham, J. trl). New York: Monthly Review Press.

Chekera, Yolanda and Nmehielle, Vincent. 2013. 'The International Law Principle of Permanent Sovereignty over Natural Resources as an Instrument for Development: The Case of Zimbabwean Diamonds'. *African Journal of Legal Studies* 6(1): 69–101.

Chigonda, Tanyaradzwa. 2018. 'More than Just Story Telling: A Review of Biodiversity Conservation and Utilisation from Precolonial to Postcolonial Zimbabwe'. *Scientifica* 8: 1–11.

Claeys, Priscilla. 2015. 'Rights to Sovereignty over Natural Resources, Development and Food Sovereignty'. *FIAN International Briefing*.

Cornwall, Andrea and Nyamu-Musembi, Celestine. 2004. 'Putting the "Rights-based Approach" to Development into Perspective'. *Third World Quarterly* 25(8): 1415–1437.

Davis, Thomas. 2009. 'The Politics of Human Rights and Development: The Challenge for Official Donors'. *Australian Journal of Political Science* 44(1): 173–192.

Easterly, William and Levine, Ross. 2012. 'The European Origins of Economic Development'. *NBER Working Paper Series Working Paper 18162*.

Elkins, Caroline and Pedersen, Susan. (Eds). 2005. *Settler Colonialism in the Twentieth Century: Projects, Practices, Legacies*. London/New York: Routledge.

Emerson, Rupert. 1965. '*Colonialism*, Political *Development and the UN*'. *International Organization* 19(3): 484–503.

Eno, Robert. 2008. 'Human Rights, Human Development, and Peace: Inseparable Ingredients in Africa's Quest for Prosperity'. PhD Thesis, University of Witwatersrand.

Fanon, Frantz. 1968. *The Wretched of the Earth* (Farington, C. trl). New York: Grove Press.

Fieldhouse, David. 1983. *Colonialism, 1870–1945: An Introduction*. London: Palgrave Macmillan.

Gbedoah, Richard. 2018. 'On Colonialism and Development—Why the Underdevelopment of the South Cannot be Delinked from the Experience of the Past'. *African Journal of Governance and Development* 7(1): 6–16.

Göcke, Katja. 2013. 'Protection and Realization of Indigenous Peoples' Land Rights at the National and International Level'. *Goettingen Journal of International* 5(1): 87–154.

Gready, Paul and Ensor, Jonathan. 2005. *Reinventing Development: Translating Rights-Based Approaches from Theory into Practice*. London: Zed Books.

Havinden, Michael and Meredith, David. 1993. *Colonialism and Development: Britain and Its Tropical Colonies, 1850–1960*. New York: Routledge.

Heldring, Leander and Robinson, James. 2012. 'Colonialism and Economic Development in Africa'. *National Bureau of Economic Research*—Working Paper 18566.

Hewitt, Vernon. 2013. 'Empire, International Development and the Concept of Good Government'. In *Empire Development and Colonialism: The Past in the Present*, edited by Duffield, Mark and Hewitt, Vernon. Pretoria: HSRC Press.

Lange, Matthew; Mahoney, James and vom Hau, Matthias. 2006. 'Colonialism and Development: A Comparative Analysis of Spanish and British Colonies'. *American Journal of Sociology* 111(5): 1412–1462.

Mamdani, Mahmoud. 1996. *Citizen and Subject: Contemporary Africa and the Legacy of Late Colonialism*. Princeton: Princeton University Press.

Maseland, Robert. 2018. 'Is Colonialism History? The Declining Impact of Colonial Legacies on African Institutional and Economic Development'. *Journal of Institutional Economics* 14(2): 259–287.

McMichael, Philip. 2012. *Development and Social Change: A Global Perspective* (5th edn). London: Sage Publication.

Memmi, Albert. 1965. *The Colonizer and the Colonized* (Greenfeld, H. trl). London: Orion Press.

Memmi, Albert. 2006. *Decolonization and the Decolonized* (Bononno, R. trl). Minneapolis: University of Minnesota Press.

Mkumbukwa, Abdallah. 2008. 'The Evolution of Wildlife Conservation Policies in Tanzania during the Colonial and Post-independence Periods'. *Development Southern Africa* 25(5): 589–600.

Mokhiber, Graig. 2001. 'Toward a Measure of Dignity: Indicators for Rights-based Development'. *Statistical Journal of the United Nations Economic Commission for Europe* 18(2–3): 155–162.

Ngang, Carol. 2018. 'Towards a Right-to-Development Governance in Africa'. *Journal of Human Rights* 17(1): 107–122.

Ngang, Carol. 2020. 'Right to Development in Africa and the Common Heritage Entitlement'. *Journal for Juridical Science* 45(1): 28–50.

Oloka-Onyango, Joe. 1999. 'Heretical Reflections on the Right to Self-determination: Prospects and Problems for a Democratic Global Future in the New Millennium'. *The American University International Law Review* 15(1): 151–208.

Reddy, Thiven. 2016. *South Africa, Settler Colonialism and the Failures of Liberal Democracy*. London: Zed Books.

Rist, Gilbert. 2008. *The History of Development: from Western Origins to Global Faith* (3rd edn, Camiller, P. trl). London: Zed Books.

Sartre, Jean-Paul. 1965. 'Introduction'. In *The Colonizer and the Colonized*, edited by Memmi, Albert and Greenfeld, Howard (trl). London: Orion Press.

Schrijver, Nicolaas. 2013. 'Self-determination of Peoples and Sovereignty over Natural Wealth and Resources'. In *Realizing the Right to Development: Essays in Commemoration of 25 Years of the United Nations Declaration of the Right to Development*, edited by UN Human Rights. Geneva: UN Publication, 95–102.

Sengupta, Arjun. 2002. 'On the Theory and Practice of the Right to Development'. *Human Rights Quarterly* 24(4): 837–889.

Sengupta, Arjun. 2004. 'The Human Right to Development'. *Oxford Development Studies* 32(2): 179–203.

Settles, Dwayne. 1996. 'The Impact of Colonialism on African Economic Development'. University of Tennessee Honors Thesis Projects.

Sheldon, Richard. 2013. 'Development, Poverty and Famines: The Case of the British Empire'. In *Empire Development and Colonialism: The Past in the Present*, edited by Duffield, Mark and Hewitt, Vernon. Pretoria: HSRC Press.

Umejesi, Ikechukwu. 2015. 'Collective Memory, Coloniality and Resource Ownership Questions: The Conflict of Identities in Postcolonial Nigeria'. *Africa Review* 7(1): 42–54.

Uvin, Peter. 2007. 'From the Right to Development to the Rights-based Approach: How "Human Rights" Entered Development'. *Development in Practice* 17(4/5): 597–606.

Part 2

Permanent sovereignty over natural resources

8 Issues of sovereignty over natural resources in Africa

Faith Kabata

Introduction

Statist views in political theory posit that the state must have control over territory to function as such.[1] From this argument the 'territorial right' accruing to the state encapsulates: jurisdictional rights to make and impose laws over the territory; right to control, extract and impose levies on resources in the territory; and the right to control the entry and exit of goods and peoples within the territory.[2] With regard to the control of natural resources, the question of who owns the resources within the territory of the state becomes more explicit.

Considerable debate surrounds the doctrine of permanent sovereignty over natural resources. While public international law attributes sovereignty over natural resources to states, international human rights law holds that it is the peoples within a state that have sovereignty over natural resources. Importantly, public international law confers on peoples the right to political self-determination, which in the post-decolonisation era has been associated with internal self-determination, the 'on-going right of peoples to govern themselves'.[3] Equally, international human rights law accords to peoples within the state the right to self-determination, which includes the right to freely dispose of their natural wealth and resources, thus introducing the concept of peoples' governance over natural resources.

Human rights treaties at the global and regional levels vest permanent sovereignty over natural resources in peoples, although ambiguities in the text of the treaties have left the debate unsettled. In relation to the rights of indigenous peoples, the international human rights system has been consistent in the finding that indigenous peoples have the right to determine the management, use and disposal of their natural resources. The question that then presents is, what about other peoples within a state—can they claim permanent sovereignty over natural resources, particularly in the context of asserting the right to development? This chapter revisits the existing problem of the inability of states to use natural resources to guarantee an adequate standard of living for their citizens and explores the right to socio-economic and cultural self-determination. The central argument made is that the right to socio-economic and cultural self-determination imposes a limitation on state's sovereignty over natural resources and grounds the peoples' claim to collective ownership rights over natural resources.

The chapter is subject to two limitations. First, it does not address itself to the debate on the concept of 'people' or 'peoples' but relies principally on the concept of people as the polity of the state, more explicitly, everyone within the territory of the state. While acknowledging and drawing from indigenous peoples' discourses on self-determination and permanent sovereignty over natural resources, the discussion is not essentially about indigenous peoples. Second, the chapter relies on the concept of self-determination as defined by human rights bodies as referring to 'peoples' pursuit of its political, economic, cultural development within the framework of the existing state or without external interference'.[4] In this sense, the right to socio-economic and cultural self-determination is conceived in the context of the right to democratic governance that accrues to all peoples within a state.[5]

The chapter proceeds as follows: the section 'Evolution of the doctrine of permanent sovereignty over natural resources' examines the evolution of the doctrine of permanent sovereignty over natural resources under public international law and international human rights law. The section 'The right to socio-economic and cultural self-determination' situates the right to socio-economic and cultural self-determination in its modern context and demonstrates that, the right entitles the people within a state to collective ownership over natural resources. The section 'Permanent sovereignty over natural resources and the right to self-determination—twin concepts or polar opposites?' grapples with the question of how to reconcile state's sovereignty over natural resources and peoples' right to socio-economic and cultural self-determination. The section 'Natural resources ownership and the right to development' makes proposals on how the right to socio-economic and cultural self-determination could be infused with the right to development to place peoples within the state at the centre of national development processes.

Evolution of the doctrine of permanent sovereignty over natural resources

Archival records of the United Nations (UN) trace the doctrine of permanent sovereignty over natural resources to the early 1950s at the UN Commission for Human Rights in relation to the established principle of state sovereignty.[6] This was in the context of decolonisation in which, the peoples in territories under colonial rule sought an equal share of the benefits accruing from exploitation of their natural resources; while newly independent states sought to take back control of natural resources as a form of socio-economic and cultural self-determination.[7]

This resulted in the adoption of UN General Assembly Resolution 523(VI) of January 1952[8] affirming the right of under-developed countries to determine freely the use of the natural resources and the duty to utilise the resources to further their national economic development interests and the expansion of the global economy.[9] Later in December 1952, Resolution 626(VII) was adopted affirming the right of each country to freely use and exploit their resources as an aspect of political independence.[10]

In 1958, drawing impetus from the ongoing debates on the drafting of the two legally binding treaties—the Covenant on Civil and Political Rights (CCPR) and the Covenant on Economic, Social and Cultural Rights (CESCR)—the UN General Assembly established the Commission on Permanent Sovereignty over Natural Resources.[11] The Commission was tasked to study the status of permanent sovereignty over natural resources in the context of the right to self-determination and make recommendations for its strengthening.[12] This culminated in the adoption in 1962 of Resolution 1803 on Permanent Sovereignty over Natural Resources, which linked the right to permanent sovereignty over natural resources to peoples.

Prior to the adoption of the 1962 Resolution, the Declaration on the Granting of Independence to Colonial Countries and Peoples asserted the right of peoples of the state to freely dispose of their natural wealth and resources with due regard to obligations arising out of international economic cooperation and based on mutual benefit and international law.[13] A subsequent Resolution by the UN General Assembly on the same day called on states to respect the sovereign rights of each state to dispose of its wealth and natural resources.[14] The conflicting positions in the Declaration and the Resolution as to who between peoples and states, permanent sovereignty over natural resources is vested, form the genesis of the contestations to date.[15]

Undoubtedly, these developments impacted the development of international human rights law on the doctrine of permanent sovereignty through the drafting of two legally binding human rights treaties. In 1966 the UN adopted the CCPR and the CESCR. The Covenants, however, do not incorporate the concept of 'permanent sovereignty over natural resources', which is only enshrined as part of the right to self-determination. Common article 1(2) reads in part:[16]

1. All peoples have the right to self-determination. By virtue of that right they freely determine their political status and freely pursue their economic, social and cultural development.
2. All peoples may, for their own ends, freely dispose of their natural wealth and resources without prejudice to any obligations arising out of international economic cooperation, based upon the principle of mutual benefit, and international law. In no case may a people be deprived of its own means of subsistence.

The provision underscores the foundational nature of the doctrine as an aspect of the right to self-determination, specifically, natural resource governance.[17] First, the inclusion of the doctrine in human rights treaties, which would outlast decolonisation underscored the importance of natural resource governance within the state.[18] In addition, the use of the term 'have' in enunciating the right to self-determination in para 1 underscores the permanency and continuing nature of peoples' sovereignty over natural resources.[19] Second, the first and second paragraphs link political and economic independence,[20] and affirm the right of peoples to economic independence, which did not automatically follow upon the attainment of political independence.[21] Third, there is the ambiguity inherent in the article which, in paragraph 1 proclaims the peoples' collective right to self-determination, but introduces a limitation in para 2, which subjects the right to self-determination to the promotion of international economic cooperation and protection of the rights of foreign investors.[22] This restricted peoples within a state from freely disposing of their resources if such disposal contravened economic cooperation and the rights of foreign investors.

The drafting history indicates that initial drafts of the article included a para 3, which expressly referred to the doctrine of permanent sovereignty.[23] It read:

> The right of peoples of self-determination shall also include permanent sovereignty over their natural wealth and resources. In no case may a people be deprived of its own means of subsistence on the grounds of any rights that may be claimed by other states.[24]

This raised heated concerns on two fronts. First, Western nations argued that it was problematic to introduce the doctrine of sovereignty over natural resources, which increasingly was viewed as a right of states, to an article dealing with a right of peoples.[25] Second, was a concern on the actual contours or scope of the right to self-determination for which it was argued that broadening the right to sovereignty over natural resources would create avenues for confiscation and expropriation of foreign property in violation of international agreements.[26]

These concerns were addressed by expunging the initial paragraph 3, which expressly provided for the doctrine of sovereignty over natural resources and by introducing limitations to the right to self-determination founded on international agreements and the protection of foreign investors.[27] The import of the limitations is that states could through the invocation of international agreements and the protection of foreign investments, violate the right to self-determination. Importantly, and after extensive debate, changes were made with a view to strengthening the peoples' right to control their natural resources.

First, para 1 which proclaims the right to political and socio-economic and cultural self-determination was rephrased to read: 'all peoples have a right of self determination', thus giving it a permanent and enduring character. Second, two other articles were introduced in the Covenants, which revisited the doctrine of permanent sovereignty over natural resources without attaching any limitations to it. Identically, article 47 of the CCPR and article 25 of the CESCR provide that nothing in the present Covenant shall be interpreted as impairing the inherent right of all peoples to enjoy and utilise fully and freely their natural wealth and resources.[28] Scholars point out that these identical articles have an absolute character and are aimed at mitigating the limitations introduced in common article 1.[29]

In addition to the UN system, the African human rights system proclaims the doctrine of permanent sovereignty over natural resources in much stronger terms, although embracing ambiguity on how the doctrine relates to state sovereignty. The African Charter on Human and Peoples' Rights (African Charter), provides in article 21 that all peoples shall have the right to freely dispose of their wealth and natural resources and that such disposal shall be done in the 'exclusive interest of the people'.[30] While the phrase seems to rise above the limitations in common article 1 of the UN Covenants, article 21(3) and (4) of the African Charter reinstates the limitation to the right in respect to international economic cooperation and principles of international law and give states power to freely dispose of wealth and natural resources in pursuit of African solidarity and unity. Paragraph 5 then obligates states to eliminate all forms of foreign economic exploitation to benefit the peoples within the state from their natural wealth and resources. Article 21 echoes the ambiguity in the UN Covenants on whether it is the people or the state that has sovereignty over natural resources and introduces limitations based on international economic cooperation and international law principles. The interpretation of article 21 of the African Charter is fully taken up in the section 'The right to socio-economic and cultural self-determination'.

Outside international human rights law, the second phase in the development of the doctrine of permanent sovereignty over natural resources was in the realm of the campaign for a new international economic order in the 1960s and 1970s.[31] Two notable developments in this phase are the adoption of Resolution 3171 in 1973 and the 1974 Charter on Economic Rights and Duties of States (CERDS).[32] This phase marked a shift in perception on the doctrine of sovereignty over natural resources. The earlier emphasis on peoples as possessing sovereignty over natural resources was shifted to the state.[33] Notably, the CERDS and Resolution 3171 refer to the rights of states in relation to sovereignty over natural resources.[34] The significance is that the doctrine in relation to peoples started to weaken and eventually, it was increasingly seen as relating to all states rather than to peoples per se.[35]

Illustratively, a number of treaties within the realm of public international law adopted after this period assert that sovereignty over natural resources is vested in states.[36] For instance, the 1982 Convention on the Law of the Sea, the 1983 Vienna Convention on Succession of States in relation to State Property, Archives and Debt and the 1993 Convention on Biological Diversity.[37]

The debate on the subject of peoples in relation to the doctrine of permanent sovereignty over natural resources re-emerged in the last two decades in the context of indigenous peoples in international human rights law.[38] The debate on indigenous peoples marks a shift in the doctrine of permanent sovereignty from the inter-state to the intra-state level.[39] It culminated in the adoption of indigenous peoples' specific human rights instruments, key among them the UN Declaration on the Rights of Indigenous Peoples (UNDRIP), which recognises indigenous peoples' rights to internal self-determination.[40] This discussion is further taken up in the section 'The right to socio-economic and cultural self-determination'.

The legal status of the doctrine of permanent sovereignty over natural resources remains unsettled. While Resolution 1803 is declaratory and thus not constituting binding treaty law, questions linger on whether the doctrine constitutes customary international law. The contestation is premised on the fact that the doctrine evolved within the UN, as a resolution of the General Assembly, meanwhile the legal status of the Assembly's resolutions is itself a subject of doctrinal controversy.[41] Nonetheless, it suffices to point out that the International Court of Justice in the *East Timor* case and in the *Democratic Republic of Congo v Uganda* case recognised the doctrine of permanent sovereignty over natural resources as customary international law.[42]

From the foregoing history of the evolution of the doctrine of permanent sovereignty and the ambiguities surrounding its conceptualisation, the question of who owns the natural resources within the state remains unanswered. As discussed, the shift in the debate to the intra-state level and drawing from the modern concept of state sovereignty, which limits authority of the state to the peoples' sovereignty, it is plausible to argue that sovereignty over natural resources within the state is vested in the peoples.

The right to socio-economic and cultural self-determination

That the debate on self-determination seems to have left behind the socio-economic and cultural or natural resource aspect is undoubted. A number of scholars have alluded to this. Reflecting on the right to self-determination, Oloka refers to economic self-determination as the 'poor second cousin' of political self-determination.[43] He attributes the neglect of economic self-determination to the socialist origins and views which treat economic, social and cultural rights as a 'Cold War hangover'.[44] Farmer also points out the overshadowing of the right to economic self-determination by other forms of self-determination, specifically secessionist self-determination.[45] Carty argues that the reduction of the right to economic self-determination to an ideological representation of the North–South confrontation by developing countries in the UN in the 1960s stripped it of its material reality.[46] Cambou and Smis allude to the neglect of economic self-determination, which she attributes to the inherent tension between the right to self-determination and the principle of state sovereignty.[47]

The genesis of the right to socio-economic and cultural self-determination can be traced to the Universal Declaration of Human Rights[48] wherein article 22 entitles

> every member of society to realization through national effort and international cooperation and in accordance to the resources and organization of each State, to the economic, social and cultural rights indispensable for his dignity and the free development of his personality.[49]

Farmer opines that article 22 distinguishes between political and socio-economic and cultural self-determination and lays down the requirement that the resources of the state are to be used for the realisation of human rights.[50] At the regional level, the African Charter provides for socio-economic and cultural self-determination in article 20 as follows:

1. All peoples shall have the right to existence. They shall have the unquestionable and inalienable right to self-determination. They shall freely determine their political status and shall freely pursue their economic and social development according to the policy they have freely chosen.

The article provides for socio-economic and cultural self-determination alongside political self-determination akin to common article 1 in the twin UN Covenants, and includes additional stronger language on its absolute and uncontestable nature. The article provides a lead to article 21, which highlights the doctrine of permanent sovereignty in the ownership of natural resources in Africa.

The right to socio-economic and cultural self-determination in common article 1 of the twin UN Covenants, the African Charter and other human rights instruments present a common theme: a repetition of the phrases 'freely pursue their economic development', 'freely dispose of their natural wealth and resources', 'disposal of natural wealth and resources for the peoples' own ends' or in the 'exclusive interest of the people', and the 'right of a people to existence or survival'. It is plausible to argue that the constitutive elements of the right to socio-economic and cultural self-determination are a people's ability to freely determine their socio-economic and cultural development, ability to freely determine the use of their natural resources, the requirement that people benefit from their natural resources and a basic minimum of the right to natural resources which must guarantee the existence of the people.[51] This position on the constitutive elements of the right of socio-economic and cultural self-determination finds support in the Human Rights Committee General Comment 12 on Common Article 1.[52]

General Comment 12 points out that the economic aspect of the right to self-determination includes the right of a people to freely dispose of their natural wealth and resources for their own needs and not to be deprived of their means of subsistence.[53] The African Commission on Human and Peoples' Rights (African Commission) in the *SERAC* case found a violation of article 21(1) of the African Charter as the Nigerian Government failed in its obligation to protect the Ogoni people from destruction of their crops and houses.[54] It points to the dignity requirement envisaged as a basic minimum of the right to socio-economic and cultural self-determination.

As discussed in the section 'The right to socio-economic and cultural self-determination', the debate on socio-economic and cultural self-determination re-emerged in two fronts. In the 1968 Tehran International Conference on Human Rights and the 1993 Vienna Conference on Human Rights,[55] and most importantly, in relation to indigenous peoples' rights to the natural resources in their ancestral lands. The 1968 Tehran and 1993 Vienna Conferences draw a link between individual rights and socio-economic and cultural self-determination by calling upon developing countries to raise the standard of living of their citizens using available resources.[56] The initiatives in the Tehran and Vienna Conferences, however, did not revive the debate considerably.[57] On the second front, with regard to indigenous peoples' rights to the natural resources on their ancestral lands, significant progress in reviving socio-economic and cultural self-determination seems to have been made. The Human Rights Committee in its Concluding Observations to a number of states indicates that article 1(2) imposes an obligation on the state to ensure that indigenous peoples' rights to control the natural resources on their lands are respected and protected.[58]

Similarly, at the regional level, the African Commission in the *Endorois* case found Kenya to have violated the right of the Endorois to freely dispose of their natural wealth and resources.[59] The African Court on Human and Peoples' Rights in 2017 also found that Kenya had violated the right of the Ogiek of their natural wealth and resources by evicting them from the Mau Forest.[60] The Inter-American Court on Human Rights in the *Saramaka* case equally found that indigenous people in Suriname have a right to the natural resources found in the lands they occupy.[61] Indigenous peoples' right to socio-economic and cultural self-determination is recognised in the Declaration on the Rights of Indigenous Peoples.[62]

While scholars have commended the re-emergence of debate on the right to socio-economic and cultural self-determination, criticism abounds. First, Gilbert argues that narrowing the application of the right to indigenous peoples raises the undesirable effect of 'compartmentalisation of human rights',[63] thus limiting the harmonious development of the right in relation to all peoples. While Gilbert acknowledges that the narrow application of the right could be justified from the perspective of natural resources being attached to indigenous peoples' lands,[64] such justification draws further criticism. Druigbo also takes issue with compartmentalisation and posits that the right belongs to all peoples within the state and views the narrowing application of the right to indigenous people as a recipe for secession.[65]

Second, that application of the discourse on the right to socio-economic and cultural self-determination to indigenous peoples is misplaced as it gives primacy to culture—that is, cultural and physical attachment to land—rather than the debate on peoples' sovereignty over their natural wealth and resources.[66] Miranda thus concludes that the discussion on indigenous peoples is not about the right to socio-economic and cultural self-determination but rather about 'cultural entitlement' to natural resources.[67]

Third, Gilbert points out that the element of free disposal of indigenous peoples' natural resources is limited to free, prior and informed consent.[68] It means that states retain absolute control over decision making on the use and management of natural resources, with indigenous peoples only giving consent after the state has exercised its decision making powers.[69] Miranda also argues that free, prior and informed consent speaks to merely procedural participatory rights while the state retains power over governance of the natural resources.[70] This argument can however be countered. The limitation to free, prior and informed consent is in the actual implementation of the norm rather than in its construction under international human rights law. Free, prior and informed consent constitute 'three interrelated and cumulative rights: the right to be consulted; the right to participate; and the right to their lands, territories and resources'.[71] Accordingly, the state's obligation to consult has been said to be a process of dialogue and negotiation with a view to obtaining consent and incorporates the right to dissent or make alternative or different proposals.[72] The process of consultation and participation is not an event but a continuous process from planning to implementation and follow-up.

The free, prior and informed consent norm is anchored on the right to self-determination. Outside the context of indigenous peoples, in relation to decolonisation, the right to be

> free and fully informed implied the right of nations and peoples to be free from any form of coercion, to a dignified existence, to be fully informed and in a position to reject or approve offers, plans, programmes and proposals that affected them and their resources.[73]

From the perspective of internal self-determination and specifically in relation to the right to socio-economic and cultural self-determination, this would imply that people within the state have the right to be free and be fully informed of development plans, programmes and projects in relation to the use and management of their natural resources and the right to reject, accept, modify or offer an alternative development agenda. It raises the question of how the right to socio-economic and cultural self-determination could be applied to all peoples within the state at the heart of national development? These questions are fully responded to in the section 'Natural resources ownership and the right to development'.

Permanent sovereignty over natural resources and the right to self-determination—twin concepts or polar opposites?

The discussions in the section 'Evolution of the doctrine of permanent sovereignty over natural resources' demonstrate that international law gives primacy to states in natural resources governance. International human rights law, on the other hand, recognises the right of peoples to control their socio-economic and cultural development through asserting the right to self-determination. This brings to the fore the tension between sovereignty over natural resources and a people's right to self-determination. Both concepts developed in the same context of decolonisation, in which sovereignty over natural resources was viewed as a corollary of the right to self-determination. Common article

1 of the human rights Covenants include the right to self-determination in permanent, continuing and absolute terms: 'All peoples *have* the right of self determination'.[74]

Gilbert argues that the permanent and continuing character of the right to self-determination in itself constitutes a limitation to states' sovereignty over natural resources.[75] Miranda notes that state sovereignty in the decolonisation period addressed sovereignty over natural resources at the inter-state level; hence seen as a right of states in relation to other states.[76] However, the inclusion of sovereignty over natural resources in human rights treaties marked a shift to the intra-state level. Identically, articles 47 and 25 of the CCPR and CESCR demonstrate the intra-state level facet of sovereignty over natural resources.[77] It connotes a shift from state's absolute authority to a duty of states in the exercise of sovereignty over natural resources.[78] Implicitly, states cannot claim absolute authority over natural resources and decision making in socio-economic and cultural development.[79]

In addition, the concept of state sovereignty has evolved into popular sovereignty, which recognises peoples as the source of power and at the core of the exercise of its authority. Wenar in his book refers to the concept of popular resource sovereignty, which draws from the idea of popular state sovereignty in modern political theory.[80] There is evidence of the co-existence of permanent sovereignty over natural resources and the right to socio-economic and cultural self-determination in the context of the right of indigenous peoples in disposing of the natural resources in their lands, in which states have a duty to ensure the free participation of indigenous people.[81]

Natural resources ownership and the right to development

The right to development has its origins in decolonisation and the search for a new international economic order,[82] akin to the right to self-determination. In the African context, the right to development is conceptualised as a collective right while at the UN level, it is conceived both as a collective and individual right.[83] The Declaration on the Right to Development (UNDRTD) provides that:

> The right to development is an alienable human right by virtue of which every person and all peoples are entitled to participate in, contribute to, and enjoy economic, social, cultural and political development, in which all human rights and fundamental freedoms can be fully realized.[84]

The African Charter is the only region-specific human rights treaty that enshrines the right to development. In article 22, the African Charter states:

1. All peoples shall have the right to their economic, social and cultural development with due regard to their freedom and identity and equal enjoyment of the common heritage of mankind.

2. States shall have the duty individually or collectively to ensure the exercise of the right to development.

Even outside of positive law, at the 1993 Vienna Conference on Human Rights, 171 states endorsed the right to development as a universal and inalienable right and as an integral part of fundamental human rights.[85] Not withstanding the universal acceptance of the right to development, there remains controversy on interpretation, obligations of states and the quantum of entitlement that accrue to the peoples to whom the right is guaranteed.[86]

In terms of interpretation of the right to development, the jurisprudence of the African Commission is informative. In the *Endorois*, the Commission adopted a two-part test for the right to development. It stated that the right to development has both a constitutive and an instrumental aspect and thus could be viewed as a means and an end.[87] The African Commission also sets out five criteria for the right to development: it must be equitable, non-discriminatory, participatory, accountable and transparent with equity and choice as over-arching themes.[88] Oka for in his insightful analysis of the right to development in the African Charter suggests that the right to development incorporates a requirement that the outcome of the process and goals of development must be enjoyment of peace; the ends and process must be guided by human rights; that no group inequities should exist in the distribution of collective benefits; peoples' participation must be viewed as an irreducible minimum; and thus the right encompasses the means, processes and outcomes of development.[89]

The UNDRTD and the African Charter place on the state a duty towards the achievement of the right to development. Article 2(3) of the Declaration provides that '[s]tates have the right and duty to formulate appropriate national development policies aimed at improvement of the well-being of the population and all individuals'.[90] A conjunctive reading of articles 3, 4 and 8 also assert the state-centric model of development which places states at the centre of the development process.[91] Similarly, the African Charter in article 22(2) places the duty to ensure the exercise of the right to development on states, implying the state-centric model of development, which is premised on the expectation that the state will, through the exploitation of the natural resources, promote economic development as it is assumed that there would be a 'trickle down' of economic gains, which will then improve the welfare of all peoples, including marginalised and vulnerable groups.[92] Miranda critiques the state-centric model of development in relation to indigenous peoples, arguing that it does not guarantee distributional equity.[93] She attributes the lack of distributional equity to the likelihood of state capture by government elites who tend to control the national development processes and outcomes for personal gains; and the possibility of the state repeating cases of historical marginalisation.[94] While this observation is made in the context of indigenous peoples, it no doubt would apply in the context of all peoples in the state. In many instances, development is conceived in the context of competing developmental goals of the state and its citizens.

This chapter is concerned with how the management of natural resources could guarantee an adequate standard of living for the peoples within the state. The central argument is that the right to socio-economic and cultural self-determination imposes a limitation on state's sovereignty over natural resources and reassigns the people within the state as right holders—a collective ownership right over natural resources. The right to socio-economic and cultural self-determination can thus be infused into the right to development to place the peoples at the centre of the national development agenda and processes. The link between the right to development and natural resources is not hard to find. At a basic level, international law instruments on sovereignty over natural resources reference socio-economic and cultural development. The UNDRTD for example, twins the right to development with full sovereignty over natural resources.[95] The African Commission ruled in the case of the *Democratic Republic of Congo v Burundi, Rwanda and Uganda* that the deprivation of the people of the Democratic Republic of Congo of the right to freely dispose of their wealth and natural resources violated the right to development.[96]

As discussed earlier, the right to development emphasises procedural and substantive rights. Equally, the constitutive elements of the right to socio-economic and cultural self-determination are a people's ability to determine their socio-economic and cultural development and to freely determine the use of their natural resources; the requirement for peoples to benefit from the natural resources; and as a basic minimum that natural resources should guarantee their survival. The procedural aspect of the right to development that speaks to participatory rights reflects the element of peoples' ability to determine their socio-economic and cultural development and to freely determine the use of their natural resources in asserting the right to socio-economic and cultural self-determination.

The participatory right as discussed in the section 'The right to socio-economic and cultural self-determination' stems from the right to self-determination, which embodies the right of peoples to be free and be fully informed of development plans, projects and programmes that affect them and their resources, including the right to reject or to approve or modify the development agenda.[97] If it is accepted that the right to socio-economic and cultural self-determination entitles peoples within the state to lay a claim to collective ownership of natural resources, then the requirement of authorisation on the use of natural resources would arise, hence the right to be free and to be informed must be accompanied by the right of the peoples to chart their socio-economic and cultural development and to freely use their natural resources.[98] The question then remains; how can the peoples within a state exercise decision making authority over their socio-economic and cultural development and the use of their natural resources?

Farmer in her analysis on the right to socio-economic self-determination suggests the need for political structures that create avenues for political participation.[99] Wenar identifies the four conditions for people to participate in decision making over natural resources as access to information about the resources, independence of the people to be free from propaganda and manipulation in their decision making, the possibility of free discussion on development policies and to dissent by expressing opinion that has an impact on state policies.[100] Further inspiration on the political structures that could create avenues for people to chart their economic development and use their natural resources could be drawn from the Extractive Industries Transparency Initiative (EITI).[101] The EITI is a non-state voluntary code of conduct that promotes open and accountable management of oil, gas and mineral resources.[102]

It is implemented by governments and involves publishing key information and data on their natural resources management such as licence allocation, taxes paid and how the revenue generated is used.[103] It comprises multiple stakeholders including governments, civil society, corporate actors and non-governmental organisations.[104] In this context, the structure could take the form of a representative political body charged with the initiation of a development agenda at the local level, which creates avenues for deliberation by the people and ensures their involvement in designing development programmes and in implementation.

The substantive aspect of the right to development compares with the requirement that all peoples benefit from their natural resources and at minimum, that natural resources guarantee subsistence to the peoples. According to Miranda, both concepts suggest distributional equity, which is tied to the notions of economic subsistence or cultural survival for indigenous peoples.[105] As discussed under the right to socio-economic and cultural self-determination, peoples within the state have collective ownership rights over natural resources. How then can these substantive rights ensure distributional equity to the notion of human dignity? She proposes a shift from a state-centric model of development to a peoples-based model in which local peoples have direct decision making authority in their economic development and the use and disposal of their natural resources.[106] Building on the EITI framework discussed above, the political structure could also have an additional role of carrying out a distributional analysis of the benefits accrued from natural resources while ensuring that vulnerable groups within the state are accommodated to meet the human dignity standard.

Conclusion

The primary argument made in this chapter is that the right to development places the state at the centre of national development processes, which often results in capture of the development agenda by state elites. The chapter revisits the right to socio-economic and cultural self-determination within the context of natural resources ownership and demonstrates that the constitutive elements of the right ground the claim of collective ownership on peoples within the state. It is argued that the doctrine of permanent sovereignty over natural resources in international human rights treaties introduced the intra-state facet of permanent sovereignty, thus imposing a duty on the state in the governance of natural resources. Based on these arguments, the chapter propounds the position that peoples should be at the centre of the national development agenda, necessitating a shift from the state-centric model of development to a peoples-based model in which peoples within the state are the ultimate decision makers in charting their socio-economic and cultural development.

Notes

1 Moore, Margaret. 2012. 'Natural Resources, Territorial Right and Global Distributive Justice'. *Political Theory* 40(1): 85.
2 *Ibid.*
3 Senarante, Kalana. 2013. 'Internal Self-determination in International Law: A Critical Third World Perspective'. *Asian Journal of International Law* 3(2): 306–307.
4 UN Committee on the Elimination of Racial Discrimination. 1996. 'General Recommendation 21'. In Note by the Secretariat, Compilation of General Comments and General Recommendations adopted by Human Rights Treaty Bodies' 8 March, U.N. Doc. HRI\GEN\1\Rev.6; *Re Secession of Quebec* 2 SCR 217; 1998 CANLII 793 (SCC).
5 Senarante, note 3, 307.
6 Schrijver, Nicolaas. 2015. 'Fifty Years Permanent Sovereignty over Natural Resources: The 1962 UN Declaration as the Opinion Iuris Communis'. In *Permanent Sovereignty over Natural Resources*, edited by Bungenberg, Marc and Hobe, Stephan. Geneva: Springer International Publishing, 16.
7 Schrijver, Nicolaas. 1997. *Sovereignty over Natural Resources: Balancing Rights and Duties*. Cambridge: Cambridge University Press, 33–81.
8 Hobe, Stephan. 2015. 'Evolution of the Principle of Permanent Sovereignty over Natural Resources'. In *Permanent Sovereignty over Natural Resources*, edited by Bungenberg and Hobe. Geneva: Springer International Publishing, 4–5.
9 UNGA Resolution 523(VI) 12 Janaury 1952.
10 UNGA Resolution 626(VII) 21 December 1952.
11 *Ibid.*
12 UNGA Resolution 1314(XIII) 12 December 1958.
13 Declaration on the Granting of Independence to Colonial Countries and Peoples, UNGA Resolution 1514(XV) 14 Dec 1960.
14 UNGA Resolution 1515(XV) 15 December 1960.
15 Gilbert, Jérémie. 2013. 'Right to Freely Dispose of Natural Resources: Utopia or Forgotten Right?' *Netherlands Quarterly of Human Rights* 31(3): 319.
16 International Covenant on Civil and Political Rights adopted 16 December 1966, 999 UNTS 171, art 1; International Covenant on Economic, Social and Cultural Rights adopted on 16 December 1966, 999 UNTS 3, art 1.
17 Gilbert, note 15, 321–322; Farmer, Alice. 2006. 'Towards a Meaningful Rebirth of Economic Self-determination: Human Rights Realization in Resource Rich Countries'. *International Law and Politics* 39(1): 430.
18 Jan Veerman, Gert and Oort van der Ploeg-van, Paulien. 1979. 'Economic Self-determination and Human Rights: Some Remarks'. In *Change and the New International Economic Order*, edited by van Lith, J.A. Dordrecht: Springer Science + Business Media, 26–27.
19 Farmer, note 17, 432–433; Gilbert, note 15, 322.
20 Gilbert, note 15, 321–322.
21 Carty, Antony. 1984. 'From the Right to Self-determination to the Right to Development: A Crisis in Legal Theory'. *Third World Legal Studies* 3: 73.

22 Gilbert, note 14, 322–323.
23 Schrivej, note 7, 49.
24 *Ibid.*
25 Gilbert, note 15, 322.
26 Cambou, Dorothee and Smis, Stefaan. 2013. 'Permanent Sovereignty over Natural Resources from a Human Rights Perspective: Natural Resources Exploitation and Indigenous Peoples' Rights in the Arctic'. *Michigan State International Law Review* 22(1): 358.
27 Gilbert, note 15, 322.
28 ICESCR art 25 and ICCPR art 47.
29 Gilbert, note 15, 323–324.
30 African Charter on Human and Peoples' Rights adopted by the Organisation of African Unity(OAU) in Nairobi Kenya on 27 June 1981, OAU Doc CAB/LEG/67/3 rev. 5, art 21.
31 Visser, Fritz. 1988. 'The Principle of Permanent Sovereignty over Natural Resources and the Nationalisation of Foreign Interests'. *CILSA* 21(1): 77.
32 *Ibid.*
33 Schrivej, note 6, 4.
34 Charter on Economic Rights and Duties of States, art 2.
35 Schrijver, note 6, 4.
36 Enyew, Endalew. 2017. 'Application of the Right to Permanent Sovereignty over Natural Resources to Indigenous Peoples: Assessment of Current Legal Developments'. *Arctic Review of Law and Politics* 8: 224–225.
37 Gilbert, note 15, 319.
38 Bungenberg, Marc and Hobe, Stephan. 2015. *Permanent Sovereignty over Natural Resources*. Geneva: Springer International Publishing, 22.
39 Miranda, Lillian. 2012. 'The Role of International Law in Intrastate Natural Resource Allocation: Sovereignty, Human Rights and Peoples' Based Development'. *Vanderbilt Journal of Transnational Law* 45(3): 795.
40 Enyew, note 36, 227; UN Declaration on the Rights of Indigenous Peoples, Resolution 61/295 adopted by the General Assembly on 13 September 2017.
41 Schrijver, note 6, 3.
42 East Timor Case (*Portugal v Australia*) [1995] ICJ Rep 90; Armed Activities in the Territory of the Congo Case (*Democratic Republic of Congo v Uganda*) [2005] ICJ Reports 251, para 244.
43 Oloka-Onyango, Joe. 1999. 'Heretical Reflections of the Right to Self-determination: Prospects and Problems for a Democratic Global Future in the New Millennium'. *American University International Law Review* 15(1): 169.
44 *Ibid*, 169–170.
45 Farmer, note 17, 419–425.
46 Carty, note 21, 73.
47 Cambou and Smis, note 26, 358–359.
48 Oloka-Onyango, note 43, 171; Farmer, note 17, 424–425.
49 Universal Declaration of Human Rights adopted 10 December 1948, UNGA Resolution 217 A(III), art 22.
50 Farmer, note 17, 425.
51 *Ibid*, 437.
52 UN Human Rights Committee. 2014. 'General Comment 12'. Note by the Secretariat, Compilation of General Comments adopted by Human Rights Treaty Bodies. 13 March, UN Doc HRI/GEN Rev. 9, para 5.
53 *Ibid.*
54 *The Social and Economic Rights Action Center and the Centre for Economic and Social Rights v Nigeria* Comm 155/9; Gilbert, note 15, 330–331.
55 Onyango-Oloka, note 43, 174–175.

56 Final Act of the International Conference on Human Rights (22 April–13 May 1968) UN Doc A/C0NF.3 2/41; Vienna Declaration and Programme of Action of the World Conference on Human Rights (14–25 June 1993) UN Doc A/CONF.157/23.
57 Onyango-Oloka, note 43, 175.
58 Cambou and Smis, note 26, 360–361.
59 *Centre for Minority Rights Development (Kenya) and Minority Rights Group International on Behalf of the Endorois Welfare Council v Kenya*, Comm 276/2003.
60 *African Commission on Human and Peoples' Rights v Republic of Kenya*, App 006/2012.
61 *Saramaka People v Suriname*, Merits, Inter-American Court of Human Rights Series Case No 172.
62 UNDRIP, note 48; Cambou and Smis, note 26, 361.
63 Gilbert, note 15, 330.
64 *Ibid.*
65 Enyew, note 36, 229.
66 Miranda, note 39, 811.
67 *Ibid.*
68 Gilbert, note 15, 329.
69 *Ibid.*
70 Miranda, note 39, 836.
71 United Nations Human Rights Council (UNHRC). 2018. 'Study of the Expert Mechanism on the Rights of Indigenous Peoples'. UN Doc A/HRC/39/62.
72 *Ibid.*
73 *Ibid.*
74 Declaration on Permanent Sovereignty over Natural Resources.
75 Gilbert, note 15, 326.
76 Miranda, note 39, 797–798.
77 *Ibid*, 800.
78 *Ibid.*
79 Miranda, note 39, 804.
80 Wenar, Leif. 2016. *Blood Oil: Tyrants, Violence and the Rules that Run the World*. Oxford: Oxford University Press.
81 Gilbert, note 15, 326; Miranda, note 39, 836.
82 Arts, Karin and Tamo, Atabongawung. 2016. 'Right to Development in International Law: New Momentum Thirty Years Down the Line'. *Netherlands International Law Review* 63(3): 225; Ngang, Carol. 2018. 'Towards a Right-to-Development Governance in Africa'. *Journal of Human Rights* 17(1): 111.
83 Arts and Tamo, note 82, 224.
84 Declaration on the Right to Development, UNGA Resolution A/RES/41/128 adopted on 4 December 1986.
85 Vienna Declaration and Programme of Action.
86 Okafor, Obiora. 2013. 'A Regional Perspective: Article 22 of the African Charter on Human and Peoples' Rights'. In *Realizing the Right to Development: Essays in Commemoration of 25 Years of the United Nations Declaration on the Right to Development*, edited by UN Human Rights, 376–377. Geneva: UN Publication; Arts and Tamo, note 82, 232.
87 Communication 276/2003, para 277.
88 *Ibid.*
89 *Ibid*, 377–378.
90 UNDRTD.
91 UNDRTD, art 3, 4 and 8; Arts and Tamo, note 82, 233; Miyawa, Maxwel. 2016. 'The International Dimension of the Right to Development: Where is the Gapping Crack of Accountability for Non-State Actors'. *The Transnational Human Rights Review* 3(1): 45.
92 Miranda, note 39, 830.
93 *Ibid*, 831.

94 *Ibid*, 831–832.
95 UNDRTD.
96 Merits Communication 227/99 African Commission on Human and Peoples' Rights (29 May 2003).
97 Study of the Expert Mechanism on the Rights of Indigenous Peoples.
98 Miranda, note 39, 836–837.
99 Farmer, note 17, 438.
100 Wenar, note 80, 235–238.
101 Miranda, note 39, 828.
102 Extractive Industries Transparency Initiative. 2019. 'Global Standard for Good Governance of Oil, Gas and Mineral Resources'. *Extractive Industries Transparency Initiative* <https://eiti.org/sites/default/files/documents/eiti_factsheet_en_oct2018_0.pdf>.
103 *Ibid*.
104 *Ibid*.
105 Miranda, note 39, 831.
106 *Ibid*, 830.

References

Arts, Karin and Tamo, Atabongawung. 2016. 'Right to Development in International Law: New Momentum Thirty Years Down the Line'. *Netherlands International Law Review* 63(3): 221–249.

Bungenberg, Marc and Hobe, Stephan. 2015. *Permanent Sovereignty over Natural Resources*. Geneva: Springer International Publishing.

Cambou, Dorothee and Smis, Stefaan. 2013. 'Permanent Sovereignty over Natural Resources from a Human Rights Perspective: Natural Resources Exploitation and Indigenous Peoples' Rights in the Arctic'. *Michigan State International Law Review* 22(1): 347–376.

Carty, Antony. 1984. 'From the Right to Self-determination to the Right to Development: A Crisis in Legal Theory'. *Third World Legal Studies* 3: 73–86.

Enyew, Endalew. 2017. 'Application of the Right to Permanent Sovereignty over Natural Resources to Indigenous Peoples: Assessment of Current Legal Developments'. *Arctic Review of Law and Politics* 8: 222–245.

Farmer, Alice. 2006. 'Towards a Meaningful Rebirth of Economic Self-determination: Human Rights Realization in Resource Rich Countries'. *International Law and Politics* 39(1): 417–473.

Gilbert, Jérémie. 2013. 'Right to Freely Dispose of Natural Resources: Utopia or Forgotten Right?' *Netherlands Quarterly of Human Rights* 31(3): 314–341.

Hobe, Stephan. 2015. 'Evolution of the Principle of Permanent Sovereignty over Natural Resources'. In *Permanent Sovereignty over Natural Resources*, edited by Bungenberg and Hobe. Geneva: Springer International Publishing, 1–11.

Jan Veerman, Gert and Oort van der Ploeg-van, Paulien. 1979. 'Economic Self-determination and Human Rights: Some Remarks'. In *Change and the New International Economic Order*, edited by van Lith, J.A. Dordrecht: Springer Science + Business Media.

Miranda, Lillian. 2012. 'The Role of International Law in Intrastate Natural Resource Allocation: Sovereignty, Human Rights and Peoples' Based Development'. *Vanderbilt Journal of Transnational Law* 45(3): 785–840.

Miyawa, Maxwel. 2016. 'The International Dimension of the Right to Development: Where is the Gapping Crack of Accountability for Non-State Actors'. *The Transnational Human Rights Review* 3(1): 39–75.

Moore, Margaret. 2012. 'Natural Resources, Territorial Right and Global Distributive Justice'. *Political Theory* 40(1): 84–107.
Ngang, Carol. 2018. 'Towards a Right-to-Development Governance in Africa'. *Journal of Human Rights* 17(1): 107–122.
Okafor, Obiora. 2013. 'A regional perspective: Article 22 of the African Charter on Human and Peoples' Rights'. In *Realizing the Right to Development: Essays in Commemoration of 25 Years of the United Nations Declaration on the Right to Development*, edited by UNHuman Rights. Geneva: UN Publication, 373–384.
Oloka-Onyango, Joe. 1999. 'Heretical Reflections of the Right to Self-determination: Prospects and Problems for a Democratic Global Future in the New Millennium'. *American University International Law Review* 15(1): 151–208.
Schrijver, Nicolaas. 1997. *Sovereignty over Natural Resources: Balancing Rights and Duties*. Cambridge: Cambridge University Press.
Schrijver, Nicolaas. 2015. 'Fifty Years Permanent Sovereignty over Natural Resources: The 1962 UN Declaration as the Opinion Iuris Communis'. In *Permanent Sovereignty over Natural Resources*, edited by Bungenberg, Marc and Hobe, Stephan. Geneva: Springer International Publishing.
Senarante, Kalana. 2013. 'Internal Self-determination in International Law: A Critical Third World Perspective'. *Asian Journal of International Law* 3(2): 305–339.
Visser, Fritz. 1988. 'The Principle of Permanent Sovereignty over Natural Resources and the Nationalisation of Foreign Interests'. *CILSA* 21(1): 76–91.
Wenar, Leif. 2016. *Blood Oil: Tyrants, Violence and the Rules that Run the World*. Oxford: Oxford University Press.

9 Rethinking natural resource ownership and the realisation of the right to development for indigenous peoples in Africa

Esther E. Njieassam

Introduction

The right to natural resource ownership as stipulated in article 21 of the African Charter on Human and Peoples' Rights has relevance for indigenous peoples. It guarantees indigenous peoples' right to freely dispose of their wealth and natural resources.[1] This right forms a fundamental basis for the development of indigenous communities, cultures, spiritual life, integrity and survival. For indigenous communities, land and natural resources are closely tied to their traditional ways of life, customs, values, knowledge and practice. One of the major demands for indigenous peoples has been that states effectively guarantee their right to live on their ancestral lands, retain their cultures and control their resources. However, states' policies often prevent them from exercising these rights and thus, deprive them of their source of livelihood.

Recognising that the right to development entails the exercise of full sovereignty over natural resources, it suffices to say that denial of the inherent and inalienable right of indigenous peoples to their traditional land and natural resources limits their ability to exercise their right to development. The worrying question is how best African countries can position their policies to ensure that indigenous peoples own and control natural resources within their territories? In answering this question, the chapter deliberates on the concept of indigenous peoples as contemplated within the African context. The section 'Natural resource ownership, indigenous peoples and the right to development' focuses on the interrelationship between natural resource ownership, indigenous peoples and the right to development. The section 'Legal systems/regimes in Africa that vest natural resource ownership rights with the state' examines some legal systems in Africa that have vested resource ownership rights with the state and how these policies have affected the development of indigenous communities. The section 'Problems arising from depriving indigenous peoples of ownership rights over their natural resources' looks at some problems encountered by indigenous peoples as a result of deprivation of their ownership rights. The concluding section calls for a possible review of these policies to grant indigenous peoples economic independence necessary for their development.

The concept of indigenous peoples within the African context

The term 'indigeneity' is a complex and controversial issue.[2] There is little or no consensus on the precise meaning of the concept in international law.[3] While it is generally considered as a political concept, the term 'indigenous peoples' may mean different things to different people depending on their historical and regional backgrounds. For instance, indigenous rights discourse is more complex on the African continent.[4] While 'indigenousness' is uncontested in the Pacific and the Americas, in Africa the notion is often misconstrued.[5] Most African states assert that the issue of indigenous peoples' rights emerged from specific and complex histories of migration, assimilation and conquest.[6] This controversy is exacerbated by the lack of a unified definition of the term by international legal instruments.[7]

However, the African Commission on Human and Peoples' Rights has developed a broader definition of the concept to put to rest the long debate of who and which groups identify as indigenous in Africa. Wang warns that in Africa the term indigenous peoples is not restricted to prior settlement.[8] Ownership rights over natural resources is crucial for advancing indigenous peoples' socio-economic conditions as well as fostering their right to development. Efforts in resolving issues surrounding the meaning of indigenous peoples led the African Commission on Human and Peoples' Rights (ACHPR) Working Group to reason that a strict definition of the concept is unnecessary. As a result, some criteria were developed as a fundamental principle in identifying indigenous peoples in Africa. This includes voluntary perpetuation of cultural distinctiveness; self-identification as a distinct collective and recognition by other groups and an experience of subjugation, marginalisation, exclusion or discrimination and dispossession as well as occupation and use of a specific territory, land and natural resources.

The African notion of indigenous peoples recognises the relationship between the people, culture, land and natural resources. Land and natural resources are prerequisite for their existence and security. Hence, the lack of access to natural resources often results in acute poverty for the affected indigenous community. It is a known fact that the inability to possess or to access their territories restricts them from making use of, and enjoying the natural resources that are needed to obtain goods crucial for their survival. It also hinders the effective development of their traditional practices such as hunting, gathering, fishing and farming methods, which ultimately poses a substantial threat to their human development.

Natural resource ownership, indigenous peoples and the right to development

Over 50 million indigenous peoples reside in different parts of the continent Africa. They are significantly over-represented among the world's poorest.[9]

Development indicators for indigenous communities are often very low compared to the broader society. This is because debates regarding the application of the right to development priorities have not captured indigenous peoples' issues.[10] Although the territories occupied by indigenous peoples are enormously rich in valuable resources, their contribution to national development processes remain insignificant. They are economically, socially and politically disadvantaged and excluded from development processes. Diergarten asserts that the commercial pressure on land has increased market demand for natural resources.[11] That approach has accelerated the displacement and disruption of indigenous communities. As such, issues of indigenous peoples' right to access land and natural resources are left unattended for state intrusion. They do not even have a right to participate or benefit from the proceeds of resources extraction, including the right to prior consultation regarding large-scale development projects in their territories.

The right to development, especially as it is conceptualised in Africa, is a process that aims at improving the well-being of all peoples. It entails the peaceful enjoyment of the right to control and own natural resources. For this to happen, it is crucial that indigenous peoples are included in development processes and are in the position to benefit from the fair distribution of its proceeds. On the contrary, almost every state on the African continent with valuable natural resources has experienced the forceful displacement or disruption of indigenous communities. While natural resources are key to indigenous peoples' existence, their territories are constantly under threats resulting from development projects such as the construction of dams, hydroelectric developments, extraction of timber and the creation of protected areas and national parks.[12] Practical examples include the forceful eviction of the Maasai indigenous community from their ancestral lands to make way for multinationals and bilateral donors like the World Bank interested in geothermal projects.[13] Similarly, the Pygmies and pastoralist groups in the Central African Republic, Cameroon and the San communities in Botswana and Namibia are facing similar challenges.

For the purpose of this chapter, Pullen's definition of natural resources is considered. He describes 'resources that are external to human beings, as distinct from the natural resources internal to human beings'.[14] He views the Earth's resources as an inherent right to be enjoyed equitably by all human beings. Equating it with the right to life, Pullen contends that every human being born on planet Earth has a right to own and use the natural resources available. Hence, the right of one person to natural resources is equal to the right of every other person to the same natural resources. Implicitly, indigenous peoples' rights to development, especially in relation to natural resource ownership are inherent and inalienable, requiring maximum state protection.

Several international instruments give relevance to the right of indigenous peoples to natural resource ownership. Most importantly, the United Nations Declaration on the Rights of Indigenous Peoples (UNDRIP) and the International Labour Organisation (ILO) Convention No 169, among other international human rights instruments are worth noting. Common article 1 of the International Covenant on Civil and Political Rights (ICCPR)[15] and the International Covenant on Economic Social and Cultural Rights (ICESCR)[16] as well as article 7 of the ILO Convention No 169[17] play a key role in shaping international efforts to enhance indigenous peoples' rights to development and resource ownership. These instruments emphasise the need for indigenous peoples among others, to assert full and complete control over their natural wealth and resources to enable them to determine their own processes for development.

Article 3 of the UNDRIP stresses the importance of the right to self-determination, which allows indigenous peoples to freely determine their political, economic, social and cultural development. For indigenous peoples, the right to self-determination is key to exercising control over their own affairs. Assurances of the right to self-determination are intended to eliminate colonial tendencies and subjugation by granting freedom to non-self-governing peoples to decide their own destinies and development priorities.[18] In principle, the enjoyment of the right to development is subject to the right to self-determination, interpreted as the ability to control, manage and own natural resources and to make informed choices on development priorities that are best for indigenous communities. Similarly, articles 21(1) and 23 of the UNDRIP guarantee to indigenous peoples, the right to decide and to develop strategies for the realisation of their right to development. It also allows them to fully participate in decisions regarding development processes that affect them directly. States are therefore obliged to ensure that all individuals are able to exercise the right to development without discrimination.

Nonetheless, indigenous peoples remain disproportionately affected by food insecurity, chronic poverty and poor health conditions as a result of forced displacements from traditional lands. Relocation away from ancestral land has had far-reaching implications for indigenous peoples' right to natural resource ownership.[19] For example, the loss of land and natural resources has been linked to poor mental and physical health conditions for indigenous communities.[20] Most African states do not adequately recognise indigenous peoples' rights to land and natural resources, thus obstructing the process of realising their right to development.

Legal systems/regimes in Africa that vest natural resource ownership rights with the state

With the arrival of colonialists in Africa in the late 1800s and the beginning of the 1900s, indigenous systems suffered great injustices.[21] Their century-long governance and economic systems infringed on the fundamental human rights of African indigenous peoples. Their cultures and unique ways of life were despised, a system that has been inherited by post-colonial African governments. This period ushered in a decentralisation process in a bid for a better forest and resource management system.[22] Since then, there has been significant pressure on indigenous peoples' territories by extractive industries interested in the exploitation of natural resources, leading to devastating consequences.

Ng'ambi and Duruigbo describe the situation as a 'pressing and perplexing' global challenge limiting indigenous peoples from accessingthe natural resources that arecrucial to improving their living standards and economic development.[23] This is despite reports that the continent holds more than half of the world's rare renewable and non-renewable resources.[24] Countries like Angola, the Democratic Republic of Congo, Ghana, Nigeria, Cameroon, Chad, Lybia and South Africa are rich in gold, diamond, timber, fossil fuel, oil and gas just to mention a few.[25] Despite the global recognition of the significance of natural resources for economic growth and development, evidence suggests that many countries in Africa are cursed by dint of an overabundance of natural resources.[26] Literature on the natural resource curse theory shows that natural resources are a driver of inequalities within societies.[27] The richly endowed resource wealth has not benefited the peoples of Africa (especially the indigenous populations), as it remains the poorest continent in the world, with an estimated 50 percent of the populations living on less than USD1.90 per day.[28] The situation is overwhelming for indigenous communities even though they are custodians of natural resources.

Most African Constitutions vest land and natural resource ownership and management rights with the state.[29] This practice is an inheritance from colonial policies that allow foreign companies from Europe and America and emerging economic powerhouses like China to largely exploit the continents'natural resources. While Sebudubudu et al feel that Botswana's success lies in the governance and efficient use of natural resources, land and minerals,[30] this chapter argues that the indigenous populations where these resources are located continue to live in want and abject poverty. Shortly after independence, land and natural resources were seen as 'strategic resources' for which ownership and control would enhance development. Chapter 2, articles 3(c), 8(1) and 9 of the Botswana Constitution can be viewed as the constitutional foundations for the regulation of the country's natural resources. Article 8(1) particularly, stipulates that 'no property of any description shall be compulsorily taken possession of, and no interest in or rights over property of any description shall be compulsorily acquired'. This provision was a means to protect and appropriate property and resources acquired by indigenous peoples.

Similarly, article 9 permits the government to acquire private property, especially where such property is 'reasonably required in the interest of defence, public safety, order, health, town and country planning', the development and utilisation of mineral resources, for the purpose of any census or in order to secure the development or utilisation of any property for a purpose beneficial to the community. This constitutional declaration is the foundation of the San indigenous struggle over their ancestral land and natural resources rights claims reached, which reached their climax with the establishment of the Tribal Land Act (TLA) of 1968 and the Tribal Grazing Land Policy (TGLP) of 1975. The objectives of these two land policies were efforts to transfer land ownership from various tribes, chiefs and individuals to the state. The TLA, Chapter 32:02, Part VI, Section 13 provides that

1. All the powers previously vested in a chief and a subordinate land authority under customary law in relation to land, including

    (a) the granting of rights to use any land;
    (b) the cancellation of the grant of any rights to use any land;
    (c) the imposition of restrictions on the use of tribal land;
    (d) authorising any change of user of tribal land; or
    (e) authorising any transfer of tribal land.

The TGLP further instituted the zooming of land categories in Basarwa's communal land. The creation of commercial leasehold ranches resulted in the displacement of many people including, the San community from their lands. During the land distribution process, 71 percent of land was declared tribal land, 25 percent was allotted to the state while the remaining 4 percent freehold land was meant to be under the control of European settlers.[31] The TLA and the TGLP did not consider the San people's right over traditionally owned lands and resources including leasehold and hunting rights. Similarly, even the Mines and Minerals Act established in 1967 entrusted ownership rights to the state. The aim was to ensure that proceeds from the exploitation of minerals benefited the entire population as opposed to a sectional interest where the minerals are discovered. This Act has greatly affected the well-being and development of San communities in Botswana. Although the discovery of diamonds has played a key role in the development and transformation of Botswana from a poor to middle-income status, the country is still grappling with the challenges of inequality and poverty, mostly affecting minority and marginalised communities like the San.

In *Roy Sesana and Others v the Attorney-General,*[32] the discovery of diamond in the Central Kalahari Game Reserve (CKGR) led to the forceful eviction of the San people.[33] The CKGR has been home for the San for several decades. It harbours one major diamond deposit now known as Ghaghoo. The government asserts that the San's presence in the area deters the execution of development programmes.[34] Aside from the resettlement, the San were denied hunting and gathering rights and access to basic social services including healthcare and water. Labelled as the longest and most expensive judgment in the history of Botswana, Justice Walia describes the process as a 'harrowing story of human suffering and despair'.[35] The Court observed that the decision to forcibly remove the San from their traditional land was unlawful and unconstitutional. The Court ordered the government to facilitate the return of the San and allow them to continue with their cultural practices including subsistence hunting and gathering. Despite the Court's ruling, government is unwilling to grant them this right. The lack of full ownership and control rights over their lands and resources makes it difficult to develop ecologically more-adapted economic activities like investment in ecotourism or entering benefit-sharing agreements over the exploitation of their resources.

Accordingly, the Constitution of Kenya, 2010, one of the relatively new constitutions in Africa is more comprehensive and covers wide-ranging issues regarding land, property and natural resource ownership. In article 61(2) land is categorised into public, community and private lands. Article 40(1) grants all Kenyan citizens the right to individually or collectively acquire and own land in any part of the country. However, article 66 allows the state to regulate the use of any land, or any interest in or right over any land for particular reasons including public safety, order, morality, health or land use for planning. In the same fashion, article 69 provides that the state shall ensure the sustainable exploitation, utilisation and management of natural resources as well as ensure that the benefits are equitably shared.[36]

In light of this provision, only the state has the capacity to manage natural resources in the country. The state's capacity for compulsory ownership of natural resources is reiterated in article 15(1) of the Community Land Act, No 6 of 2012. It provides that the state reserves ownership of land located on the surface of the earth and sub-surface of the rock including, the natural resources contained in or under the surface and in the air space above the surface. In addition, Part II, section 5(1) and 19 of the National Land Commission Act of 2019, Part IV, section 14(1) of the Petroleum Act of 2019 and the Local Content Bill of 2018 all share the same view. Ten percent of the proceeds derived from the exploitation of these resources are allocated to the population that resides where the resources are found. This can be interpreted to mean that in spite of the classification of land into different categories, the state retains ownership rights over land and natural resources. This practice has jeopardised the living conditions of the indigenous communities in Kenya.

This was the position in the landmark case of *Centre for Minority Rights Development (CEMIRIDE) (on behalf of the Endorois) v Kenya*.[37] In this case, the CEMIRIDE accused the government of Kenya of violating articles 21 and 22 of the African Charter on Human and Peoples Rights, after forcefully evicting the Endorois people from their traditional homelands for development purposes. The African Commission on Human and Peoples Rights found that the Kenyan government had not only violated the Endorois peoples' right to enjoy and freely dispose of their natural resources but also prevented them from exercising their right to development. Although the above decision illustrates indigenous peoples' rights to natural resources and its relevance to development, regrettably most African states continue to retain ownership rights over surface and sub-surface resources.[38] Similar to most African countries and Kenya in particular, mining activities and modern development schemes are considered the preferred means of development.[39] Paragraph 25 of the draft Kenyan Policy shows that indigenous peoples' economic pattern of survival is undermined.

Nigeria like most African countries is endowed with mineral resources. These resources are located in traditional lands occupied by the indigenous

Ogoni people in the Niger Delta. Scholars describe them as a distinct indigenous minority residing in an area of approximately 1,000 square kilometres in the South-South of Nigeria as it is commonly known.[40] The production of crude oil began in 1953 with the arrival of Shell Petroleum Development Company (SPDC) in Ogoniland, River State, Nigeria.[41] It is estimated that since the discovery of petroleum, over 100 billion worth of USD has been extracted from the area.[42] The local communities of Ogoniland do not benefit from the proceeds; rather they are confronted with environmental pollution including food insecurity and other health-related complications.[43] The economy of Nigeria depends on revenue from petroleum for survival. It is reported that most state governments use proceeds from petroleum to pay the salaries of civil servants. Irrespective of the significance of petroleum to the Nigerian economy, the Niger Delta State where these resources are found remains impoverished and underdeveloped.[44] Efforts to prevent the over-exploitation of these resources have resulted in frequent violence between the Ogoni people, government and multinational corporations.

The global trend of economic and regional development resulted in the enactment of the legislation that grants resource ownership rights to the government. Section 44(3) of the Nigerian Constitution[45] stipulates that the control of all minerals, mineral oils and natural gas in, under or upon any land in, under or upon the territorial waters and Exclusive Economic Zone (EEZ) of Nigeria is vested with the state. Section 1 of the Petroleum Act of 1990 vests ownership and control of all petroleum in, under or upon any land on the Federal government of Nigeria. Similarly, section 1 of the Mining and Minerals Act of 2007 confers ownership of all mineral resources found in, under or upon any land in Nigeria with the state. This implies that land found in commercial quantities shall immediately become the property of the state in line with the provisions of the Land Use Act of 1978. Section 3(1) of the Mining and Minerals Act stipulates that no person shall search, exploit or impound minerals in Nigeria except as provided by law. This clearly means communities residing in areas where these minerals are located do not have ownership or control rights over them.

While section 2 of the EEZ[46] confers ownership of natural resources of the seabed, sub soil and superjacent waters of the EEZ on the Federal Republic of Nigeria, section 1 of the Land Use Act equally vests all land on the government of each state of the Federation of Nigeria. Section 2 gives total control and management of land to the advisory bodies of the state. The main objective of these laws is to ensure the sustainable use, control and management of land and mineral resources. It deprives indigenous communities of their ancestral rights over land and resources on which they rely for subsistence. The Mineral Laws of Nigeria have prompted the displacement and relocation of most indigenous populations from the Ogoniland who are now living in regret for originating from an area that is rich in petroleum resources. The endowed resources have rather subjected them to a disadvantaged position.

Ikeya asserts that the relocation policy and sedentary lifestyle imposed on indigenous populations have not only affected their livelihood but also limited their control over land and natural resources.[47] The world's genetic resources are located in territories occupied by indigenous peoples where specialised livelihood practices and traditional ecological knowledge have a significant contribution to sustainable development.[48] Unfortunately, increased pressure on these resources has put new strains on indigenous peoples' ability to own the resources that can facilitate their development.[49]

In Ghana, article 1 of the Minerals and Mining Act, 703 of 2006 also bestows ownership rights of all minerals in its original state with the president in trust for the people. However, article 5 mandates the Minister to grant Mineral rights in the form of Mineral licences to private individuals wishing to exploit mineral resources. This process can be challenging to the five self-identified indigenous groups that always encounter multiple challenges in accessing basic rights. The bureaucratic nature surrounding this process disadvantaged indigenous communities that encounter various forms of inequalities in all spheres of life. In addition, in Uganda the Koogere indigenous land was transformed into Queen Elizabeth's National Park. While these national parks tend to attract millions of tourists and generate income, the indigenous communities do not benefit from the proceeds that accrue from them. Anghie suggests that in protecting developing states' ability to achieve economic growth, efforts should be made to ensure that they benefit from the proceeds that derived from the commercialisation of their natural resources.[50] Although, the right to development is rooted in principles of international cooperation, transparency, accountability and non-discrimination, indigenous peoples' concerns are often not taken into consideration. It is obvious that development projects initiated by governments do not respect the fundamental rights of indigenous peoples. The next section zooms into some of the problems that arise as a result of indigenous peoples' inability to own and control their resources.

Problems arising from depriving indigenous peoples of ownership rights over their natural resources

Food insecurity

The well-being of indigenous peoples depends on the relationship they share with their natural environment. They are sustained by the natural resources within their communities. The lack of access to land and other resources has altered their food systems, resulting in acute hunger and malnutrition.[51] Food plays a key role in the well-being and mental development of the entire populations globally. Children and especially pregnant women become very vulnerable if they are unable to access the quality food needed for their development. Altering or destroying indigenous peoples' traditional means of livelihood places a huge burden on their existence.

There is a widening gap of income disparity among indigenous and non-indigenous communities in Africa. Those confronted with forced evictions find it difficult to secure a healthy lifestyle. The majority of them are unable to afford meals with a balanced diet as they need money to do so. Living in their communities, they depend on wild fruits and animals, which provide high protein levels vital for good health and mental development. Their right to food is intimately linked to access to land and natural resources. Hence, it is imperative to protect and revitalise indigenous peoples' local food systems to ensure a healthy and active lifestyle for them.

Poverty

Despite the lack of accurate data on the situation of indigenous peoples at national levels, studies suggest that they are 'at the bottom of socio-economic scale'.[52] It is reported that indigenous peoples make up 5 percent of the world's population and 15 percent of the poorest group in the world.[53] Although they own and occupy about one-quarter of the world's land and resources, constituting 80 percent of the world's biodiversity, the pressures exerted on their lands severely threaten their existence and render them impoverished.[54] The main goal of the 2015 Sustainable Development Goals is to eradicate extreme poverty in all its forms by 2030.[55] However, poverty, hunger and diseases continue to affect the most vulnerable groups in Africa.[56] Indigenous peoples suffer disproportionately, as they have little or no access to basic social services including education, health and employment systems. Their impoverishment is exacerbated by forceful expulsion from ancestral lands, loss of control over natural resources including state policies of assimilation into the mainstream society. Lee and Neves discern that most of the world's poor are people who live in rural areas, especially as they only depend on agriculture and natural resources for survival.[57] The absence of these resources instigates issues of food insecurity and poor health.

Accordingly, issues of financial poverty are increasingly disturbing in indigenous communities. They live in isolation in forested areas that are ranked at the bottom of the income scale, employment and other indices of poverty. Irrespective of their intuitive relationship with nature, they are denied ownership rights over the resources they have conserved for many years. Sustainable economic development processes can only be realised if the scope and degree of indigenous peoples' decision-making power and control over natural resources are enhanced.

Non-representation in state affairs

Land and natural resources form the basis of indigenous peoples' spiritual, cultural and political identity.[58] The United Nations Inter-Agency Support Group notes that lack of access to land and natural resources exposes indigenous peoples to multiple human rights abuses, including violation of their right to development.[59] Clavero affirms that any development assistance policy negotiated without their full consent and participation, especially those that have the potential to negatively affect their livelihood is illegal.[60] Regrettably, indigenous peoples have constantly been targeted for opposing acts of land expropriation and resource extraction perpetrated within their territories without their consent.[61]

The right to participate in all decision-making processes concerning indigenous peoples is recognised in a number of international instruments including the ICCPR, the ILO Convention No 169 and the UNDRIP. These instruments grant indigenous peoples the opportunity to determine and to develop priorities and strategies for the development or use of their land and resources. It obligates state governments to include indigenous peoples in the formulation and implementation of development policies that have a direct bearing on their well-being. This is because by being part of the decision-making process, indigenous peoples can easily object to development projects that can alter their traditional ways of life.[62] Colchester, Eversole, Mcneish and Cimadamore opine that national policies and institutional structures still exclude indigenous peoples from participating in all aspects of decision-making processes.[63]

As such, indigenous peoples face impediments in having adequate representation, accessing land and natural resources within their territories. With the inability to access natural resources essential for their survival, actualising the right to development will almost be impossible. Worse is the fact that states reserve the right to expropriate or permit development activities on the pretext of serving public interest, notwithstanding the adverse impact that such projects usually have on the livelihood of the custodians of these territories.

Conclusion

This chapter looked at how indigenous peoples' inability to own and control natural resources limits the process of actualising their right to development. Making this analysis entailed delving into the African concept of indigenous peoples. A study by the International Work Group on Indigenous Populations (IWGIP), presents criteria in identifying indigenous populations in Africa, including occupation and attachment to land and natural resources, cultural distinctiveness and marginalisation, exclusion and expulsion from ancestral lands among others.

In examining the connection between natural resource ownership, indigenous peoples and the right to development, it is observed that despite the amount of resources endowed in their territories, national policies generally allow states to retain full ownership of surface and sub-surface resources in Africa. This often results in indigenous peoples not benefitting from the proceeds that accrue from the exploitation of those resources. This is the reason why indigenous peoples across Africa have remained destitute, unable to develop their traditionally acquired skills and indigenous knowledge that can aid their personal development.

By exploring some legal systems in Africa that vest natural resource ownership rights with the state, it is observed that that was the basis for depriving indigenous peoples of their right to own, control and freely dispose of their natural resources as they deem appropriate. This hinders the global campaign of achieving the Sustainable Development Goals by the year 2030, which is anchored on the promise of leaving no one behind. Hence, domestic legal frameworks in Africa need to be reviewed to enable African governments to implement a strict model to ensure that indigenous peoples exercise control inmanaging their natural resources. This will enable them to negotiate deals with extractive companies to ensure the sustainable exploitation of these resources. This process will not only guarantee a fair share in the proceeds from exploitation but also enable them to preserve these resources for the benefit of future generations.

The challenges encountered in these circumstances are state government's reluctance to grant indigenous peoples' formal recognition at the national level. With respect to the special rights allocated to indigenous peoples in a number of international and regional legal instruments, it is vital for African state governments to adopt and implement a strict model that allows indigenous peoples the right to control, own and manage the natural resources that they traditionally own from time immemorial.

Notes

1 African Charter on Human and Peoples' Rights, adopted by the Organisation of African Unity in Nairobi on 27 June 1981, OAU Doc CAB/LEG/67/3 rev. 5; 1520 UNTS 217.
2 Kingsburg, Benedict. 1998. '"Indigenous Peoples" in International Law: A Constructivist Approach to the Asian Controversy'. *American Journal of International Law* 92(3): 414–457.
3 Weaver, Hilary. 2001. 'Indigenous Identity: What is it, and Who Really Has It?' *American Indian Quarterly* 25(2): 240–255.
4 Bojosi, Kealeboga and Wachira, Mukundi. 2006. 'Protecting Indigenous Peoples in Africa: An Analysis of the Approach of the African Commission on Human and Peoples' Rights'. *African Human Rights Law Journal* 6: 382–406.
5 Barume, Albert. 2010. *Land Rights of Indigenous Peoples in Africa: With Special Focus on Central, Eastern and Southern Africa*. Denmark: Copenhagen.
6 IWGIA. 2019. *Indigenous World 2019*. Denmark: International Work Group for Indigenous Affairs.
7 ILO Convention No 169 Concerning Indigenous and Tribal Peoples in Independent Countries, adopted by the General Assembly Resolution 1650 UNTS 383, 27 June 1989; United Nation Declaration on the Rights of Indigenous Peoples, adopted by General Assembly Resolution A/RES/61/295, 13 September 2007.
8 Wang, Linzhu. 2015. 'The Definition of Indigenous Peoples and its Applicability in China'. *International Journal on Minority and Group Rights* 22(2): 232–258.
9 International Work Group for Indigenous Affairs. 2020. <www.iwgia.org/regions/africa>; UN Office of the High Commissioner for Human Rights. 'The Right to Development and Indigenous Peoples: Threats to the Rights of Indigenous Peoples'. <www.ohchr.org/Documents/Issues/RtD/RTD_IndigenousPeoples.pdf>.
10 Gibbs, Meredith. 2005. 'The Right to Development and Indigenous Peoples: Lessons from New Zealand'. *World Development* 33(8): 1365–1378.

11 Diergarten, Yorck. 2019. 'Indigenous or Out of Scope? Large-Scale Land Acquisition in Developing Countries, International Human Rights Law and the Current Deficiencies in Land Rights Protection'. *Human Rights Law Review* 19(1): 193–198.
12 Southalan, John; Culotta, Kenneth and Fallon, Archie. 2011. 'Indigenous Peoples and Resources Development: A Rapidly Changing Legal Landscape'. *Oil, Gas and Energy Law Journal* 9(4): 1–22.
13 Koissaba, Ben. 2016. 'Elusive Justice: The Maasai Contestation of Land Appropriation in Kenya: A Historical and Contemporary Perspective'. In *African History and Modernities Kenya After 50: Reconfiguring Historical, Political, and Policy Milestones*, edited by Kithinji, Michael; Mwanzia, Koster and Rotich, Jerono. London: Macmillan, 180–215.
14 Pullen, John. 2013. 'An Essay on Distributive Justice and the Equal Ownership of Natural Resources'. *The American Journal of Economic and Sociology* 72(5): 1044–1074.
15 International Covenant on Civil and Political Rights (ICCPR) adopted by UN General Assembly on 16 December 1966, 2200A (XXI), art 1.
16 International Covenant on Economic, Social and Cultural Rights (ICESCR) adopted by UN General Assembly Resolution on 16 December 1966, A/RES/2200, art 1.
17 ILO Convention No 169, note 7.
18 Demir, Ebru. 2017. 'The Right to Internal Self-Determination in Peace Building Processes: A Reinterpretation of the Concept of Local Ownership from a Legal Perspective'. *The Age of Human Rights Journal* 8: 18–48.
19 Hagen, Renée and Minter, Tessa. 2020. 'Displacement in the Name of Development: How Indigenous Rights Legislation Fails to Protect Philippine Hunter-Gatherers'. *Society and Natural Resources: An International Journal* 33(1): 65–82.
20 Anderson, Ian *et al.* 2016. 'Indigenous and Tribal Peoples' Health (The Lancet-Lowitja Institute Global Collaboration): A Population Study'. *The Lancet* 388(10040): 131–157; Valeggia, Claudia and Snodgrass, Josh. 2015. 'Health of Indigenous Peoples'. *Annual Review of Anthropology* 44: 117–135.
21 Koissaba, note 13, 189–210.
22 German, Laura; Karsenty, Alain and Tiani, Anne-Marie. 2009. 'Governance and Decentralisation in Africa: Linking Local, Regional and Global Dialogues'. In *Governing Africa's Forest in a Globalised World*, edited by German, Laura; Karsenty, Alain and Tiani, Anne-Marie. United Kingdom: Earthscan, 1–28.
23 Duruigbo, Emeka. 2006. 'Permanent Sovereignty and Peoples' Ownership of Natural Resources in International Law'. *George Washington International Law Review* 38: 33–100; Ng'ambi, Sangwani. 2015. 'Permanent Sovereignty over Natural Resources and the Sanctity of Contracts, from the Angle of Lucrum Cessans'. *Loyola University Chicago International Law Review* 12(2): 153–172.
24 International Institute for Democracy and Electoral Assistance. 2017. 'Information Brief: Africa and West Asia Programme'. <www.idea.int/sites/default/files/publications/enhancing-natural-resource-governance-in-africa.pdf>.
25 Choma, Hako. 2009. 'The Impact of Mining Rights on the Local Community in Limpopo Province, South Africa: A Comparative Study'. *US-China Law Review* 6(2): 8–15.
26 Van der Ploeg, Frederick. 2011. 'Natural Resources: Curse or Blessing?' *Journal of Economic Literature* 49(2): 366–420.
27 Wegenast, Tim and Schneider, Gerald. 2017. 'Ownership Matters: Natural Resources Property Rights and Social Conflict in sub-Sahara Africa'. *Erschienen in Political Geography* 61: 110–122.
28 The World Bank. 2020. 'Poverty in a Rising Africa'. *The World Bank* <www.worldbank.org/content/dam/Worldbank/document/Africa/poverty-rising-africa-poverty-report-main-messages.pdf>.
29 Botchway, Francis and Rukuba-Ngaiza, Nightingale. 2015. 'The Constitutional Regime for Resource Governance in Africa: The Difficult March toward Accountability'. In *The World Bank Legal Review Volume 6: Improving Delivery in Development: The Role of Voices, Social Contract, Accountability*, edited by Wouters, Jan *et al.* Washington DC: The World Bank.

30 Sebudubudu, David *et al.* 2014. 'Governance of Land and Natural Resources for Sustainable development in Botswana: Possible Lessons for the Agricultural and Tourism Sectors'. *International Journal of African Development* 2(1): 81–100.

31 Adams, Martin; Kalabamu, Faustin and White, Richard. 2003. 'Land Tenure Policy and Practice in Botswana: Governance Lessons for Southern Africa'. *Austrian Journal of Development Studies* 19(1): 55–74.

32 *Roy Sesana and Others v the Attorney-General* [2002] 1 BLB 452 (HCR).

33 Winters, Olivia. 2014. 'The Botswana Bushmen's Fight for Water and Land Rights in the Central Kalahari Game Reserve'. *The Journal of Sustainable Development* 13(1): 285–300.

34 Hitchcock, Robert. 2002. 'We are the First People: Land, Natural Resources and Identity in the Central Kalahari, Botswana'. *Journal of African Studies* 28(4): 797–824.

35 Dinokopila, Bonolo. 2011. 'The Right to Water in Botswana: A Review of the Matsipane Mosetlhanyane Case'. *African Human Rights Law Journal* 11: 282–295.

36 The Constitution of Kenya, 2010, Sect 69.

37 *Centre for Minority Rights Development (Kenya) and Minority Rights Group International on Behalf of Endorois Welfare Council v Kenya* Comm 276/2003 *(Endorois).*

38 Ulfstein, Geir. 2004. 'Indigenous Peoples' Right to Land'. *Netherlands: Max Planck Yearbook of United Nations Law* 8: 1–48.

39 Ngugi, Joel. 2001–2002. 'The Decolonisation-Modernisation Interface and the Plight of Indigenous Peoples in Post-Colonial Development Discourse in Africa'. *Wisconsin International Law Journal* 20: 297–352.

40 Ojide, Makuachukwu. 2015. 'Socioeconomic Baseline Study of Ogoniland in the Niger Delta Region'. *Research Journal of Economics* 3(1): 8–15.

41 Bodo, Tombari. 2019. 'Deep Issues behind the Crisis in the Niger Delta Region: The Case of Oil Exploration in Ogoniland River State, Nigeria'. *Asian Journal of Geographical Research* 2(1): 1–12.

42 Bodo, Tombari. 2018. 'Community Understanding of the Environmental and Socio-Economic Consequences of Petroleum Exploitation in Ogoni, River State, Nigeria'. *International Journal of Advanced Research and Publications* 2(11): 51–55.

43 Bodo, Tombari and David, Kepka 2018. 'The Petroleum Exploitation and Pollution in Ogoni, Rivers State, Nigeria: The Community Perspective'. *European Scientific Journal* 14(32): 197–212; Bodo, note 42, 51.

44 Ezekwe, Ugwo; Chinedu, Umeh and Ochuba, Daniel. 2019. 'Analysis of the Impact of Crude Oil Export and Economic Growth in Nigeria (1980-2017): An Approach of Time Series Econometric Model'. *Economics and Social Sciences Academic Journal* 1(2): 12–19; Uwakonye, Nga; Osho, Gbolahan and Anucha, Hyacinth. 2006. 'The Impact of Oil and Gas Production on the Nigerian Economy: A Rural Sector Econometric Model'. *International Business and Economic Research Journal* 5(2): 61–62.

45 Constitution of the Federal Republic of Nigeria, 1999.

46 Exclusive Economic Zone Decree No 28 of 5 October 1978.

47 Pemunta, Vitalis. 2019. 'Forest Conservation, Wildlife Legislation and the Baka Pygmies of Southeast Cameroon'. *GeoJournal* 84(3): 1035–1055; Awuh, Harrison. 2015. 'Adaptive Livelihood Strategies in Conservation-Induced Displacement: The Case of the Baka of East Cameroon'. *African Studies Review* 58(2): 135–156.

48 Inter-American Commission on Human Rights. 2010–2011. 'Indigenous and Tribal Peoples' Rights over their Ancestral Lands and Natural Resources: Norms and Jurisprudence on the Inter-American Human Rights System'. *American Indian Law Review* 35(2): 285.

49 Gilbert, Jérémie. 2017. *Land Grabbing, Investments and Indigenous Peoples' Rights to Land and Natural Resources: Case Studies and Legal Analysis.* Copenhagen: International Work Group for Indigenous Affairs, 26.

50 Anghie, Anthony. 2004. *Imperialism, Sovereignty and the Making of International Law.* London: Cambridge University Press.

51 Lemke, Stefanie and Delormier, Treena. 2017. 'Indigenous Peoples' Food Systems, Nutrition, and Gender: Conceptual and Methodological Considerations'. *Maternal and Child Nutrition* 13(S3): 1–12.

52 Cobo, Martínez. 1987. 'Study of the Problem of Discrimination against Indigenous Populations: Final Report Submitted by the Special Rapporteur: Conclusions, Proposals and Recommendation'. UN Publications V.

53 Hall, Gillette and Gandolfo, Ariel. 2016. 'Poverty and Exclusion among Indigenous Peoples: The Global Evidence'. *World Bank Blogs* <https://blogs.worldbank.org/voices/poverty-and-exclusion-among-indigenous-peoples-global-evidence>.

54 Minority Rights Group International. 2016. 'Minority and Indigenous Cultures in Africa Facing Serious Threat: New Global Report' State of the World's Minority and Indigenous Peoples'. <https://minorityrights.org/2016/07/12/minority-and-indigenous-cultures-in-africa-facing-serious-threat-new-global-report/>.

55 Transforming Our World: The 2030 Agenda for Sustainable Development adopted by the UN General Assembly, September 2015.

56 The Sustainable Development Goals Report. 2019. <https://unstats.un.org/sdgs/report/2019/The-Sustainable-Development-Goals-Report-2019.pdf>.

57 Lee, David *et al.* 2009. 'Rural Poverty and Natural Resources: Improving Access and Sustainable Management'. *ESA Working Paper No 09-03*, 1–130.

58 Northcott, Heather. 2012. 'Realisation of the Right of Indigenous Peoples to Natural Resources under International Law through the Emerging Right to Autonomy'. *The International Journal of Human Rights* 16(1): 73–79.

59 UN Inter-Agency Support Group. 2014. 'Thematic Paper on Lands, Territories and Resources'. Thematic paper towards the preparation of the 2014 World Conference on Indigenous Peoples'. *United Nations* <www.un.org/en/ga/69/meetings/indigenous/pdf/IASG%20Paper%20_%20Lands%20territories%20and%20resources%20-%20rev1.pdf>.

60 Clavero, Bartolomé. 2005. 'The Indigenous Rights of Participation and International Development Policies'. *Arizona Journal of International and Comparative Law* 22(1): 41–43.

61 United Nations. 2018. 'Lands, Natural Resources Represent Life for Indigenous Peoples, not Mere Commodities'. Speakers Stress as Permanent Forum begins Session, Economic and Social Council, Permanent Forum on Indigenous Issues. Seventeenth Session, 16 April. <www.un.org/press/en/2018/hr5387.doc.htm>.

62 Inter Parliamentary Union. 2014. 'Implementing the UN Declaration on the Rights of Indigenous Peoples: Handbook for Parliamentarians'. <http://archive.ipu.org/PDF/publications/indigenous-en.pdf>.

63 Colchester, Marcus. 2004. 'Conservation Policy and Indigenous Peoples'. *Environmental Sciences and Policy* 7: 145–153; Eversole, Robyn; McNeish, John-Andrew and Cimadamore, Alberto. (Eds). 2005. *Indigenous Peoples and Poverty: An International Perspective*. London: Zed Books.

References

Adams, Martin; Kalabamu, Faustin and White, Richard. 2003. 'Land Tenure Policy and Practice in Botswana: Governance Lessons for Southern Africa'. *Austrian Journal of Development Studies* 19(1): 55–74.

Anderson, Ian *et al.* 2016. 'Indigenous and Tribal Peoples' Health (The Lancet-Lowitja Institute Global Collaboration): A Population Study'. *The Lancet* 388(10040): 131–157.

Anghie, Anthony. 2004. *Imperialism, Sovereignty and the Making of International Law*. London: Cambridge University Press.

Awuh, Harrison. 2015. 'Adaptive Livelihood Strategies in Conservation-Induced Displacement: The Case of the Baka of East Cameroon'. *African Studies Review* 58(2): 135–156.

Barume, Albert. 2010. *Land Rights of Indigenous Peoples in Africa: With Special Focus on Central, Eastern and Southern Africa*. Denmark: Copenhagen.

Bodo, Tombari and David, Kepka 2018. 'The Petroleum Exploitation and Pollution in Ogoni, Rivers State, Nigeria: The Community Perspective'. *European Scientific Journal* 14(32): 197–212.

Bodo, Tombari. 2018. 'Community Understanding of the Environmental and Socio-Economic Consequences of Petroleum Exploitation in Ogoni, River State, Nigeria'. *International Journal of Advanced Research and Publications* 2(11): 51–55.

Bodo, Tombari. 2019. 'Deep Issues behind the Crisis in the Niger Delta Region: The Case of Oil Exploration in Ogoniland River State, Nigeria'. *Asian Journal of Geographical Research* 2(1): 1–12.

Bojosi, Kealeboga and Wachira, Mukundi. 2006. 'Protecting Indigenous Peoples in Africa: An Analysis of the Approach of the African Commission on Human and Peoples' Rights'. *African Human Rights Law Journal* 6: 382–406.

Botchway, Francis and Rukuba-Ngaiza, Nightingale. 2015. 'The Constitutional Regime for Resource Governance in Africa: The Difficult March toward Accountability'. In *The World Bank Legal Review Volume 6: Improving Delivery in Development: The Role of Voices, Social Contract, Accountability*, edited by Wouters, Jan *et al.* Washington DC: The World Bank, 149–167.

Choma, Hako. 2009. 'The Impact of Mining Rights on the Local Community in Limpopo Province, South Africa: A Comparative Study'. *US-China Law Review* 6(2): 8–15.

Clavero, Bartolomé. 2005. 'The Indigenous Rights of Participation and International Development Policies'. *Arizona Journal of International and Comparative Law* 22(1): 41–51.

Colchester, Marcus. 2004. 'Conservation Policy and Indigenous Peoples'. *Environmental Sciences and Policy* 7: 145–153.

Cobo, Martínez. 1987. 'Study of the Problem of Discrimination against Indigenous Populations: Final Report Submitted by the Special Rapporteur: Conclusions, Proposals and Recommendation'. UN Publications V.

Demir, Ebru. 2017. 'The Right to Internal Self-Determination in Peace Building Processes: A Reinterpretation of the Concept of Local Ownership from a Legal Perspective'. *The Age of Human Rights Journal* 8: 18–48.

Diergarten, Yorck. 2019. 'Indigenous or Out of Scope? Large-Scale Land Acquisition in Developing Countries, International Human Rights Law and the Current Deficiencies in Land Rights Protection'. *Human Rights Law Review* 19(1): 193–198.

Dinokopila, Bonolo. 2011. 'The Right to Water in Botswana: A Review of the Matsipane Mosetlhanyane Case'. *African Human Rights Law Journal* 11: 282–295.

Duruigbo, Emeka. 2006. 'Permanent Sovereignty and Peoples' Ownership of Natural Resources in International Law'. *George Washington International Law Review* 38: 33–100.

Eversole, Robyn; McNeish, John-Andrew and Cimadamore, Alberto. (Eds). 2005. *Indigenous Peoples and Poverty: An International Perspective*. London: Zed Books.

Ezekwe, Ugwo; Chinedu, Umeh and Ochuba, Daniel. 2019. 'Analysis of the Impact of Crude Oil Export and Economic Growth in Nigeria (1980–2017): An Approach of Time Series Econometric Model'. *Economics and Social Sciences Academic Journal* 1(2): 12–19.

German, Laura; Karsenty, Alain and Tiani, Anne-Marie. 2009. 'Governance and Decentralisation in Africa: Linking Local, Regional and Global Dialogues'. In *Governing Africa's Forest in a Globalised World*, edited by German, Laura; Karsenty, Alain and Tiani, Anne-Marie. United Kingdom: Earthscan, 1–28.

Gibbs, Meredith. 2005. 'The Right to Development and Indigenous Peoples: Lessons from New Zealand'. *World Development* 33(8): 1365–1378.

Gilbert, Jérémie. 2017. *Land Grabbing, Investments and Indigenous Peoples' Rights to Land and Natural Resources: Case Studies and Legal Analysis*. Copenhagen: International Work Group for Indigenous Affairs.

Hagen, Renée and Minter, Tessa. 2020. 'Displacement in the Name of Development: How Indigenous Rights Legislation Fails to Protect Philippine Hunter-Gatherers'. *Society and Natural Resources: An International Journal* 33(1): 65–82.

Hitchcock, Robert. 2002. 'We are the First People: Land, Natural Resources and Identity in the Central Kalahari, Botswana'. *Journal of African Studies* 28(4): 797–824.

Inter-American Commission on Human Rights. 2010–2011. 'Indigenous and Tribal Peoples' Rights over their Ancestral Lands and Natural Resources: Norms and Jurisprudence on the Inter-American Human Rights System'. *American Indian Law Review* 35(2): 263–496.

Kingsburg, Benedict. 1998. '"Indigenous Peoples" in International Law: A Constructivist Approach to the Asian Controversy'. *American Journal of International Law* 92(3): 414–457.

Koissaba, Ben. 2016. 'Elusive Justice: The Maasai Contestation of Land Appropriation in Kenya: A Historical and Contemporary Perspective'. In *African History and Modernities Kenya After 50: Reconfiguring Historical, Political, and Policy Milestones*, edited by Kithinji, Michael; Mwanzia, Koster and Rotich, Jerono. London: Macmillan, 180–215.

Lee, David *et al.* 2009. 'Rural Poverty and Natural Resources: Improving Access and Sustainable Management'. ESA Working Paper No 09-03, 1–130.

Lemke, Stefanie and Delormier, Treena. 2017. 'Indigenous Peoples' Food Systems, Nutrition, and Gender: Conceptual and Methodological Considerations'. *Maternal and Child Nutrition* 13(S3): 1–12.

Ng'ambi, Sangwani. 2015. 'Permanent Sovereignty over Natural Resources and the Sanctity of Contracts, from the Angle of Lucrum Cessans'. *Loyola University Chicago International Law Review* 12(2): 153–172.

Ngugi, Joel. 2001–2002. 'The Decolonisation-Modernisation Interface and the Plight of Indigenous Peoples in Post-Colonial Development Discourse in Africa'. *Wisconsin International Law Journal* 20: 297–352.

Northcott, Heather. 2012. 'Realisation of the Right of Indigenous Peoples to Natural Resources under International Law through the Emerging Right to Autonomy'. *The International Journal of Human Rights* 16(1): 73–79.

Ojide, Makuachukwu. 2015. 'Socioeconomic Baseline Study of Ogoniland in the Niger Delta Region'. *Research Journal of Economics* 3(1): 8–15.

Pemunta, Vitalis. 2019. 'Forest Conservation, Wildlife Legislation and the Baka Pygmies of Southeast Cameroon'. *Geo Journal* 84(3): 1035–1055.

Pullen, John. 2013. 'An Essay on Distributive Justice and the Equal Ownership of Natural Resources'. *The American Journal of Economic and Sociology* 72(5): 1044–1074.

Sebudubudu, David *et al.* 2014. 'Governance of Land and Natural Resources for Sustainable development in Botswana: Possible Lessons for the Agricultural and Tourism Sectors'. *International Journal of African Development* 2(1): 81–100.

Southalan, John; Culotta, Kenneth and Fallon, Archie. 2011. 'Indigenous Peoples and Resources Development: A Rapidly Changing Legal Landscape'. *Oil, Gas and Energy Law Journal* 9(4): 1–22.

Ulfstein, Geir. 2004. 'Indigenous Peoples' Right to Land'. Netherlands: Max Planck *Yearbook of United Nations Law* 8: 1–48.

Uwakonye, Nga; Osho, Gbolahan and Anucha, Hyacinth. 2006. 'The Impact of Oil and Gas Production on the Nigerian Economy: A Rural Sector Econometric Model'. *International Business and Economic Research Journal* 5(2): 61–62.

Valeggia, Claudia and Snodgrass, Josh. 2015. 'Health of Indigenous Peoples'. *Annual Review of Anthropology* 44: 117–135.

Van der Ploeg, Frederick. 2011. 'Natural Resources: Curse or Blessing?' *Journal of Economic Literature* 49(2): 366–420.

Wang, Linzhu. 2015. 'The Definition of Indigenous Peoples and its Applicability in China'. *International Journal on Minority and Group Rights* 22(2): 232–258.

Weaver, Hilary. 2001. 'Indigenous Identity: What is it, and who Really Has It?' *American Indian Quarterly* 25(2): 240–255.

Wegenast, Tim and Schneider, Gerald. 2017. 'Ownership Matters: Natural Resources Property Rights and Social Conflict in sub-Sahara Africa'. *Erschienen in Political Geography* 61: 110–122.

Winters, Olivia. 2014. 'The Botswana Bushmen's Fight for Water and Land Rights in the Central Kalahari Game Reserve'. *The Journal of Sustainable Development* 13(1): 285–300.

10 Ownership of natural resources and the right to development for African indigenous peoples

Ebun Abolarin

Introduction

The perspectives articulated in this chapter are human rights-based, structured in four main parts and a conclusion in addition to this introductory section. The first part expounds the collective rights of indigenous peoples in Africa as owners of the natural resources based on the provisions of relevant international human rights treaties and the African Charter on Human and Peoples' Rights. The discussion culminates in the jurisprudence of the African Commission on Human and Peoples' Rights on the principle of inclusion and the right to development, both derived from property rights in natural resources as it concerns African indigenous peoples. The second part covers the the matic enquiry; who owns the natural wealth and resources needed to achieve the right to development and, to what extent is full sovereignty over natural resources achievable for the peoples to whom the right to development is guaranteed? This enquiry occasions another question with regards to whether African governments protect indigenous peoples' right to development as a universally acknowledged entitlement. The fourth part explains the mechanisms for redress available to African indigenous peoples, nationally and regionally, in the event of violations of any of the stipulated rights under international and regional human rights treaties.

Natural resources ownership—collective right of indigenous peoples in Africa

The right to natural resources is tied to land ownership and is culturally symbolic to indigenous peoples and their communities. The fact that indigenous peoples' land rights have been in existence prior to the new system of land ownership of African states, is undisputable[1] and as Barume points out, the communities claim a collective right over the natural resources pertaining to these lands, which range from hunting and gathering to grazing areas, grasslands, forests, mixed savannah, wetlands, mountain sides, lakes, rivers, coastal areas, fishing grounds and so on.[2]

Individualism is not an acceptable cultural value of indigenous peoples; hence the right to land and natural resources is vested in the entire community and envisaged to be used collectively in a non-exclusive way.[3] Traditionally, indigenous peoples operate and live a communal lifestyle, usually as one extended family, with close-knit connections to one another, which Chen describes as 'a network of deeply committed horizontal relationships'.[4] The collective position in making and executing decisions constitutes the basis for the exercise and enjoyment of individual member's rights within the community and thus, as Chen further asserts, is essential for securing their cultural survival.[5]

International human rights instruments generally do not make provision for collective or group rights, and also do not define the concept of 'peoples'. Nevertheless, the repeated use of the term 'all peoples' in the common article 1 of the International Covenant on Civil and Political Rights (ICCPR) and the International Covenant on Economic, Social and Cultural Rights (ICESCR), both of 1976, which provide that: 'all peoples have the right to self-determination', and 'all peoples may, for their own ends, freely dispose of their natural wealth and resources', infer not 'single' but 'collective' persons, bringing indigenous peoples into the fold of 'all peoples' whose right to self-determination, and the right to freely dispose of their natural wealth and resources must be protected by state parties. The collective rights of indigenous peoples are brought into clearer focus with the adoption of specific instruments under international law, that guarantee to them special protection.

The International Labour Organisation Convention No 169 on Indigenous and Tribal Peoples Convention, 1989 (ILO C169), is the primary treaty under international law that guarantees absolute protection of indigenous peoples' collective right to their natural resources, especially pertaining to their ancestral lands. It expressly provides, in article 15, that indigenous peoples are entitled to participate in the use, management and conservation of their natural resources. This provision has been expanded in article 26 of the United Nations Declaration on the Rights of Indigenous Peoples (UNDRIP), which stipulates that

1. Indigenous peoples have the right to the lands, territories and resources which they have traditionally owned, occupied or otherwise used or acquired.
2. Indigenous peoples have the right to own, use, develop and control the lands, territories and resources that they possess by reason of traditional ownership or other traditional occupation or use, as well as those which they have otherwise acquired.[6]

These are explicit international law provisions protecting the collective rights of indigenous peoples to their lands and natural resources, including the right to development and use. Prior to adopting the UNDRIP, regional human rights institutions such as the Inter-American Human Rights Commission and the African Commission on Human and People's Rights had recognised indigenous peoples' collective rights to property by relying on the two international human rights covenants and the ILO C169.

The 2001 decision of the Inter-American Court on Human Rights in the *Mayagna (Sumo) Awas Tingni v Nicaragua* (*Awas Tingni* case)[7] represents a classic juridical interpretation of property as 'the right of members of the indigenous communities within the framework of communal property'.[8] The African Commission relied on the *Awas Tingni* case in the *Centre for Minority Rights Development (Kenya) and Minority Rights Group (On Behalf of Endorois Welfare Council) v Kenya* (*Endorois* case)[9] in granting recognition to the Endorois claim to communal property rights. The interpretation of the 'concept of property in indigenous communities' was given by the Inter-American Court as 'a communitarian tradition regarding a communal form of collective property of the land, in the sense that ownership of the land is not centred on an individual but rather on the group and its community'.[10] This interpretation distinguishes the individualistic approach to the right to property.

With its decision in the *Endorois* case, the African Commission's dynamism in promoting indigenous peoples' rights became pronounced where it emphasised that, 'normatively, the African Charter is an innovative and unique human rights document'[11] which places particular emphasis on the rights of 'peoples', and making substantial departure from 'the narrow formulations of other regional and universal human rights instruments by weaving a tapestry which includes the three "generations" of rights: civil and political rights; economic, social and cultural rights; and group and peoples' rights'.[12] The African Commission held that in the African context, the term 'indigenous' is not 'intended to create a special class of citizens, but rather to address historical and present-day injustices and inequalities' and that in the notion of 'peoples' in the African Charter, it is closely related to collective rights.

The African Charter therefore makes provision for peoples to retain rights as collectives,[13] and as expounded in this section, the property right to land and natural resources is recognised as a traditional right guaranteed to African indigenous peoples. The exercise of state sovereignty over natural resources must thus include the peoples concerned as the African Commission concluded.

The African Commission's jurisprudence on the right to development

Principle of inclusion

The African Charter on Human and Peoples' Rights provides in article 21(1) that: 'all peoples shall freely dispose of their wealth and natural resources. This right shall be exercised in the exclusive interest of the people. In no case shall a people be deprived of it'. The African Commission's interpretation of this provision in the *Front for the Liberation of the State of Cabinda v Republic of Angola* (*Cabinda*) case as it pertains to natural resources lays emphasis on the principle of inclusion[14] and also provides context to the term 'peoples' contained in the African Charter.

The African Commission emphasised that the term 'peoples' in article 21 of the African Charter could mean, either the entire people of a given state or 'a peoples' within the state, and that the right enshrined in the article imposes a duty on the state to involve representatives of its peoples in decisions concerning the management of national wealth and natural resources.[15] The Commission thereby instituted the principle of inclusion (of the peoples) for the purpose of management of their national wealth and natural resources. The indication by Angola that elected representatives of the people of Cabinda were in the Angolan Parliament, which is mandated to exercise oversight supervision of the management of natural resources in the country,[16] satisfied the prescribed principle of inclusion for the national benefit of all the people of Angola.

In reaching its conclusion in the *Cabinda* case, the Commission noted that article 21 of the African Charter presupposes that the right to natural resources is 'held in trust for the people'.[17] To be 'held in trust' means 'a property interest held by one person for the benefit of another'.[18] Simplified, a thing held in trust means the holder is not the owner but those to whom the thing is entrustedare the owners. Owners cannot be left behind but rather, must be engaged in compliance with the principle of inclusion concerning the thing entrusted to them. Hence, although states hold in trust, ownership of natural resources ultimately rests with African peoples and thus, all forms of exploitation of the natural resources must involve and include them.

In the *Social and Economic Rights Action Centre and Centre for Economic and Social Rights (CESR) v Nigeria* (*Ogoni*) case,[19] the African Commission explained the origin of article 21(1) thus:

> The origin of this provision may be traced to colonialism, during which the human and material resources of Africa were largely exploited for the benefit of outside powers, creating tragedy for Africans themselves, depriving them of their birth right and alienating them from the land. The aftermath of colonial exploitation has left Africa's precious resources and people still vulnerable to foreign misappropriation. The drafters of the [African] Charter obviously wanted to remind African governments of the continent's painful legacy and restore co-operative economic development to its traditional place at the heart of African Society.[20]

From the Commission's explanation, grave injustice was committed by colonial powers against the peoples of Africa through the act of dispossession. This act has been sustained by African governments which are supposed to protect the people but chose to collaborate with foreign exploiters and multinational corporations as in the *Ogoni* case. The question that begs for an answer is why the dispossession persists even after independence, and why do African States still continue to exploit the peoples' natural resources, which amounts to violation of the peoples' right to property since they are neither involved nor included, as noted in the *Ogoni* case wherein the African Commission observed that

> The government did not involve the Ogoni communities in the decisions that affected the development of Ogoniland. The destructive and selfish role played by oil development in Ogoniland, closely tied with repressive tactics of the Nigerian Government, and the lack of material benefits accruing to the local population may well be said to constitute a violation of article 21.[21]

Dispossession and deprivation are human rights violations that ought to be reversed and redressed to make way for the manifestation of the principle of inclusion. Still on the *Ogoni* case, the African Commission emphasised the duty of governments to protect their citizens not only through appropriate legislation and effective enforcement, but also from damaging acts perpetrated by private parties.[22] The dispossession by private parties recorded in the *Ogoni* case was also reiterated in the *Endorois* case as violation committed against the indigenous peoples. In adjudicating the *Endorois* case, the Commission relied on the decision in the *Ogoni* case in stating that 'the right to natural resources contained within their traditional land was vested in the indigenous people' and that 'the right to freely dispose of natural resources is of crucial importance to indigenous peoples and their way of life'.[23] To the Endorois claim, the African Commission reaffirmed its decision in the *Ogoni* case that the right to natural resources found within their traditional lands is vested in the indigenous people and as a people inhabiting a specific region within a state, they could make claim to the right under article 21 of the African Charter.[24] The Commission thereby maintained its jurisprudential consistency about indigenous peoples' ownership of natural resources in Africa, which invokes the principle of inclusion.

The position of the African Commission on article 21 of the African Charter that 'all peoples shall freely dispose of their wealth and natural resources' has been analysed to emphasise the principle of inclusion for the exploitation, development and management of the natural resources for the benefits of the African indigenous peoples, which can justifiably facilitate the realisation of the right to development.

The right to development

The right to development as an integral part of property right in natural resources is guaranteed by the African Charter, which states in article 22(1) that 'all peoples shall have the right to their economic, social and cultural development with due regard to their freedom and identity and in the equal enjoyment of the common heritage of mankind'. Article 22(2) provides that '[s]tates shall have the duty, individually or collectively, to ensure the exercise of the right to development'.

According to Stamatopoulou, the right to land and natural resources is inextricably linked to the survival of indigenous peoples and hence, deprivation from their land is tantamount to depriving them of identity and culture.[25]

Deprivation is a violation of the fundamental right of indigenous peoples as provided for in article 1(2) of both covenants on human rights that 'In no case may a people be deprived of its own means of subsistence'. This provision is synonymous with article 21(1) of the African Charter on Wealth and Natural Resources in terms of which states are mandated to protect their peoples from all forms of deprivation, more especially from foreign economic exploitation. Articles 21 and 22 of the African Charter, read in conjunction with article 5 of the Declaration on the Right to Development (DRTD) make it mandatory for states to protect the right to natural resource ownership and the right to development of their peoples.

Three African Commission decisions relating to the deprivation of indigenous peoples of their natural resources, and by so doing, denying them the right to development, are discussed in this section. In the *Democratic Republic of the Congo v Burundi, Rwanda and Uganda* (*DRC*) case,[26] it was reported that the armies of Burundi, Rwanda and Uganda violated the fundamental principles that govern friendly relations between states as enshrined in the United Nations Charter and the Charter of the Organisation of African Unity (OAU), resulting in disrespect for the sovereignty and territorial integrity and violation of the inalienable right to self-determination of the peoples of the DRC.[27]

The African Commission held in its decision that the illegal exploitation and looting of the natural resources of the DRC by Burundi, Rwanda and Uganda were in contravention of article 21 of the African Charter, which according to reports, was carried out through different means including confiscation, extraction, forced monopoly and price-fixing, with the first two reaching 'proportions that made the war in the Democratic Republic of the Congo a very lucrative business'.[28] The facts of this case and the subsequent decision of the Commission gave credence to the principle of state sovereignty over natural resources inthe United Nations General Assembly (UNGA) Resolution 1803(XVII),[29] which provides in article 7 that 'violation of the rights of peoples and nations to sovereignty over their natural wealth and resources is contrary to the spirit and principles of the Charter of the United Nations'. The case exemplifies the interpretation and understanding of how the principle ought to be applied to uphold the sovereignty of states without violating the right of peoples, holding that

> the deprivation of the right of the people of the Democratic Republic of Congo, in this case, to freely dispose of their wealth and natural resources, has also occasioned another violation—*their right to their economic, social and cultural development, and of the general duty of States to individually or collectively ensure the exercise of the right to development,* guaranteed under *article 22* of the African Charter.[30] (author's emphasis)

In summary, the atrocities committed by Burundi, Rwanda and Uganda violated the individual and collective right to development of people of the DRC.

In the *Sudan Human Rights Organisation and Centre on Housing Rights and Evictions (COHRE) v Sudan (Sudan)* case,[31] which recorded gross, massive and systematic violations of human rights against the indigenous Black African tribes in the Darfur region, in particular members of the Fur, Marsalit and Zaghawa tribes (Western Sudan) by the Republic of Sudan, the violations included large-scale killings, forced displacement of populations, destruction of public facilities, properties and disruption of life through bombing by military fighter jets in densely populated areas.[32]

The African Commission found that the attacks and forced displacement of Darfurian people denied them the opportunity to engage in economic, social and cultural activities, and the activities of the Sudanese military constituted a violation of not only the economic, social and cultural rights, but other individual rights of the Darfurian people, in violation of article 22 of the African Charter.[33] In addressing the violations committed against the people of Darfur, the Commission described them in their collective capacity as 'a people' as provided for under article 19, which states that 'all peoples shall be equal; they shall enjoy the same respect and shall have the same rights. Nothing shall justify the domination of a people by another'. The people of Darfur therefore do not deserve to be dominated by a people of another race in the same state which might be responsible for the claim for unequal treatment that arose from alleged underdevelopment and marginalisation.[34]

In the *Centre for Minority Rights Development (Kenya) and Minority Rights Group (on Behalf of Endorois Welfare Council) v Kenya*, mentioned earlier, the Endorois alleged among others, violation of article 22 of the African Charter on the right to development. The violation resulted from the Republic of Kenya's failure to adequately involve the indigenous Endorois communities in the development process to ensure their continued improvement and well-being. The Endorois were evicted from their lands and forced to give way for the conservation of the area as a game reserve, which accordingly denied them access to Lake Bogoria, the salt licks and their usual pastures and thus, caused their cattle to die in large numbers. The Endorois claimed that they were consequently unable to pay their taxes, which made the Kenyan authorities take away more of their cattle. The eviction from the game reserve diminished the set of choices and capabilities available to the Endorois who argued that the lack of choices for the community contradicts the guarantee of the right to development expected to be provided by the Kenyan government in accordance with the provisions of the African Charter. On the right to development in the *Endorois* case the African Commission established the following principles:

(a) **Consultation**: Adequacy of consultations is an essential part of development. States are obligated to adequately consult with indigenous people and to secure their participation in development initiatives within their community or on development projects that concern them.
(b) **Free, prior and informed consent**: In addition to consultation, states must obtain the free, prior and informed consent of indigenous peoples

before carrying out any development projects within their communities. The UN Special Rapporteur on the Situation of Human Rights and Fundamental Freedoms of Indigenous Peoples' notes with regard to the *Endorois* case that free, prior and informed consent is essential for the protection of human rights of indigenous peoples in relation to major development projects.[35] The African Commission made reference to the recommendations of the Committee on the Elimination of Racial Discrimination that, the prior informed consent of communities must be sought when major exploitation activities are planned in indigenous territories and that equitable sharing of benefits must be ensured and derived from such exploitation.[36]

(c) **Participation**: Participation is inextricably connected to the right to development. The African Commission made reference to article 2(3) of the DRTD by reiterating that the right to development includes 'active, free and meaningful participation in development'. For participation to be meaningful, it must be carried out through consultations done in good faith and through culturally appropriate procedures with intent to reach an agreement. This entails constant communication between the parties.[37]

(d) **Benefit sharing**: The right to development would be considered violated when a particular development decreases the well-being of the community. The African Commission considered that the concept of benefit sharing serves as an important indication of compliance with property rights.[38] The Commission further notes that in 1990, 'the African Charter on Popular Participation in Development and Transformation' set benefit sharing as key to the development process.[39] Thus, in the *Endorois* case,

> the right to obtain 'just compensation' in the spirit of the African Charter 'translates into a right of the members of the Endorois community to reasonably share in the benefits made as a result of a restriction or deprivation of their right to the use and enjoyment of their traditional lands and of those natural resources necessary for their survival'.[40]

Pursuant to the spirit of the African Charter, benefit sharing may be understood as a form of reasonable equitable compensation resulting from the exploitation of traditionally owned lands and of those natural resources necessary for the survival of the Endorois community. Hence, the government of Kenya was required to ensure mutually acceptable benefit sharing with the Endorois community.

(e) **Capabilities and choices**: The freedom of choices, values and capabilities were lost by eviction from the land. Human development values being integrated into the right to development aresymbolic of the all-inclusiveness of the right to development. The Endorois 'faced significant losses in choice since their eviction from the land'[41] and thus, lost capabilities, opportunities, values and choices that would have made their lives better. Capabilities and choices for the Endorois[42] are values responsible

> for their human development, which must improve so that the right to development, which incorporates civil and political as well as economic, social and cultural rights can be fully realised.

African states should, therefore, change their irresponsiveness towards indigenous peoples' yearnings for their economic, social and cultural rights to development, and the need to enjoy their natural wealth and resources. The irresponsiveness was registered in the deliberate avoidance of the Kenyan government to implement the African Commission's *Endorois* decision of 2009[43] and subsequent failure of the government to be represented at the September 2013 workshop organised by the African Commission.[44] Prior to that decision, the Nigerian government was irresponsive by its failure in 1996 to respond to complaints in the *Ogoni case* and the subsequent non-implementation of the 2001 decision.[45] Such irresponsiveness compromises human rights generally and the right to development in particular on the African continent.

Thematic summarisation

The two thematic questions are given consideration in this section in reaffirming the fact that the right to development and ownership of natural resources are the rights of indigenous peoples.

Indigenous ownership of natural resources

From the purview of this chapter, the answer has been laid in the body of arguments presented above to the effect that indigenous peoples indeed own the natural resources located in their territories and communities.To corroborate the argument, the DRTD explicates on natural resources ownership in article 1(2) which stipulates that

> the human right to development also implies the full realization of the right of peoples to self-determination, which includes, subject to the relevant provisions of both International Covenants on Human Rights, the exercise of their inalienable right to full sovereignty over all their natural wealth and resources.

It clarifies doubts about ownership, which must not be confused with the principle of sovereignty over natural resources. The UN's efforts at resolving this dilemma are documented in the Final Report of the Special Rapporteur, on 'Indigenous peoples' permanent sovereignty over natural resources' which affirmed that

> The interest in the application of this principle to indigenous peoples follows from the similarity of their circumstances to the situation of the peoples to whom the principle was first applied. [...] And as a result, it has

> become clear that meaningful political and economic self-determination of indigenous peoples will never be possible without indigenous peoples having the legal authority to exercise control over their lands and territories.[46]

As the Special Rapporteur explains, the term 'sovereignty' in reference to indigenous peoples does not diminish or contradict the 'sovereignty' of the state.[47] The intrinsic value of this understanding is that

> Indigenous peoples' permanent sovereignty over natural resources might properly be described as a collective right by virtue of which the State is obligated to respect, protect, and promote the governmental and property interests of indigenous peoples (as collectivities) in their natural resources.[48]

The principles for actualising the right to development for indigenous peoples in Africa, which include their involvement, participation and inclusion and above all, recognition of their human right to property, which have been reiterated in the jurisprudence of the African Commission earlier discussed, support the argument that indigenous peoples own the natural wealth and resources on their lands and communities. This is the people-centred and rights-based approach to development that has increasingly been advanced.

Upholding this approach for indigenous peoples is intended to reaffirm their inalienable rights to the natural resources that qualify them for participatory development and benefit sharing that are obtained from the exploitation of the resources. The guarantees to indigenous peoples to exercise and enjoy their rights to land, natural resources and socio-economic and cultural development is, according to the Special Rapporteur, to ensure that indigenous peoples who own the natural resources have the permanent right to control and dispose of the resources in a manner that is appropriate to them; free from subjugation or unfairness of any kind, including the often unequal and unjust conditions of the private market place.[49]

The non-recognition of indigenous peoples' right over their natural resources and the right to participate in the governance and be involved in the processes for development are the factors driving the agitation of indigenous peoples globally and in Africa. At the core of indigenous peoples' activism is the fundamental liberty and legitimate assertion against being excluded or side-lined; marginalised or dispossessed of the inherent right to the natural resources that are required for the realisation of their right to development. By claiming sovereign ownership of their natural resources, indigenous peoples in Africa are not contesting for territorial sovereignty with the states but for recognition of their right to permanent sovereignty over the natural resources, which they traditionally own.

Realising the right to development for indigenous peoples

Asserting the right to full sovereignty over natural wealth and resources is in principle achievable to the extent that African state governments have ratified various treaties that guarantee to indigenous peoples the right to sovereign ownership over their natural resources and the right to development, which converge in article 1(2) of the DRTD. The full implementation of the DRTD is important to indigenous peoples' full enjoyment of all human rights and fundamental freedoms as reaffirmed in the Vienna Declaration and Programme of Action adopted by the World Conference on Human Rights in 1993. It should be understood that the DRTD constitutes an integral link between the Universal Declaration of Human Rights and the Vienna Declaration by way of its 'elaboration of a holistic vision integrating economic, social and cultural rights with civil and political rights' as recognised in article 3 of UN General Assembly Resolution 53/155.[50]

This chapter advances the view that Africa as a continent should adopt a regional development strategy that specifically responds to the needs of African peoples, including, in particular, indigenous peoples. The industrialised nations of the world have held sway over developing and underdeveloped countries while building and improving their own economies. This has been to the detriment of developing countries, particularly those in Africa, which have not benefitted from the principle of state sovereignty that promotes international cooperation for the socio-economic development and autonomy of their countries as contained in Resolution 1803.[51]

Africa can, therefore, no longer rely on global agendas that are unsuitable for the needs of her peoples because of the unique development needs of the African peoples, who over the centuries have demonstrated that they are distinct peoples. Although Agenda 2063 keys into this purpose, it, however, totally ignores the most vulnerable and marginalised—the indigenous peoples. African governments seem unwilling to embrace and protect indigenous peoples on the continent and their entitlements to the wealth and resources that naturally belong to them. The next section explores how African governments could potentially guarantee indigenous peoples' right to development.

Obligation to protect indigenous peoples' right to development

Unlike the 2030 Agenda for Sustainable Development,[52] which diligently and explicitly mentions indigenous peoples' concerns in different and specific areas of need, the African 2063 Agenda[53] focuses on women, youth and children but makes no mention of indigenous peoples as a group with peculiar needs. Paragraph 49 anticipates that by 2063, development in Africa would be people-oriented and caring. It puts children first, seeks to empower women and the youths but makes no mention of the measures intended for indigenous peoples. Paragraph 48 states that 'all the citizens of Africa will be actively involved in decision making in all aspects of development, including social, economic, political and environmental' and Aspiration 6 deals with 'An Africa whose development is people-driven, relying on the potential of African people, especially its women and youth, and caring for children'. Both of these provisions leave out indigenous peoples.

It is a grave oversight that indigenous peoples are continuously left out in both the drafting and implementation of programmes of the African Union, which makes Agenda 2063 exclusively elitist and not all-inclusive as expected. It is the case with the Common African Position (CAP) on the post-2015 Development Agenda, which relied heavily on the wealth of information collected and collated from national and regional stakeholders comprising of the executive and legislative arms of governments, private sector, civil society organisations, youth associations, women's groups, trade unions and academia, but excluded indigenous peoples.[54]

The African attitude towards indigenous peoples is contrary to the African Union Assembly decision to be part of the global drive through the UN and other multilateral organisations, to find multilateral approaches to humanity's most pressing concerns.[55] The UN 2030 Agenda variously includes indigenous peoples in its plans as affirmed in its preamble, relying on the UN Charter, which states that

> 'We the peoples', whereby embodying the inclusiveness of the 2030 Agenda in its profound pronouncement that 'we the peoples' who are embarking on the road to 2030 include—'Governments as well as parliaments, the United Nations system and other international institutions, local authorities, indigenous peoples, civil society, business and the private sector, the scientific and academic community—and all people, [...] and this, we believe, will ensure its success.[56]

Implementation of the 2030 Agenda requires commitments at the global, regional and national levels and hence, a systemic follow-up and review over the 15 years' timeframe. Indigenous peoples are given significance, particularly because of their close proximity to nature and their rich knowledge of natural resources and biological diversity, which cannot be ignored in the processes for sustainable development.

This observation of the 2030 global agenda on development is crucial as a means to point out the shortcomings of the 2063 African agenda for failing to include or recognise the role that indigenous peoples can and should play in advancing sustainable development on the continent, which of course amounts to a denial of their right to contribute to development. If Africa is to make substantive use of the natural resources on the continent as a means to accelerate development, it is imperative to review its policies with regard to the exploitation of natural resources, particularly within the extractive sector with vast operations in indigenous territories and communities. It requires factoring in indigenous peoples' concerns, necessitating a committed resolve by the African Union (AU) Commission in taking measures to discontinue arrant dispossession of their lands and natural resources by extractive industries—an initiative that is already being piloted by the World Bank Group.[57]

It is anticipated that the governments of resource-rich countries in Africa would yield to the concern that the DRTD is insufficiently disseminated, and thus, take appropriate steps for bilateral and multilateral cooperation programmes, national development strategies and policies and the activities of international organisations to create an economic, social, cultural as well as the enabling political and legal environment for the African people to achieve socio-economic and cultural development,[58] particularly with indigenous peoples in focus. In the event that rights in general and the right to development in particular are violated, opportunity for redress avails to all peoples, not excluding indigenous peoples asdiscussed in the next section.

Redress mechanisms

It is pertinent to affirm that instruments of international human rights law are legally binding with laid-down obligations which state parties are bound to respect and to fulfil. International treaties are expected to apply universally to the benefit of all peoples without discrimination and thus, their implementation and enforcement ought to be internationally guided, monitored and accompanied by effective redress mechanisms in accordance with states' duties to respect, to protect and to fulfil human rights. The obligation to respect means that states must refrain from interfering with or curtailing the enjoyment of human rights; the obligation to protect requires states to protect individuals and groups against human rights abuses while the obligation to fulfil obligates states to take positive action to facilitate the enjoyment of basic human rights.[59]

Effective implementations of human rights at the national level depend solely on states fulfilling these obligations, although each country decides to what extent or how its obligations are met. The ratification of international human rights treaties prompts governments to put in place 'domestic measures and legislation compatible with their treaty obligations and duties, thereby enabling the domestic legal system to make provision for legal protection of the human rights guaranteed under international law. Domestic application becomes possible after a ratified treaty is transformed into domestic law through legislative enactment for direct application within the domestic legal order'.[60]

At the regional level, the African Charter[61] embraces the economic, social and cultural rights in international law and is required to make African people economically fulfilled. All African countries have ratified the African Charter and are expected to have domesticated same as national law, making it applicable in their national courts. All African peoples, including indigenous peoples, can further seek remedy at the African Commission or the African Court on Human and Peoples' Rights when all domestic remedies have been exhausted and failed to redress alleged violations. Communications for violations have generally been regulated by the rules of procedure of the African Commission for the purpose of redress, but, following the creation of the African Court, new rules came into effect in 2010,[62] and the Court has in effect adjudicated one case involving the Ogiek indigenous community in Kenya.[63]

Conclusion

Under international human rights law, human rights protection of individuals and groups are inexhaustible.[64] The same international principle applies to the African Charter which has absorbed most of the relevant international human rights provisions for civil, political, economic, social, cultural and environmental rights that have been applied in the jurisprudence of the African Commission. African states as parties to these human rights instruments are under obligation to promote and protect the rights of their peoples, not excluding indigenous peoples. Thus, African indigenous peoples have the opportunity to seek redress nationally or regionally when their rights, particularly the right to development are violated.

With reference to the African Commission's decisions in the cases discussed in this chapter, indigenous peoples are legitimately entitled to be involved and consulted with regard to any development that concerns them or has to do with their lands and natural resources. This is the universal doctrine of the right to development, which lies at the core of the global Sustainable Development Goals,[65] and thus, also needs to reflect in the African development aspirations. Recognition of the property right of indigenous peoples as core participants in the processes for the realisation of their right to development is the solution for economic buoyancy in Africa.

Notes

1 Barume, Albert. 2014. *Land Rights of Indigenous Peoples in Africa* (2nd edn). Copenhagen: International Work Group on Indigenous Affairs, 51.
2 *Ibid*, 203.
3 *Ibid*.
4 Chen, Cher. 2017. 'Indigenous Rights in International Law'. *Oxford Research Encyclopaedia of International Studies*, 14.
5 *Ibid*.
6 UN Declaration on the Rights of Indigenous Peoples adopted by the General Assembly on 13 September 2007 (144 states voted in favour, 4 against and 11 abstentions).
7 *Mayagna (Sumo) Awas Tingni v Nicaragua* [2001] Inter-Am Ct H R (Ser C No 79) (*Awas Tingni*).
8 *Ibid*, para 148; *Centre for Minority Rights Development (Kenya) and Minority Rights Group (On Behalf of Endorois Welfare Council) v Kenya* ACHPR Comm 276/03 (2009) (*Endorois*) para 190.
9 *Endorois*, note 8.
10 *Awas Tingni*, note 7, para 149.
11 *Endorois*, note 8, para 149.
12 *Ibid*, para 149.
13 *Ibid*, paras 149–150; African Charter on Human and Peoples' Rights adopted by the OAU in Nairobi Kenya on 27 June 1981, OAU Doc CAB/LEG/67/3 rev.5, arts 20–24.
14 *Front for the Liberation of the State of Cabinda v Republic of Angola* Comm 328/06 ACHPR (2013).
15 *Ibid*, paras 130–131.
16 *Ibid*, para 132.
17 *Ibid*, para 132.

18 Merriam-Webster Dictionary.
19 *Social and Economic Rights Action Centre (SERAC) and Centre for Economic and Social Rights v Nigeria* Comm 155/96 ACHPR (2001) (*Ogoni).*
20 *Ibid*, para 56.
21 *Ibid*, para 55.
22 *Ibid*, para 57.
23 *Endorois*, note 8, para 121.
24 *Ibid*, para 255.
25 Stamatopoulou, Elsa. 1994. 'Indigenous Peoples and the United Nations: Human Rights as a Developing Dynamic'. *Human Rights Quarterly* 16(1): 58–81.
26 *Democratic Republic of the Congo v Burundi, Rwanda and Uganda* Comm 227/99 ACHPR (2003).
27 *Ibid*, paras 8, 77.
28 *Ibid*, paras 93–94.
29 UN General Assembly Resolution A/RES/1803 (XVII) on Permanent Sovereignty over Natural Resources adopted in 1962.
30 *DR Congo*, note 26, para 95.
31 *Sudan Human Rights Organisation and Centre on Housing Rights and Evictions (COHRE) v Sudan* Comm 279/03-296/06 ACHPR (2009).
32 *Ibid*, paras 2–3.
33 *Ibid*, paras 223–224.
34 *Ibid*, para 223.
35 *Endorois*, note 8, para 293.
36 *Ibid*, para 296.
37 *Ibid*, paras 282, 283, 289.
38 *Ibid*, para 294.
39 *Ibid*, para 295.
40 *Ibid*, para 295.
41 *Ibid*, para 297.
42 *Ibid*, para 283.
43 African Commission on Human and Peoples' Rights. 2015. 'Monitoring the Situation of Indigenous Peoples on the Continent'. Intercession Report of the Working Group on Indigenous Populations/Communities in Africa.
44 Final Communiqué of the Workshop on the Status of the Implementation of the Endorois Decision of the African Commission on Human and Peoples' Rights organised by the African Commission's Working Group on Indigenous Populations/Communities in collaboration with the Endorois Welfare Council, Nairobi, Kenya on 23rd September 2013 <wwwachpr.org/news/viewdetail?id=139>.
45 *Ogoni*, note 19, para 30.
46 UN Economic and Social Council–E/CN.4/Sub.2/2004/30—Final Report of the Special Rapporteur, Erica-Irene, A. Daes 'Prevention of Discrimination and Protection of Indigenous Peoples—"Indigenous peoples" Permanent Sovereignty over Natural Resources'. (2004), paras 6, 8.
47 *Ibid*, para 30.
48 *Ibid*, para 40.
49 *Ibid*, para 47.
50 UN General Assembly Resolution A/RES/53/155 on the Right to Development 1999.
51 A/RES/1803 (XVII), note 29, preamble.
52 UN General Assembly Resolution A/RES/70/1 Transforming our World—The 2030 Agenda for Sustainable Development: A Call for Action to Change our World.
53 AU Commission. 2015. 'Agenda 2063: The Africa We Want' adopted in September.
54 Common African Position (CAP) on the Post-2015 Development Agenda, African Union, March 2014 para 2.

55 Agenda 2063, note 53, para 66(c).
56 Transforming Our World, note 52, para 52.
57 The World Bank Group and Extractive Industries. 2004. 'Striking a Better Balance: The Final Extractive Industries Review'. *World Bank Group Management Response*, xi.
58 UN General Assembly Resolution A/RES/55/108 on the Right to Development 2001, preamble.
59 The Foundation of International Human Rights Law. <www.un.org/en/sections/universal-declaration/foundation-international-human-rights-law/index.html>.
60 Shaw, Malcolm. 2003. *International Law* (5th edn). Cambridge: Cambridge University Press, 129.
61 Adopted in Nairobi, Kenya in June 1981 and entered into force on 21 October 1986. All 55 African countries have ratified the Charter.
62 These rules were approved by the African Commission on Human and Peoples' Rights during its 47th Ordinary Session held in Banjul (The Gambia) from 12–26 May and entered into force on 18 August 2010.
63 *African Commission on Human and Peoples' Rights v the Republic of Kenya* Appl 006/2012.
64 Other treaties on human rights protection include the following: International Convention on the Elimination of All Forms of Racial Discrimination, New York, 7 March 1966; Amendment to article 8 of the International Convention on the Elimination of All Forms of Racial Discrimination, New York, 15 January 1992; Optional Protocol to the International Covenant on Economic, Social and Cultural Rights, New York, 10 December 2008; Optional Protocol to the International Covenant on Civil and Political Rights, New York, 16 December 1966; Second Optional Protocol to the International Covenant on Civil and Political Rights, aiming at the abolition of the death penalty, New York, 15 December 1989; Agreement establishing the Fund for the Development of the Indigenous Peoples of Latin America and the Caribbean, Madrid, 24 July 1992.
65 United Nations. 2019. 'Human Rights'. <www.un.org/en/sections/issues-depth/human-rights/>.

References

Adopted in Nairobi, Kenya in June 1981 and entered into force on 21 October 1986. All 55 African countries have ratified the Charter.

African Charter on Human and Peoples' Rights adopted by the OAU in Nairobi Kenya on 27 June 1981, OAU Doc CAB/LEG/67/3 rev. 5; 1520 UNTS 217.

African Charter on Human and Peoples' Rights adopted by the OAU in Nairobi Kenya on 27 June 1981, OAU Doc CAB/LEG/67/3 rev.5.

African Commission on Human and Peoples' Rights v the Republic of Kenya Appl 006/2012.

African Union Commission. 2015. 'Agenda 2063: The Africa We Want' adopted in September.

Barume, Albert. 2014. *Land Rights of Indigenous Peoples in Africa* (2nd edn). Copenhagen: International Work Group on Indigenous Affairs.

Centre for Minority Rights Development (Kenya) and Minority Rights Group (On Behalf of Endorois Welfare Council) v Kenya ACHPR Comm 276/03 (2009).

Chen, Cher. 2017. 'Indigenous Rights in International Law'. *Oxford Research Encyclopaedia of International Studies*, 1–29.

Declaration on the Right to Development Resolution A/RES/41/128 adopted by the UN General Assembly on 4 December 1986.

Democratic Republic of the Congo v Burundi, Rwanda and Uganda Comm 227/99 ACHPR (2003).

Front for the Liberation of the State of Cabinda v Republic of Angola Comm 328/06 ACHPR (2013).

Mayagna (Sumo) Awas Tingni v Nicaragua [2001] Inter-American Court of Human Rights (Ser C No 79).

Shaw, Malcolm. 2003. *International Law* (5th edn). Cambridge: Cambridge University Press.

Social and Economic Rights Action Centre (SERAC) and Centre for Economic and Social Rights v Nigeria Comm 155/96 ACHPR (2001).

Stamatopoulou, Elsa. 1994. 'Indigenous Peoples and the United Nations: Human Rights as a Developing Dynamic'. *Human Rights Quarterly* 16(1): 58–81.

Sudan Human Rights Organisation and Centre on Housing Rights and Evictions (COHRE) v Sudan Comm279/03-296/06 ACHPR (2009).

The World Bank Group and Extractive Industries. 2004. 'Striking a Better Balance: The Final Extractive Industries Review' *World Bank Group Management Response.*

United Nations Declaration on the Rights of Indigenous Peoples adopted by the General Assembly on 13 September 2007.

United Nations Economic and Social Council–E/CN.4/Sub.2/2004/30—Final Report of the Special Rapporteur, Erica-Irene, A. Daes. 'Prevention of Discrimination and Protection of Indigenous Peoples—"Indigenous peoples" Permanent Sovereignty over Natural Resources'. (2004).

United Nations General Assembly Resolution A/RES/1803 (XVII) on Permanent Sovereignty over Natural Resources adopted in 1962.

United Nations General Assembly Resolution A/RES/70/1 Transforming our World—The 2030 Agenda for Sustainable Development: A Call for Action to Change Our World.

11 Assessing the impact of water on the right to development (RTD) within the tourism industry in Africa

Furaha Joy Sekai Saungweme

Introduction

Natural resources in tourism are integral to sustainable development and economic diversification for African economies. However, the mere existence of natural resources is not indicative of access, use and ownership by the majority, especially poor local communities inhabiting tourism-rich destinations. Water is one such natural resource which is the focus of this chapter in assessing how the availability, affordability and quality of water in tourist communities impact on the right to development. This chapter contends that deepening inequalities and social conflict in tourism-impacted communities point towards inequitable distribution of water.[1] It is argued that water is a crucial catalyst for the realisation of the right to development as recognised in national jurisprudence and regional and international instruments.

The Declaration on the Right to Development (UNDRTD) is discussed as an important international law instrument whose text explicitly places humans at the centre of development processes through its emphasis on the need to ensure 'the constant improvement of the well being of the entire population and of all individuals on the basis of their active, free and meaningful participation in development and in the fair distribution of benefits resulting therefrom'.[2] The principles derived from the UNDRTD therefore support the assertion that where the use of natural resources such as water in the tourism sector develops and advances in a manner that marginalises and excludes a particular section of the population whether intentionally or unintentionally, such development is in conflict with the fundamental principles of the UNDRTD.

The tourism sector has not reduced the socio-economic inequalities that create access barriers to water.[3] It is argued that there is an obligation to do so on the part of the private sector but primarily on the government. The tourism sector is often projected as an economic driver and job creator for workers,[4] yet it is the workers who benefit the least.

An analysis of tourism policies from selected countries is examined and it is demonstrated that tourism policies in many developing countries contradict the standards established by various socio-economic and cultural development goals, leading to economic, social and political marginalisation of communities impacted by tourism.This includes coastal communities that are wholly dependent on water for survival, income generation and sustainability of livelihood activities. Human development is water dependent. However, because water is misappropriated and often distributed in a discriminatory manner, on the basis of class, social status, gender, sex, and even on tribal grounds, it fosters inequality amongst communities and creates hostilities.

Racially motivated demarcations of socio-economic resources were addressed under South Africa's 1997 White Paper on National Water Policy,[5] as was the case in many African countries under colonialism. Although the division of access to resources in present-day Africa may not be as openly divided along racial lines, it is certainly divided along ethnic and political lines and on the basis of socio-economic status. For example, water demands in the tourism industry sharpen inequalities because the water needs of tourists have priority over the needs of the locals whose buying power is the determining factor.

The analysis also intersects international human rights to the right to development, water and the tourism sector and examines national realities that drive water policies. Tourism-rich coastal destinations suffer tensions between the needs of tourism and host communities causing strained fragile coastal resources.[6] Examples of tourism-rich coastal destinations examined in this chapter are Zanzibar and The Gambia. The impact of court decisions on the right to water is discussed in the last section.

International and regional approaches to water

This section outlines key legal positions under international law and African regional policies relating to water. It will then expound on the relationship between the water and the tourism industry, its impact on local communities and the undermining of the right to development.

The African Union (AU) Agenda 2063 framework for development[7] adopted in 2015 serves as an example of an African-driven policy model, which at least, on paper, acknowledges the critical role that water plays in development. One of its priority areas is to improve on basic quality services by reducing the number of those without access to safe drinking water by 95 percent and to ensure that at least 5 percent of national budgets is allocated to water and sanitation and further that at least 70 percent of the population enjoy access to quality basic services (water, sanitation, electricity, transpiration, internet connectivity). The above goals were to be achieved before 2020 in addition to expectations that water policy sector reforms would be implemented between 2016 and 2018 under specific action plans leading to 2023.[8] According to a German Development Cooperation (GIZ) review report, which looks at water sector reforms in Burkina Faso, Kenya, Tanzania, Uganda and Zambia, the Millennium Development Goals (MDGs) for water were not achieved, raising concern about whether water equity is possible under the Sustainable Development Goals (SDGs).[9]

The main international human rights instrument relevant to the right to water is the Covenant on Social, Economic and Cultural Rights (ICESCR), which has been ratified by at least fifty African states, thus creating clear obligations. Although the text of the Covenant is silent on the question of water rights, the ICESCR is often used to justify the existence of the right to water as an extension of several socio-economic rights, including the right to an adequate standard of living recognised in the Covenant.[10] The Optional Protocol to the ICESCR is another international instrument that has been used to monitor states' compliance or violations of the rights contained in the Covenant. This Optional Protocol is administered by the United Nations (UN) Committee on Social, Economic and Cultural Rights, which plays a significant role in clarifying the parameters and application of socio-economic rights. In this capacity, the committee has specifically addressed the right to water within a legal context under General Comment 15 on how water in its normative content consists of freedoms and entitlements, formulated as follows:

> The freedoms include the right to maintain access to existing water supplies necessary for the right to water, and the right to be free from interference, such as the right to be free from arbitrary disconnections or contamination of water supplies. By contrast, the entitlements include the right to a system of water supply and management that provides equality of opportunity for people to enjoy the right to water.[11]

The purpose of General Comment No 15 has been to place the right to water as a human right in relation to key provisions of the ICSECR such as the right to health and the right to adequate housing and food.[12] The Committee notes the interconnectedness between the right to water and other basic rights, notably the rights to life and human dignity.[13] The parameters of the right to water are further defined within the context of availability, accessibility and quality; adding that water accessibility is to be understood in terms of physical and economic access, non-discrimination and information accessibility.[14] These four principles are relevant to addressing the equality disparities found in the tourism sector and relevant to the right to development of local communities in tourism destinations because many of the contentions on water arise from physical and economic access; physical in the sense that barriers are created to prevent local communities from accessing water sources and economic in the sense that local communities often lack the means and buying power to, for instance, drill boreholes or purchase water as a commercial commodity. Non-discrimination is also a relevant consideration in terms of its recognition internationally as a principle that underpins equality and dignity and remains key to all human rights values.

The UN Committee on Social, Economic and Cultural Rights under General Comment20,[15] drawing from the ICESCR, deals specifically with the issue of discrimination and articulates the fact that 'non-discrimination and equality are fundamental components of international human rights law and essential to the exercise and enjoyment of economic, social and cultural rights'. General Comment 20 defines discriminations as

> any distinction, exclusion, restriction or preference or other differential treatment that is directly or indirectly based on the prohibited grounds of discrimination and which has the intention or effect of nullifying or impairing the recognition, enjoyment or exercise, on an equal footing, of Covenant rights.[16]

The committee further advocates for any formal and substantive discrimination to be prohibited entirely and observed that formal anti-discrimination is enshrined in laws, and policies in contrast to substantive forms of discrimination that result from actual discriminatory practices often targeting historically disadvantaged groups. Formal anti-discrimination policies and laws cannot be effective without addressing substantive discrimination. The committee also identified systemic discrimination noting:

> discrimination against some groups is pervasive and persistent and deeply entrenched in social behaviour and organisation, often involving unchallenged or indirect discrimination. Such systemic discrimination can be understood as legal rules, policies, practices or predominant cultural attitudes in either the public or private sector which create relative disadvantages for some groups, and privileges for other groups.[17]

It is submitted that local communities impacted by tourism development often encounter such systemic discrimination as described by the Committee and that although water distribution in Africa may no longer be advanced along racial lines, it is certainly advanced by the socio-economic realities of local communities in tourism areas.

The Committee therefore expresses concern in General Comment No 15 that it is a privileged group that enjoys unfettered access to water and urged state parties to invest in 'services and facilities that benefit a far larger part of the population'.[18] The Committee holds the view that state parties should appreciate the special obligation to provide water to particularly vulnerable groups such as women, children, indigenous peoples, asylum seekers, internally displaced persons, refugees, prisoners and detainees, and minority groups.[19] It is unfortunate however, that the Committee fails to address tourism-impacted local communities as a vulnerable group given the specific water challenges these communities face, triggered by high levels of inequality and poverty that characterise tourist destinations.

The International Convention on the Elimination of All Forms of Discrimination against Women (CEDAW) requires state parties to the Convention to guarantee to women the right to 'enjoy adequate living conditions, particularly in relation to [...] water supply'.[20] In so doing, the CEDAW reflects the obvious gender dynamics of water recognised in Africa's regional conventions, for instance, the Protocol to the African Charter on Human and Peoples Rights on the Rights of Women in Africa,[21] requiring state parties to 'provide women with access to clean drinking water'.[22] The International Convention on the Rights of Children (CRC) illustrates this pattern as well in calling on state parties to combat disease and malnutrition 'through the provision of adequate nutritious foods and clean drinking water'.[23] The poverty narrative is expressed in a World Bank report,[24] describing Africa in terms of absolute poverty, defined as severe deprivation of basic socio-economic rights such as food, safe drinking water, sanitation facilities, health, shelter, education and basic social services.

In 2010, the United Nations General Assembly (UNGA) adopted a non-binding resolution reaffirming the importance of the right to water as a fundamental human right and called on states to provide 'financial resources, build capacity and transfer technology, particularly to developing countries in scaling up efforts to provide safe, clean, accessible and affordable drinking water and sanitation for all'.[25] There were 122 votes in favour of the resolution whilst 41 member states abstained from voting while none denounced it. Those states that abstained did so for varying reasons; whilst some held the view that the legal implications of a declared right to water had not been fully considered in the Assembly or in Geneva, others expressed concern that the resolution seemed to establish that there was a right without setting out a scope.[26] This non-binding resolution is relevant to the discourse on the right to development because it is consistent with the narrative that water is an essential element of the right to development and the duty of states to ensure its realisation.

Some of the member states that voted in favour of the resolution did so for varying reasons. Hungary viewed water and sanitation as part of a right to an adequate standard of living; Egypt appreciated the fact that the resolution did not create new rights or sub-categories of rights other than those contained in international human rights instruments; Brazil considered that the right to water and sanitation was intrinsically connected to the rights to life, health, food and adequate housing and that it was the states' responsibility to ensure this right.[27] These varying positions reflected by member states in the process leading to the adoption of the resolution provide insight into the possible application of the right to water by member states.

Civil society activism has helped to create awareness on water within a human rights context and relying on international guidelines such as the International Tourism Partnership (ITP). The ITP sets out the vision for how hotel companies can work to meet certain human rights objectives including access to water.[28] Under the ITP Goals for 2030 the core commitments to a human rights-centred form of tourism are summarised as follows: 'to support increased water use efficiency, sustainable withdrawals and supply of fresh water to address water scarcity and reduce the number of people affected by water scarcity by embedding water stewardship programmes across hotel portfolios'.[29]

The ITP juxtaposed the lavish use of water consumption in the hotel industry per guest against its use by local communities and argues that such use reflects an inequitable and unsustainable pattern of water access given that the demand for water could outstrip supply by 40 percent in 2030.[30] The ITP position reflects the Sustainable Development Goal on water, which seeks to reduce water scarcity and enable all sectors to have more efficient water systems in a sustainable, effective and holistic manner.[31] The Global Sustainable Tourism Council sets minimum standards of compliance for the tourism sector regarding the provision of water to neighbouring communities, thereby placing obligations on tourism businesses.[32] It emphasises the importance of good governance and proper regulatory frameworks as essential elements for the monitoring and enforcement of strategies to ensure availability and access to water.

A report conducted by Tourism Concern identified nine principles that must be established in order to guarantee water equity in the tourism industry.[33] The nine principles include a call on governments to undertake positive measures to ensure that the three tenets of quality, availability and accessibility are not compromised by tourism development. These measures may be accomplished through regulations which ensure 'sustainable and equitable water and tourism management'.[34] Governments are also required to undertake water resource assessments before engaging in land use and tourism planning. This has the implication that potential water scarcity in an area targeted for tourism development or expansion may not proceed until the water concerns pertaining to sustainability and accessibility for all the impacted stakeholders are resolved. The nine principles also serve as an affirmation of the core principles enunciated in the UNDRTD, namely that development prospects should take cognisance of the fact that the 'human person is the central subject of development and should be the active participant and beneficiary of the right to development'.[35]

The human rights obligations placed on tourism businesses mirror the corporate responsibility under the Guiding Principles on Business and Human Rights to respect human rights. Respecting human rights means to 'avoid infringing on the human rights of others' and to 'address adverse human rights impacts with which they are involved'.[36] Both the Guiding Principles and the nine principles developed by Tourism Concern underscore the importance of upholding pro-poor tourism policies through the equitable use and access to water.

National realities on the right to water

This section demonstrates how access to water remains a source of conflict for tourism-affected local communities as a result of national policies and practices. National water policies should involve preventing undue interference with access to water as well as forging a right to participatory development, allowing people to pursue social and economic development on the basis of freely chosen policies.[37] It means ensuring that local communities are part of the decision-making processeson water distribution. However, this is hardly ever the case as demonstrated by a UN Environment Programme report noting the competing interests between the tourism industry and local communities causing pressure on domestic resources to the detriment of these communities as noted in this except:

> There are examples of communities overrun by large numbers of visitors, commercialization of traditions and threats to cultural survival from unplanned and unmanaged tourism. Tourism destinations are occasionally built by outsiders (usually with government approval) in areas that indigenous or traditional communities consider to be theirs, and where the development was neither desired nor locally validated [...] these situations could lead to conflicts that make cooperation and mutual benefits nearly impossible to achieve, and instil animosities that negatively affect the local communities and the tourism destination. In countries where resources are limited and communities are vulnerable to external shocks, the cultural issues could overlap or even be aggravated by environmental concerns and problems related to access to water, coastal resources and wildlife.[38]

The survival of the tourism sector is heavily reliant on its ability to pander to the demands of its clients, much of which depends on water consumption and water-dependent technologies. However, the failure to balance competing interests has compromised the basic water rights of tourism-impacted communities, thus jeopardising their food security and accelerating deforestation and land degradation as noted in specific locations in countries that are tourist attractions like Zanzibar to start with.

Zanzibar[39] is a tourism-rich area that prioritises leisure and tourism over the water access rights of the locals. Despite significant economic growth, it is reported that local communities live below the poverty line and the availability and quality of water is poor.[40] Yeshanew Sisay argues against the narrative that economic growth is indicative of fulfilment of the right to development, observing that '[w]hile it is true that economic growth generates income which in turn may be used for further development many argue for a wider definition of development emphasizing the quality and distribution of the growth and income effects'.[41]

Access to water in the Zanzibar region is regulated by the Water Act,[42] which describes the various licensing procedures required to extract, use or store water and how such licensing may be terminated.[43] Even though there are such procedures determining ownership and use, the Act goes on to state in section 20 that water belongs to the government. It suggests that any right of claim to access water and the grievances arising from such procedures should be directed to the government, which bears the final responsibility regarding water distribution. The lack of access to water in Zanzibar is a violation of the three tenets prescribed under General Comment No 15 pertaining to the quality, accessibility and affordability of water.

Zanzibar's coastal region boasts of about 325 hotels with its water consumption described as follows:

> The hotels use a lot of water for bathing, swimming pools and laundry. Community uses an average of 93 litres per day per household whereas the average consumption per room varies 686 litres per day in guest houses up to 3195 litres per room per day for five star hotels with average of 1482 litres per room per day.[44]

The issue of excessive use of water is addressed under the Zanzibar Water Act and ensures that licence holders will have their water licences suspended or varied where excessive use is causing scarcity for those using it or causing poor quality of the water. However, it is not clear which affected peoples this applies to as it broadly states that the provision applies where the water becomes 'insufficient for the needs of the people using it on account of excessive use, drought or an emergency',[45] given that scarcity and poor-quality water is the status quo for the local communities.

According to the Act, an order of preferential use and users may be directed in cases of water scarcity. On the list of six, tourism development is in the fourth category for preferential use when water is running low.[46] So, although tourism development is mentioned in the Act, it is again silent in respect to tourism-impacted communities. Excessive and inequitable use of water remains a challenge in the popular resort villages of Kiwengwa and Nungwi in the East of Zanzibar where it is reported that tourists use '16 times more fresh water a day per head than locals', making it clear that there is an obvious and unjustifiable discrepancy between water demand and water consumption. The challenges around water scarcity, water borne diseases and poor health infrastructure[47] raise questions about why the Zanzibar Tourism Act[48] is silent on the issue of water.[49]

Whilst the Zanzibar Tourism Policy of 2017[50] in theory discourages tourism that fails to conserve and improve the welfare of local people, it is also silent on addressing the barriers to local participation and enjoyment of the benefits of local tourism including access to water. The water challenges in Zanzibar range from pollution, loss, climate change and poor use of rainwater to high rates of water wastage.[51] The social conflict resulting from the water and tourism nexus are also reflective and perpetuate existing power imbalances between the rich and powerful tourism industry against poor tourism-impacted communities. For instance, hoteliers and the resort sector have the machinery to drill for water and build boreholes that allow them to determine local community access to water. The hoteliers become de facto water distributaries as a result of the government's inability to do so.

The gender dimension to water is largely due to the fact that women's household responsibilities and income-generating opportunities are intrinsically tied to water accessibility, including in tourism-rich destinations. It is argued that gender equality and access to water are co-dependents, where without one, the other cannot survive. Sustainable Development Goal 6 guarantees that water and sanitation cannot be achieved without attaining Sustainable Development Goal 5, which focuses on gender equality. In The Gambia for example, where public water infrastructure is badly managed, women are forced to rise early and queue in line for water collection. Micro enterprises run by women fruit sellers identify the lack of access to water as a barrier to their development. Frequent power cuts also affect water access and incapacitates water-producing electric pumps.[52]

The water scarcity experienced by locals is in direct contrast to the abundance of water enjoyed by tourists in The Gambia, a country in which tourism has become the fastest growing sector. Such inequalities conflict with the equality of opportunity provision guaranteed under the UNDRTD regarding access to a series of enumerated rights, which fall under the umbrella of socio-economic rights.[53] Tourism in The Gambia has been cited as the prime foreign exchange earner and job creator in both the formal and informal sectors.[54] Yet its legislative and policy framework is challenged on two fronts; firstly, the tourism industry is poorly regulated and secondly, water sources are increasingly constrained by a weak regulatory policy framework.[55] Country-specific assessments of water policies including The Gambia reveal a lack of alignment between water and tourism. Lack of alignment means that tourism policies develop without incorporating water access realities; similarly, water policies are developed without reference to tourism.

The Gambia's Water Policy[56] acknowledges water as a common right held in trust by the government on behalf of the people. However, it has been criticised for being poorly implemented and lacking enforcement capacity as it is not supported by legislation. In the most cursory way, the policy does reference the water needs of various sectors including the tourism sector and notes the importance of utilising water in a sustainable manner. Although the policy incorporates an inclusive approach to stakeholders and women, it does not go far enough in addressing water scarcity in poor communities; neither does it advance any solutions. The Gambia's National Tourism Policy[57] is also silent on the issue of water and as such perpetuates water inequities by its silence while the tourism industry of The Gambia is highly dependent on water. The policy, however, does highlight the need to promulgate new laws empowering the Ministry of Tourism and its relevant agencies to administer the tourist industry.[58]

The intense interest in tourism development in coastal areas and islands like Zanzibar, and The Gambia is propelled by tourist demands; yet at the same time, the water demands of local communities have been undermined.

Appraisal of selected jurisprudence

This section considers the impact of national jurisprudence on the right to development for local communities in tourism-impacted communities in both coastal and landlocked areas.

Coastal areas

Local communities in coastal areas are highly dependent on water: The United Nations World Tourism Organisation (UNWTO) Code of Ethics on Tourism notes that

> [s]pecial attention should be paid to the specific problems of coastal areas and island territories and to vulnerable rural or mountain regions, for which tourism often represents a rare opportunity for development in the face of the decline of traditional economic activities.[59]

In the *Coastal Links Langebaan v Minister of Agriculture, Forestry and Fisheries* case,[60] the applicants were a community of fishermen who claimed customary fishing rights on the basis that they and their ancestors had lived and worked on these lands for several generations with access to the lagoon waters and marine resources.[61] They experienced exclusion when the lagoon was proclaimed to be part of the Laangebaan National Park in 1985 under the National Parks Act and placed under the management of the South African National Parks Board (SANP) and their customary fishing rights were ignored.

Proceedings were launched in the Equality Court on behalf of traditional artisanal fishermen and their communities, including the Laangebaan net fishers alleging that the government's failure to provide subsistence and artisanal fishers with access to marine resources violated a number of their constitutional rights, specifically the right to equality, the right to not be discriminated against unfairly, the right to engage in a trade and occupation and the right to access sufficient food. In response, the Court granted an order to develop a policy for traditional fishers with due regard to their socio-economic rights and equitable access to marine resources.[62] This led to the development of the Small Scale Fishing (SSF) Policy, which was gazetted in June 2012 and sought to redress wrongs committed against traditional fishing communities as a result of being dispossessed of their lands that were adjacent to the coasts during the apartheid and colonial periods.[63] However, they were denied access to zone B on the basis that net fishing was destructive and should only be allowed in very limited circumstances.[64] In contravention of the new restrictive conditions, the applicants were caught fishing in zone B and this led to criminal prosecutions.

The Court was of the view that permitting some fishermen to exercise their fishing rights in the same lagoon in which others were being prohibited on the basis of conservation and ecological reasons was arbitrary and irrational.[65] It noted that although the restrictive imposition appeared to be based on geographical divisions, there were clear racially discriminatory undertones that motivated the distinction. The Court concluded that as a result, the restrictive condition was unfair and unconstitutional.[66] What is required is balancing the interests of local communities and the tourism industry as is the case with the coastal part of Spain where local communities reap the benefits of water-based tourism:

> These benefits include the following: the improvement of access infrastructure and signage; the active involvement of local communities in the conservation of natural ecosystems where water is a prominent feature in order to continue to derive advantages from the responsible use of this resource—an impulse for economic, social, and local development; increases in the sales of local products, industries, and services.[67]

The tourism industry in African countries needs toalso gravitate towards mutually beneficial water policies to facilitate the right to development for tourism-impacted communities.

Landlocked territories

Access to water in landlocked territories poses significant challenges for local communities inhabiting tourist-rich areas. Pederson notes that 'in Africa, more than 80% of foreign visitors visit national parks for wildlife watching with up to 5 million visitors per year in protected areas alone'.[68] Such large numbers of tourists are more than likely to influence water consumption patterns and strain scarce water resources located within the nature reserves to the detriment of the local communities.

In Botswana, the indigenous people called the Basarwa filed an application with the courts seeking access to their land for the purpose of water appropriation to support their livelihood. A relocation policy confined them to a settlement outside the Central Kalahari Game Reserve (CKGR) for the purpose of ensuring that the CKGR would be used solely for purposes of conservation of wildlife. The Basarwa people were aggrieved by the fact that they no longer had access to water due to a dismantling of the only borehole in the area[69] and alleged that such denial was unlawful and unconstitutional. The court found in favour of the Basarwa people by ordering that the applicants had a right to re-commission the borehole in the CKGR and to sink one or more boreholes at the site.[70]

The *Basarwa* case is reflective of a pattern of exclusionary policies and practices denying local communities access to natural resources. Under General Comment No 15, the Committee categorised indigenous peoples as a special group in need of protective measures with regard to access to water by requiring states to ensure that '[i]ndigenous peoples' access to water resources on their ancestral lands is protected from encroachment and unlawful pollution. States should provide resources for indigenous peoples to design deliver and control their access to water'.[71]

In the case of *Mazibuko*,[72] the Constitutional Court of South Africa held that 'the achievement of equality, one of the founding values of [the South African] constitution, will not be accomplished whilst water is abundantly available to the wealthy, but not to the poor'.[73] This principle is relevant for local communities in tourism-rich destinations.

Water as a minimum core obligation

The *Mazibuko* case also brought to the forefront the issue of minimum core obligations, a principle established under General Comment No 3 which stipulates that '[a] minimum core obligation to ensure the satisfaction of, at the very least, minimum essential levels of each of the rights is incumbent upon every state party'.[74] The minimum core obligation imposes a responsibility on states to ensure the provision of basic goods and services because they are perceived to be fundamental rights that flow from the 'very basic interest people have in survival and the socio-economic goods required to survive'.[75] The court, however, rejected the notion of minimum core in favour of a reasonableness approach in line with prior Constitutional Court judgments:

> [it] is clear that the right does not require the state upon demand to provide every person with sufficient water without more; rather it requires the state to take reasonable legislative and other measures progressively to realise the achievement of the right of access to sufficient water, within available resources.[76]

Thus, the Court rejected the applicants' argument that it should determine an appropriate minimum water quantity for the community in question. The Court further rejected the argument that the constitutional provision in question conferred a right to claim sufficient water from the state immediately.[77] In so doing, the Constitutional Court rejected the international standard set by the CESCR, and the ruling of its own lower courts, which had found in favour of the applicants. On the aspect of policy, the Court noted that it was a constitutional requirement for the state to 'adopt reasonable measures to progressively realise the right of access to water'.[78] The Court assessed the policy using the reasonableness criteria and found that the policy was reasonable and compliant with section 27 of the Constitution as the city had progressively sought to increase access to water for larger households that were prejudiced by the 6 kilolitre limit.[79] The *Mazibuko* decision is more aligned to the concept of progressive realisation, enshrined in the ICESCR[80] and in constitutions such as those of South Africa and Kenya. It allows states to progressively facilitate the realisation of socio-economic rights over time, as opposed to immediately.

The question is whether the anti-minimum core obligation approach adopted by the court conflicts or aligns with the principles of the UNDRTD. The minimum core obligation has been interpreted by the CESCR to mean that states are required to ensure the fulfilment of minimum access to socio-economic rights. In a similar thread, the UNDRTD urges states to 'formulate appropriate national development policies that aim at the constant improvement of the well-being of the entire population and all individuals'.[81] There is room to argue that the minimum core obligations approach closely relates to the principles embedded in the UNDRTD as opposed to the reasonableness approach adopted and preferred by South African courts. The application of the minimum core obligation to the tourism sector is a viable option for the realisation of water rights in tourism-impacted communities.

Conclusion

This chapter has analysed the right to water as a core component of the right to development, particularly for tourism-impacted communities. Reference is made to the main international law instruments, including the UNDRTD which articulates socio-economic rights within the framework of the right to development. The availability, accessibility and quality of water is used as a standard to assess the level of enjoyment of water for tourism-impacted communities and their right to development enshrined in international law instruments as well as in the decisions rendered by the courts in the cases discussed in this chapter. It is argued that a country's economic growth is not sufficient to guarantee the enjoyment of basic socio-economic rights as the right to water and therefore, additional measures must be undertaken to ensure the realisation of essential livelihood sustainability rights. This begins with the recognition that the right to development, which incorporates the right to water, is a fundamental right, which must be reflected in water policies within the tourism sector with assurances of penalties in the event of lack of compliance.

Notes

1 Spenceley, Anna; Ashley, Caroline and de Kock, Melissa. 2009. *Tourism-Led Poverty Reduction Programme: Core Training Module*. International Trade Centre, 41–42.
2 Declaration on the Right to Development Resolution 41/128 adopted by the UN General Assembly on 4 December 1986.
3 Bolwell, Dain and Weinz, Wolfgang. 2008. 'Reducing Poverty through Tourism'. *ILO Working Paper Series Sectoral Activities Programme.*
4 UN Conference on Trade and Development (UNCTAD) Sustainable Tourism. 2013. 'Contribution to Economic Growth and Sustainable Development'. Expert Meeting on Tourism's Contribution to Sustainable Development: Geneva, Item 3 of the provisional agenda, 14–15 March. The UNCTAD secretariat noted that the link between tourism, economic growth and poverty reduction was not an obvious one, special measures had to be undertaken.
5 Department of Water Affairs and Forestry. 1997. 'White Paper on a National Water Policy for South Africa'. April 1997 <www.dwaf.gov.za/Documents/Policies/nwpwp.pdf>, 2.
6 For additional discussions on water scarcity in tourism destinations, see Cole, Stroma. 2012. 'A political Ecology of Water Equity and Tourism: A Case Study from Bali'. *Annals of Tourism Research* 39(2): 1221–1241.
7 Agenda 2063 First Ten-Year Implementation Plan. 2014. <https://au.int/sites/default/files/documents/33126- doc-01_background_note.pdf>.
8 *Ibid.*
9 GIZ 2018 Synthesis Report Part 1. *Access to Water and Sanitation in Sub-Saharan Africa* Review of Sector Reforms and Investments, Key Findings to Inform Future Support to Sector Development <www.oecd.org>. The GIZ report focuses more on water access in urban populations, the same principles and concerns that apply to the tourism sector are also found in urban areas.
10 International Covenant on Social, Economic and Cultural Rights, art 11(1).
11 UN Committee on Economic, Social and Cultural Rights, Twenty-Ninth Session Geneva, 11–29 November 2002 Agenda item 3 E/C.12/2002/11 20 January 2003.
12 General Comment No 15: The Right to Water (arts 11 and 12 of the Covenant) adopted at the Twenty-Ninth Session of the Committee on Economic, Social and Cultural Rights, on 20 January 2003 Contained in Document E/C.12/2002/11, para 3.

13 *Ibid*, para 3.
14 *Ibid*, para 12.
15 UN Committee on Economic, Social and Cultural Rights. 2009. General Comment 20, Forty-Second Session Geneva, on *Non-Discrimination in Economic, Social and Cultural Rights* (art. 2, para. 2) UNDoc. E/C.12/GC/20 (2009).
16 *Ibid*.
17 General Comment 20, note 15, para 12.
18 General Comment No 15, note 12, para 14.
19 *Ibid*, para 15.
20 Convention on the Elimination of All Forms of Discrimination against Women, art 14(2).
21 Protocol to the African Charter on Human and Peoples Rights on the Rights of Women in Africa adopted in Maputo, Mozambique on 11 July 2003.
22 *Ibid*, art 15(a).
23 Convention on the Rights of Children, art 24(2).
24 Garcia, Marito; Pence, Alan and Evans, Judith. (Eds). 2008. *Africa's Future, Africa's Challenge Early Childhood Care and Development in Sub-Saharan Africa*. Washington, DC: The World Bank, 57.
25 UN General Assembly Resolution on the Right to Water, Sixty-Fourth Session Agenda item 48 Integrated and coordinated implementation of and follow-up to the outcomes of the major UN conferences and summits in the economic, social and related fields A/64/L.63/Rev.1.
26 Sixty-Fourth General Assembly Plenary 108th Meeting Comments from USA.
27 Resolution on the Right to Water, note 25.
28 International Tourism Partnership (ITP). 'ITP's Goals for 2030 Unite the Hotel Industry for a Sustainable Future'. <www.tourismpartnership.org/itp-goals/>.
29 *Ibid*.
30 ITP, note 28.
31 UN Sustainable Development Goal 6.4. <www.un.org/sustainabledevelopment/water-and-sanitation/>.
32 Global Sustainable Tourism Council. <www.gstcouncil.org>.
33 Tourism Concern. <www.tourismconcern.org.uk>.
34 *Ibid*, principle 2.
35 UNDRTD, note 2, art 2.
36 UN Guiding Principles on Business and Hunan Rights, principle 11. </www.ohchr.org/documents/publications/guidingprinciplesbusinesshr_en.pdf>. The UNGP is an instrument consisting of 31 principles for implementing the UN 'Protect, Respect and Remedy' Framework.
37 Yeshanew, Sisay. 2013. *The Justiciability of Economic, Social and Cultural Rights in the African Regional Human Rights System: Theories Laws, Practices and Prospects*. Cambridge: Intersentia, 261–268.
38 UN Environment Programme. 2011. *Towards a Green Economy: Pathways to Sustainable Development and Poverty Eradication: A Synthesis for Policy Makers*. Nairobi: United Nations Environment Programme.
39 Located 25 km from the coast of mainland Tanzania.
40 UNEP, note 38.
41 Yeshanew, note 37, 261–268.
42 Zanzibar Water Act No 4 of 2006.
43 *Ibid*, sects 20–25.
44 Revolutionary Government of Zanzibar. 2015. *Zanzibar Research Agenda 2015–2020*. Zanzibar Government.
45 Zanzibar Water Act, note 42, sect 28.
46 *Ibid*, sect 26–27.

47 Tourism Concern Research Report. 'Water Equity in Tourism, A Human Right-A Global Responsibility'. <www.tourismconcern.org.uk/wp.../10/Water-Equity-Tourism-Report-TC.pdf>.
48 Zanzibar Tourism Policy Act No 06 of 2009 passed in the House of Representatives of Zanzibar on the 16th day of October 2009.
49 *Ibid*, sect 27(1) only refers to HIV/AIDS and drugs awareness campaigns.
50 <www.zanzibartourism.go.tz/images/joomlart/documents/ZanzibarTourism_Policy.pdf>.
51 Revolutionary Government of Zanzibar, note 44, 24.
52 Tourism Concern Research Report, note 47.
53 UNDRTD, note 2, art 8.
54 Tourism Concern Research Report, note 47.
55 Latchford, Rebecca and Niang, Dauda. 2012. *Case Study Summary of Water Equity in Tourism in the Gambia*. Tourism Concern.
56 The Gambia Water Policy adopted in 2008.
57 Responsible Tourism Policy for Gambia. <www.accessgambia.com/information/gambia-tourism-authority.html>.
58 *Ibid*, 14.
59 UN World Tourism Organisation Code of Ethics on Tourism.
60 *Coastal Links Langebaan v Minister of Agriculture, Forestry and Fisheries* Case No 11907/13.
61 *Ibid*, para 14.
62 *Ibid*, para 23.
63 *Ibid*, para 35.
64 *Ibid*, para 31.
65 *Ibid*, para 80.
66 *Ibid*, para 80–85.
67 Folgado-Fernández, José *et al.* 2019. 'Water Tourism: A New Strategy for the Sustainable Management of Water-Based Ecosystems and Landscapes in Extremadura (Spain)'. *Land* 8(2): 1–18.
68 Pederson, Charnette. No date. 'The South African Tourism Policy with Regards to Sustainable Tourism'. *Development and International Relations*, 2.
69 *Matsipane Mosetlhanyane v the Attorney General of Botswana Court of Appeal*, CALB -074-10 para 6.
70 *Ibid*, para 25(1.1).
71 *Ibid*, para 16(d).
72 *Mazibuko and Others v City of Johannesburg and Others* 2010 (4) SA (1) (CC).
73 *Ibid*, para 2.
74 Committee on Economic, Social and Cultural Rights, General Comment No 3 (Fifth Session, 1991) UN, Doc E/1991//23.
75 Bilchitz, David. 2003. 'Towards a Reasonable Approach to the Minimum Core: Laying the Foundations for Future Socio-Economic Rights Jurisprudence'. *South African Journal of Human Rights* 19(1): 9–10.
76 *Government of the Republic of South Africa and Others v Grootboom and Others* 2001 (1) SA 46(CC).
77 Sect 27(1)(b) of the Constitution of South Africa.
78 *Grootboom*, note 76, para 94.
79 *Ibid*, para 97.
80 ICESCR, note 10, art 2(1).
81 UNDRTD, note 2, art 2(3).

References

Bilchitz, David. 2003. 'Towards a Reasonable Approach to the Minimum Core: Laying the Foundations for Future Socio-Economic Rights Jurisprudence'. *South African Journal of Human Rights* 19(1): 9–10.

Bolwell, Dain and Weinz, Wolfgang. 2008. 'Reducing Poverty through Tourism'. *ILO Working Paper Series Sectoral Activities Programme.*

Cole, Stroma. 2012. 'A political Ecology of Water Equity and Tourism: A Case Study from Bali'. *Annals of Tourism Research* 39(2): 1221–1241.

Folgado-Fernández, José *et al.* 2019. 'Water Tourism: A New Strategy for the Sustainable Management of Water-Based Ecosystems and Landscapes in Extremadura (Spain)'. *Land* 8(2): 1–18.

Garcia, Marito; Pence, Alan and Evans, Judith. (Eds). 2008. *Africa's Future, Africa's Chall enge Early Childhood Care and Development in Sub-Saharan Africa.* Washington DC: The World Bank.

Latchford, Rebecca and Niang, Dauda. 2012. *Case Study Summary of Water Equity in Tourism in the Gambia.* Tourism Concern.

Pederson, Charnette. n.d. 'The South African Tourism Policy with Regards to Sustainable Tourism'. *Development and International Relations.*

Spenceley, Anna; Ashley, Caroline and de Kock, Melissa. 2009. *Tourism-Led Poverty Reduction Programme: Core Training Module.* International Trade Centre.

UN Environment Programme. 2011. *Towards a Green Economy: Pathways to Sustainable Development and Poverty Eradication: A Synthesis for Policy Makers.* Nairobi: United Nations Environment Programme.

Yeshanew, Sisay. 2013. *The Justiciability of Economic, Social and Cultural Rights in the African Regional Human Rights System: Theories Laws, Practices and Prospects.* Cambridge: Intersentia, 261–268.

Part 3

Collective benefit-sharing of natural resources in sustaining the right to development

12 Sustainable Development Goals and the bumpy road to 2030 in Ethiopia

The challenges of land grabbing

Jean-Claude N. Ashukem

Introduction

The Sustainable Development Goals (SDGs) and land grabbing could be considered as panaceas for some of the 21st-century challenges. While land grabbing has been found to play an indisputable important role in addressing some of the concerns of the SDGs, such as those exemplified by food and energy,[1] at the same time the negative impacts of the phenomenon inexorably pose an existential threat to the attainment of the SDGs. Land grabbing also has a direct adverse effect on the right to development (RTD). As an inherent natural resource, land ownership, as stipulated in article 1(2) of the Declaration on the Right to Development, is central to the realisation of the RTD. Concerns over land grabbing thus relate to core issues of the RTD, including the right to sovereign ownership over land and natural resources, and the resultant developmental needs of local communities. Secured land rights are pivotal to and play a catalytic role in economic growth, poverty alleviation and inclusive social development. Thus, the issues discussed in this chapter relate to the central theme of the RTD.

This chapter evaluates and critiques some of the SDGs, including goals 1, 2 and, 7, in the context of land grabbing and argues that the phenomenon, which is prevalent in Africa—wrongly conceived to have vast portions of empty and abundant land[2]—is counter-intuitive to the SDGs and provides a potential difficulty for their achievement. Using Ethiopia as a case study, I examine various examples of the (negative) impact of land grabbing on the SDGs in the country to substantiate this claim. The focus on Ethiopia is because the country has become a major site of development enclosures, evident by the increasing expansion of agribusiness activities that is dramatically re-forming the country's social and physical landscape. The premise of this argument is twofold. Firstly, land is the anchor for the realisation of most, if not all, of the SDGs for which any negative impact on land (as a resource), as in the context of land grabbing, could have corresponding negative impact on other goals.

Efforts to achieve either Goal 1 on zero poverty, Goal 2 relating to food or Goal 13 on climate change/renewable energy could, for example, have profound negative impact on other goals and targets, such as those relating to land, water, the environment, and poverty. Logically, efforts to attain any of the SDGs in the instance where land grabbing takes place could have serious unintended negative impact on the achievement of other goals. Secondly, although a critical reading of the SDGs confirms that they revolve around basic human rights, including the right to land, land grabbing is characterised as having profound negative impact on some of these rights, including the forceful eviction of local communities from their land, water and food sources, a degraded environment, poor working conditions, gender discrimination and forceful migration.[3] Collectively, these negative impacts pose fundamental epistemological and ontological challenges to human rights, and as a consequence, could greatly thwart the overall purpose of the SDGs.

The section 'Conceptualising land grabbing' of this chapter provides a contextual background to land grabbing and how it links to the SDGs. The section 'Interface between the SDGs, land grabbing and the RTD' provides an overview of the SDGs. The section 'Difficulties in achieving the SDGs in the context of land grabbing in Ethiopia' provides an analysis of the negative impact of land grabbing and the resultant difficulties of achieving the SDGs in Ethiopia. The last part is the conclusion.

Conceptualising land grabbing

There is no unanimously agreed definition of land grabbing. The phenomenon has been defined differently by different commentators.[4] Generally, land grabbing denotes the purchase of large swathes of land by transnational companies in developing countries, *including Ethiopia*, for various purposes. Land grabbing is driven by a combination of factors, and a generally acceptable understanding is that, in the context of a globalised world, food and energy and the surge in global demand for agricultural commodities represents the underlying forces behind the phenomenon. The 2007–2008 global crises—regarding food, financial systems and energy—ignited the global rush for agricultural land.[5] In the immediate aftermath of these crises, land grabbing became prevalent in developing countries generally, and in Ethiopia specifically, as the next best alternative to address the global food and energy concerns.[6] In this regard, land grabbing needs to be understood within the context of power—the desire to capture land and its associated resources in order to control the benefit of its use. Land grabbing epitomises a contemporary ongoing and accelerated practice of the 21st century that has fundamentally changed the power dynamics of access to, and use of, land and its resources. Such changes unequivocally have significant underlying epistemological challenges to land rights that are detrimental to rural peasants' rights. This aptly explains the normative weight often used to describe the process: land grabbing. Irrespective of its size, the phenomenon raises serious ethical concerns about the violation of human rights, dispossession of the commons, environmental impacts and the overall power imbalance that characterises the negotiation process.

Land grabbing is connected to two interrelated transformations. These include a small number of companies and other actors that control large concentrations of extended land, and the global agrarian practice that involves a significant shift from traditional, local, small-scale systems of production to a much larger, intensive-driven and commercially oriented agricultural model. Land acquisitions in the Global South usually come with greater productivity and cost-efficiency than the lost domestic production capacity that they compensate for. In a more general sense, land grabbing is a particularly invasive form of food import strategy with strong colonialist tendencies.

The parties to land grabbing typically include a foreign investor (which can be a foreign government or multinational company) and a host country/government. While issues such as food imports could be particularly important for the foreign investor, land availability, underpinned by weak land governance and absence of local population land rights protection, often emerges as the necessary prerequisite and fundamental pre-condition for investment in the host country.[7] The host country, in turn, relies on infrastructural development, employment opportunities and know-how transfer commitments.[8]

Land grabbing occurs in two ways: when host governments solicit foreign investors to invest in their countries in order to boost agricultural sector and increase economic growth and development needs, and also when governments forcefully appropriate land from local communities and lease them to foreign investors as a means of creating development.[9] In either of these cases, the state assumes the role of land broker and their role has raised significant governance and regulatory concerns,[10] in part as land deal contracts are non-transparent and are not inclusive.[11]

These governance examples are predicated on how different actors frame the same phenomenon differently, based on their political ideologies and interests. What, for example, might form a critical perspective by one actor, and be portrayed as a much-needed development opportunity for the host country, could be depicted as a case of dispossession of local communities' land rights by another actor. This is evident from the discussions on land governance arrangements that have been organised by United Nations agencies, such as the United Nations Food and Agricultural Organisation, the International Fund for Agricultural Development, the United Nations Conference on Trade and Development and the World Bank Group. These agencies have greatly promoted voluntary approaches to large-scale land investment, including the Voluntary Guidelines on the Responsible Governance of Tenure of Land, Fisheries and Forests, and the Voluntary Certification Schemes for Sustainability Standards. These guidelines were meant to ensure proper governance of natural resources, including land in the broader framework of development activities. However, the idea that corporate self-regulation and codes of conduct could substantially promote good land governance in the context of land grabbing has been heavily contested as part of a neoliberal agenda that threatens human rights, food security and the livelihoods and well-being of peasant farmers and traditional users of land and natural resources.[12]

Interface between the SDGs, land grabbing and the RTD

Understanding the SDGs

The SDGs were adopted in 2015 by a United Nations General Assembly Resolution entitled *Transforming Our World: The 2030 Agenda for Sustainable Development*[13] as the successor framework to the Millennium Development Goals (MDGs). They are meant to shape, guide and direct the global development agenda for the next 15 years in ways that would address some of the most pressing contemporary challenges of the 21st century. The preamble of Agenda 2030 projects a globally shared trajectory and vision for global development that promotes 'a plan of action for people, planet and prosperity [...] to free [the] human race from the tyranny of poverty and want and to heal and secure our planet'. Evidently, the SDGs are cast in aspirational language and this seeks to emphatically convey the world's determination to improve people's lives; protect the planet; ensure that everyone enjoys a prosperous and fulfilled life; foster a peaceful, just and inclusive society that is free from fear and violence; and revitalise the global partnership for sustainable development that could be instrumental in achieving the SDGs.[14]

The SDGs represent commitments that states and relevant stakeholders are encouraged to work towards and implement. In this regard, they represent a much more comprehensive and ambitious initiative that includes, in multitudinous ways, other pertinent global concerns, such as climate change as an inherent feature of its 2030 agenda.[15] Unlike the MDGs that were mostly applicable only to developing countries, the SDGs are global and they are applicable to both developed and developing countries. To this end, they extend the focus of international development beyond poverty to sustainability. Although termed global goals, the fact that the SDGs are development-induced, as they strive to propel global development efforts for the next 15 years, suggests that they represent a more viable potential to already financially rich and privileged countries. This constitutes countries of the Global North that are often the principal actors of land grabbing against underprivileged but resource-sufficient countries of the Global South.

The Global South harbours most of the development projects for the benefit of the Global North but is still battling with serious endemic developmental challenges. As global goals, the SDGs take an all-inclusive and possibly, diverse approach to the achievement of its objectives, in both the developed and developing countries. The SDGs consist of 17 ambitious goals and 169 targets, and their underlying agenda would be commendable insofar as they help to eradicate global poverty, among other ills.[16] There are over 232 global indicators that were developed by the Inter-Agency and Expert Group Indicators (IAEG-SDGs) adopted by Resolution A/RES/71/313 at the 48th session of the United Nations Statistical Commission of 2017.[17] Although these indicators act as a catalyst in assessing the results of the SDGs made by states, commentators have argued that, because they are not well-designed, they may lead to ambiguous or confusing results relating to the assessment of the SDGs.[18]

Intersection with the RTD

The United Nations Declaration on the Right to Development of 1986 (UNDRTD) is the anchor of the RTD. The Declaration outlines the relevance of the right to the extent that any derogation constitutes both a violation of the right and other rights-based entitlements that are associated to the right, as stated by the African Commission on Human Rights in the *Centre for Minority Rights Development and Minority Rights Group International on Behalf of Endorois Welfare Council v Kenya* (*Endorois*) case.[19]

Article 1 of the Declaration entitles everyone and all peoples to the RTD, portrayed as the right to participate in, contribute to and enjoy economic, social, cultural and political development. Article 22 of the African Charter on Human and Peoples' Rights guarantees to all people the right to economic, social and cultural development that takes into account their freedom and identity as well as equality in the enjoyment of the common heritage of mankind, the obligation on states to 'individually or collectively [...] ensure the exercise of the right to development'.[20] Clearly, the human person is the subject of the RTD to the extent that only the human person can participate in and could be considered a beneficiary of the right.[21] Given the centrality of land as a resource for development, it is apposite that only the human person ought to be the beneficiary of land rights and its resources in a way that improves their socio-economic and cultural life. Hence, they must have the right to use, own and control the development that is undertaken on their land. It is in this context that the interface between land grabbing, the RTD and the SDGs must be understood and conceived, which reposes squarely on the right and ability of people to exercise ownership and control over their land and its resources. However, the central element of land, the RTD and human-induced development, including the SDGs, suggests that an impact on land and its resources, as in the context of land grabbing could greatly undermine both the purpose of the RTD and the objectives of the SDGs.

Difficulties in achieving the SDGs in the context of land grabbing in Ethiopia

Ending poverty

Goal 1 of the SDGs establishes a link between poverty alleviation and access to natural resources, including land. In terms of the Goal, there is an urgent need to end extreme poverty in all its forms by 2030,[22] since endemic poverty is associated with, and inimical to, people's socio-economic advancement. Goal 1 seeks to uplift the socio-economic situation of everyone, particularly the poor and the vulnerable. Over 41 percent of the population in Africa and in Ethiopia still lives and continues to survive on less than USD1.25 per day, and the prevalence of extreme poverty has been perceived as being close-knit with the non-development of its rural areas.[23]

Meeting this Goal could be more challenging in this era of land grabbing, as the practice has further exacerbated the living conditions of local communities in Ethiopia, instead of uplifting it. Even where land grabbing activities provide employment prospects for local communities, they are often poorly paid, and this does not in any way raise their living standards. This is because the activity does not take into account the plight of local communities, nor does it institutionalise this within the framework of land grabbing contracts since they are most often exclusionary.

The deprivation of local communities from their traditionally owned lands through land grabbing has a direct correlation with and negative implication on their ability to sustain themselves and to curb poverty. In Ethiopia, land epitomises a source of wealth, culture, power and ancestral lineage to the rural population of approximately 85 percent of the total population.[24] By this, it meant that losing access to and control over land (such as through land grabbing) is linked to multifaceted socio-cultural and economic consequences that impair the ability to curb poverty. Agriculture is also a major source of income generation for most households and the eviction of locals from their land deprives them from generating the much-needed income and thereby impairs their ability to curb poverty.[25] Since the implementation of the Saudi Star PLC rice project in the Gambella region, over 70.000 people have been displaced from their land under the villagisation programme.[26]

Land grabbing increases village-to-city migration, a situation that adds further constraints on Goal 11 on sustainable cities. Loss of income represents a clear example of how land grabbing activities could impact on the livelihood of local communities.[27] Land grabbing directly contradicts the socio-economic and cultural rights of local communities, and this has a direct bearing on their ability to sustain themselves and to address poverty-related concerns. Along this path, it is logical that practices of land grabbing could severely affect the attainment of Goal 1, as they have the potential to impoverish the livelihoods of local communities in Africa. This does not detract from the argument that, without land grabbing, local communities in Africa will not be poor or live on less than USD1 a day. Rather, it emphasises the view that practices of land grabbing exacerbate their living conditions and are an anathema to Goal 1.

Zero hunger

Poverty and hunger are deeply intertwined. Poverty has a direct bearing on one's ability to have access to good-quality, sufficient and nutritious food. Goal 2 is distinctively restrictive to addressing zero hunger by 2030. Goal 2 is accompanied by seven targets and these targets epitomise the overwhelming global desire to curtail hunger by 2030 by ensuring specifically that everyone has access to safe, nutritious and sufficient food,[28] that there will be an end to all forms of malnutrition,[29] that there will be an increase in agricultural productivity[30] and that sustainable food production will be ensured,[31] among others.

Yet, practices of land grabbing are counter-intuitive to the realisation of Goal 2. Generally, and in Ethiopia in particular, small-scale farming characterises one of the most appropriate ways of food crop production, and local populations depend heavily on local land and natural resources for self-sustenance.[32] However, land grabbing has dramatically altered the agrarian landscape while impacting negatively on local communities and means of livelihood. Land grabbing creates food insecurity and famine, as most of the food produced is exported back to the investor's country. This is the case with the Gambella rice project, where over 75 percent of the rice produced was exported to the Middle East.[33] Such situations irrefutably create and (possibly) exacerbate food security in the host countries and is absolutely inharmonious with Goal 2, and do not take into consideration dire socio-economic consequences associated with food security, such as massive displacement.

The disproportionate distribution of food during land grabbing between the foreign investor and the host country pushes upward the curve of the number of hungry people to over 185 million, with a concomitant increase in malnutrition.[34] Even if one were to argue that hunger prevailed prior to land grabbing, it must be borne in mind that over the years, the practice has worsened the situation in Ethiopia and Africa as a whole, harbouring the vast majority of the over 800 million people suffering from hunger.[35]

Equitable access to land

As a significant resource and cross-cutting issue, land plays a vital role in the realisation of the SDGs agenda. This importance of land suggests the need to properly monitor land-related indicators so as to promote and ensure responsible land governance that protects land rights. Increased attention to, and protection of (local communities') land rights could be evidence of the critical role that land tenure security and access to land could play in pushing the move to achieve economic growth in an inclusive and sustainable way.

Target 1.4 of Goal 1 seeks to ensure that by 2030, all men and women, and particularly the poor and the vulnerable, have equal rights to economic resources, including land and access to basic services, own and control land and natural resources among other things. Relatedly, indicator 1.4.2 aims to measure the proportion of the total adult population with secure land tenure rights in the form of legally recognised documentation issued under a formalised land tenure system, and who perceive their right to land as being secured. These indicators in principle provide suitable cognitive measures for advancing the concept and spirit of the continuum of tenure rights, with the hope that states would strive to ensure the protection of people's rights to land at the domestic levels by ensuring, specifically, that land grabbing does not trample upon the human right to land ownership.

Clearly, therefore, measures to secure land tenure rights, such as those advocated by the SDGs, could potentially help to improve people's livelihoods and address the sustainable management of natural resources, while concomitantly striving to promote responsible foreign agricultural investments that could help to eradicate poverty and secure food production. Secure land rights could also have the potential to guarantee the protection of land rights through, for example, the application of legal remedies whenever they are violated or abused, the prevention of arbitrary deprivation of local communities' rights from their land, and the creation of land markets that could unlock the potential of land as an asset and encourage efficient allocations and transactions,[36] during land grabbing, for example.

Nonetheless, the realisation of this Goal in this era of land grabbing is illusory and highly elusive since the activity is heavily characterised by the deprivation of local communities' rights to own and control land and its resources. It remains to be seen how governments and the international community at large might move the 'needle' of land governance forward without a 'gauge', or without agreements as to what exactly the gauge will measure. Although the Inter-Agency and Expert Group on Sustainable Development Goals Indicators have provided a three-tier ranking that determines which indicator could possibly be implemented, it must be noted that all three land-related indicators were stuck in the lowest category—Tier III.[37] With the exception of indicator 5a.1, which measures the percentage of people having secure rights to agricultural land and the share of women who were elevated to Tier III, indicators 1.4.2 and 5a.2 are still mired in Tier III.[38]

In Africa as a whole, most people still hold land rights under undocumented forms of customary land rights regimes, and are easily deprived of those rights under this form of tenure when land grabbing occurs.[39] Given that over 90 percent of land rights in Ethiopia, particularly in the Gambella region, are not covered by formal property ownership rights,[40] it is logical that the lack of a certificate of title as conclusive evidence of ownership places local communities and other vulnerable groups in a difficult and precarious position during land grabbing operations, to the extent that their land rights are violated. In other words, insecure land rights, premised on a lack of certification of ownership, constrains valuable opportunities for millions of local communities and vulnerable populations in Ethiopia to own and control land. Regrettably, such displacement often occurs without prior consultation, but forceful evictions of local communities from their land as confirmed by a villager of Gambella 'we were not told that our land will be given to foreign investors'.[41] The Oakland Institute succinctly captures this lack of consultation and narrates that

> Prior to relocation, no community consultation was carried out, either by Saudi Star or the government. Villagers only knew that their land had been given to investors once the bulldozers began clearing the area. When they expressed concern to the government about the clearing of their ancestral lands, government officials reportedly replied; 'You don't have any land, only government has land'.[42]

Land grabbing generally happens at the expense of the customary land rights of local communities, which is further exacerbated by constant evictions and displacements of local communities from their homes and farmlands to the extent that such displacements have become potential limitations to the RTD of local peoples. Land grabbing, whether for food or biofuel production,results in the degeneration of entire communities through displacement and forced migration into cities, and most local communities'members who lack certificates of title as is the case in Gambella, would have no redress in the form of compensation under the Ethiopian land tenure system.[43] Such displacement increasingly poses serious threats and problems to the realisation of Goal 11. Although Helmut and Du Plessis[44] are of the view that SDG 11 appears to be the thickening layer of international normative guidance on the question of good urban governance, or put more colloquially, the urban turn of governance, I argue that the forceful eviction and displacement of local communities from their homes to cities due to land grabbing obscures the fundamental issue, meaning and importance of urban governance because cities themselves would no longer be safe and inclusive, resilient and sustainable for the 2030 development agenda.

It must be pointed out that formal land tenure systems, which recognise the registration of land rights as proposed by the SDGs, comprise a Western style of land administration that is entirely novel and contrary to the systems of land holding in sub-Saharan Africa. To be sure, significant tension exist between the two. The Western form of land administration (though embraced by domestic legislation in Africa) has potential limitations, including the fact that it is more costly, time and capacity demanding and it does not seem to serve millions of Africans, whose form of land holding—customary form of tenure—is social rather than legal.[45] The security of land rights would under formal land administration provides the certainty that one's right to land will be recognised and protected in cases of dispossession through land grabbing.

The SDGs focus more on promoting security of formal land tenure, and not customary land tenure. Instead of bridging the dichotomy between them, the SDGs seem to intensify the divide. An understanding that land grabbing is characterised by eviction and deprivation of local communities from their traditionally owned and occupied lands, indicates that the activity serves to promote poor land governance (that does not take into account customary forms of tenure), and this is contrary to what the SDGs seek to achieve. In essence, if the global aspiration to promote a land governance regime that protects land rights and ensures economic growth in a sustainable manner is anything to go by, then much needs to be achieved in terms of improving the state's land governance systems, both statutory and customary. This should be achieved through making the land governance systems more robust against the exogenous forces of land grabbing and should promote and ensure homogeneity with the land-related SDGs to the extent that adequate safeguards are implemented to ensure the security of land rights on a large scale. Without this, attempts to achieve this goal and the general aim of the SDGs would be illusory.

In Ethiopia, as elsewhere in Africa, land is not only an asset or commodity, but also a form of identity and recognition, as it represents a spiritual and cultural bond between sacred land sites and local inhabitants. Yet, practices of land grabbing incontestably have the potential to destroy the ancestral lineage, spiritual bonds and cultural identity between the Anuak people of Gambella and their lands.[46] In this regard, it is hard to imagine how the SDGs could, in the face of continuous land grabbing, address the issue of land security within the 15 years time frame. Perhaps a more pragmatic approach would have been to include comprehensive targets/indicators relating to the recognition of customary land rights within the framework of the SDGs and of the need to safeguard customary land rights generally and in the context of development activities, not excluding land grabbing.

Nevertheless, the SDGs and particularly the one relating to land, only refer to lofty concerns to promote land governance by 2030, without setting out a clear-cut procedure on how this might be achieved. If the SDGs are truly committed to ensuring the land rights of everyone, and particularly of the poor and the vulnerable, attention might have to be paid to their land rights, as these are not documented under any formal land administration. The glaring absence of such commitment in any land-related targets of the SDGs supports the contention that the SDGs have systemically paved the way for the continuous deprivation of local communities of their land rights in order to promote the much-needed global development, which in principle seems to largely favour developed countries. By ignoring the protection of customary land rights, the SGDs place theirholders in precarious positions of vulnerability, if one considers that legal rights/documentation goes beyond mere land ownership rights to include legally enforceable documentation of user rights. This adds further complexity to the plight of local communities in Africa which, as indicated earlier, survive on less than USD1 a day.

Land grabbing, thus, appears as a potential barrier to the achievement of the SGDs, at least in Africa where most of the local people still hold undocumented land rights that are characterised by far-reaching ramifications, such as those mentioned above. Consequently, the problem with land rights creates interwoven problems of poverty, food security, gender issues, city migrations and sustainable management of natural resources—issues which the ambitious SDGs have sought to resolve by 2030. The problem of land rights, and presumably customary land rights, is further exacerbated by the fact that the SDGs do not place land and property rights at the forefront of development strategies; instead, they are included as sub-goals within various broader goals.[47] It suggests that the SDGs do not provide any bold stance in addressing systemic poverty and inequality that might arise from the lack of protection from activities such as land grabbing, and which result from the customary land rights of local communities, women and indigenous people.

Supposedly, the concomitant impacts of land grabbing on the land rights of local communities could continue in perpetuity, since local communities are unable to practically defend their customary land rights and to successfully compete with the exogenous forces that perpetuate land grabbing. Hence, land grabbing leads to the development and implementation of systematic ways to deprive locals of their land rights in pursuit of socio-economic development that benefits rich bourgeoisie countries. In a sense, the negative impacts of land grabbing on land rights actually support, justify and legitimise the potential limits of achieving the SDGs.

It is important to bear in mind that land is indispensable for agricultural production and this indispensability of land is emblematic of many contemporary sustainability issues, ranging from impoverished livelihoods of local communities in sub-Saharan Africa, who largely depend on their land for survival and income generation. The reason is that the eviction of local communities from their land through land grabbing for, inter-alia, the production of food crops, significantly affects Goal 1 of the SDGs, which requires the taking of concerted effort to eradicate—and not merely reduce—poverty by 2030. It has been reported that prolonged droughts—one of the impacts of climate change—are gaining a negative grip on agriculture and this severely affects a source of livelihood and income.[48]

Effective, climate-smart agriculture could require, for example that attention be given to inter alia, gender and social inclusion.[49] Yet, land grabbing significantly impacts on the gender dimension, and women who constitute the bulk of rural agricultural workers in Africa[50] continue to face unequal and unsecure rights to land and lack of meaningful participation in decision-making concerning land rights. This contradicts indicator 5.a.2, which measures the proportion of countries whose legal framework guarantees women's equal rights to land ownership and control. Conceivably, it would be potentially difficult to achieve this Goal and its targets within the context of land grabbing.

Renewable energy

Recognising the increasing global challenge of energy security, Goal 7 promotes the renewable energy agenda and commits to ensure that by 2030, everyone would have access to affordable, reliable and modern energy.[51] In alignment with the specific targets, there is a need to substantially increase the share of renewable energy in the global energy mix,[52] double by 2030 the global rate of improvement in energy efficiency,[53] enhance international co-operation with the hope of facilitating access to clean energy research and technology, ensure energy efficiency and promote investment in energy and clean energy technology.[54] Evidently, the SDGs seemingly unlock the potential for the Global North to continuously advance economically and environmentally by propagating the renewal energy agenda, as exemplified by their active engagements in land grabbing activities in the Global South.

In furtherance of the proposition that energy security is one of the drivers of land grabbing, it is remarkable that both the SGDs and land grabbing are interrelated and deeply concerned with energy security, to the extent that investments in renewable energy, in the guise of biofuel crops, have become predominant. Land grabbing for purposes of renewable energy has incontestably triggered negative ramifications that are potential barriers to the attainment of other SDGs. For example, the production of biofuel crops could significantly impair food production. This relates to situations where land that was previously used for the production of food crops is being converted for the production of biofuel crops, such as palm oil, jatropha and sugar cane.[55]

In other parts of Africa such as Uganda and Cameroon, foreign investors (mostly from the Global North) have invested more resources in the production of these biofuel crops, than in food crops.[56] Such investments further highlight the complexity and difficulty of achieving Goal 2 within the context of land grabbing.

Conclusion

It is noted that a noble intention accompanied the SDGs, namely that of moving the world forward within a time frame of 15 years through measures that would address contemporary global challenges, and that they have been hailed as representing the most successful anti-poverty movement of the 21st century. More realistically, perhaps, the SDGs are considered to have provided a common vision and contributed to making some remarkable progress. However, this chapter demonstrates the impracticability of achieving the overall aims of the SDGs in the context of land grabbing in Ethiopia. The magnitude of land grabbing and its negative impact on the rights and means of livelihoods of local communities evidently demonstrate the implicit controversial role played by the land grabbing phenomenon in the realisation of the global development agenda.

Any negative impact of land grabbing creates a negative chain effect on people's fundamental rights and on the global Earth systems. These effects are inimical to the vision and aim of the SDGs and their sustainability concerns. It was argued that while the SDGs privileged the Global North, who are mostly investors in land grabbing, to advance for the next 15 years, on the other hand, it has hindered the progress and developmental agenda of underprivileged countries—the Global South—where land grabbing is predominant, with profound socio-cultural and environmental consequences, such as those mentioned above. It is affirmed that land grabbing is implicated in hindering the attainment of the SDGs in Ethiopia. The SDGs do not prohibit activities such as land grabbing that are truly destructive to the global Earth systems and the ideals of the SDGs, nor do they consider the underlying causes of poverty and gender equality, or the underlying environmental issues that are related to the human development agenda and the improvement of the human conditions. It was argued that because the pursuit of any of the SDGs would have profound adverse implications on other goals in the context of land grabbing, it does not suffice for any of the SDGs to be attained at the expense of another.

If the SDGs are indeed the principal roadmap for global development, they represent a lost opportunity (at least in Ethiopia and Africa in general where land grabbing is prevalent) as they would most likely continue to advance socio-economic growth at the expense of global ecological considerations. Accordingly, a critical evaluation of the SDGs through the lens of land grabbing does not in any way represent the envisaged suitable roadmap for the 2030 development agenda.

In retrospect, just like the MDGs, the SDGs are not likely to achieve much. Consequently, the 2030 development agenda is only ephemeral and should be viewed as otiose in the context of land grabbing, taking into account that the world still sits with enormous challenges and will be forced to adopt yet another set of global aspiration to address its needs. Presumably, any (intended) new global aspirations should provide measures on how to address some of the hurdles that might constrain its objectives with respect to foreseeable challenges like land gabbing. There is a need to take into account African development priorities, as well as the socio-economic and cultural settings of African societies and the power imbalances in that context. Unless this is done, global efforts willforever be moving in circles, with mere aspirations that realistically will never be translated into reality.

Notes

1 Ashukem, Jean-Claude. 2020. 'The SDGs and the Bio-Economy: Fostering Land Grabbing in Africa'. *Review of Africa Political Economy* 47(164): 275–290.
2 Deininger, Klaus. 2011. 'Challenges Posed by the New Wave of Farmland Investment'. *Journal of Peasant Studies* 38(2): 223; Borass, Saturnino and Franco, Jennifer. 2012. 'Global Land Grabbing and Trajectories of Agrarian Change: A Preliminary Analysis'. *Journal of Agrarian Change* 12(1): 37.
3 Ashukem, Jean-Claude. 2016. 'A Rights-Based Approach to Foreign Agro-Investment Governance in Cameroon, Uganda and South Africa'. LLD Thesis, North-West University, 76–83.
4 De Schutter, Olivier. 2011. 'The Green Rush: The Global Race for Farmland and the Rights of Land Users'. *Harvard International Law Journal* 52(2): 504; Cotula, Lorenzo *et al.* 2009. *Land Grab or Development Opportunity? Agricultural Investment and International Land Deals in Africa*. Rome: Food and Agricultural Organisation, 17.
5 Cotula, Lorenzo. 2014. 'Addressing the Human Rights Impacts of Land Grabbing'. *European Parliament—Policy Department*, 11–13.
6 Narula, Smita. 2013. 'The Global Land Rush: Markets, Rights, and the Politics of Food'. *Stanford Journal of International Law* 103; Zoomers, Annelies. 2010. 'Globalisation and the Foreignisation of Space: Seven Processes Driving the Current Global Land Grab'. *Journal of Peasant Studies* 37(2): 429–447.
7 Cotula, note 5; Ashukem, note 3, 44–59.
8 Dell'Angelo, Jampel; D'Odorico, Paulo and Rulli, Maria. 2007. 'Threats to Sustainable Development Posed by Land and Water Grabbing'. *Current Opinions in Environmental Sustainability* 26(27): 121.
9 Ashukem, note 3, 5; Ashukem, Jean-Claude. 2019. 'A Rights-Based Approach to Foreign Agro-Investment in Cameroon: Enhancing the Protection of Local Communities' Rights'. *Journal of African Law* 63(2): 166.
10 Wolford, Wendy *et al.* 2013. 'Governing Global Land Deals: The Role of the State in the Rush for Land'. *Development and Change* 44(2): 189–210.

11 Ashukem, Jean-Claude. 2019. 'A Human Rights-Based Approach to Foreign Agricultural Investment in Uganda'. *African Journal of International and Comparative Law* 27(2): 272.
12 Dell'Angelo, Paulo and Maria, note 8, 122.
13 UN General Assembly 'Transforming Our World: The 2030 Agenda for Sustainable Development' A/RES/701 of 21 October 2015.
14 *Ibid.*
15 Kotzé, Louis. 2018. 'The Sustainable Development Goals: An existential critique alongside three new-millennial analytical paradigms'. In *Sustainable Development Goals: Law, Theory and Implementation*, edited by French, Duncan and Kotzé, Louis. London: Edward Elgar Publishing, 42.
16 Jaiyesimi, Rotimi. 2016. 'The Challenge of Implementing the Sustainable Development Goals in Africa: The Way Forward'. *African Journal of Reproductive Health* 20(3): 13.
17 'SDG Indicators: Global Indicator Framework for the Sustainable Development Goals and Targets of the 2030 Agenda for Sustainable Development'. <https://unstats.un.org/sdgs/indicators-list/>.
18 Janoušková, Svatava; Hák, Tomášand Moldan, Bedřich. 2018. 'Global SDGs Assessments: Helping or Confusing Indicators?' *Sustainability* 10(5): 1–14.
19 African Commission (2009) 276/2003.
20 African Charter on Human and Peoples' Rights adopted by the Organisation of African Unity (OAU) in Nairobi Kenya on 27 June 1981, OAU Doc CAB/LEG/67/3 rev.5, arts 22.
21 Sengupta, Arjun. 2001. 'Right to Development as a Human Right'. *Economic and Political Weekly* 36(27): 2528.
22 See Target 1.1.
23 Hufe, Paul and Heuermann, Daniel. 2017. 'The Local Impacts of Large-Scale Land Acquisitions: A Review of Case Study Evidence from Sub-Saharan Africa'. *Journal of Contemporary African Studies* 35(2): 168–189.
24 Abegaz, Tekle. 2020. 'Displacement from Land as a Limit to the Realisation of the Right to Development in Ethiopia'. *Journal of Juridical Science* 45(1): 62.
25 Ashukem, Jean-Claude. 2020. 'Land Grabbing and Customary Land Rights in Uganda: A Critical Reflection of the Constitutional and Legislative Right to Land'. *International Journal on Minority and Group Rights* 27(1): 137.
26 Narula, note 6.
27 Davis, Kyle; D'Odorico, Paulo and Rulli, Maria. 2014. 'Land Grabbing: A Preliminary Quantification of Economic Impacts on Rural Livelihoods'. *Population and Environment* 36: 190.
28 Target 2.1 of the SGDs.
29 Target 2.2 of the SDGs.
30 Target 2.3 of the SDGs.
31 Target 2.4 of the SDGs.
32 Dell'Angelo, Jampel *et al.* 2017. 'The Tragedy of the Grabbed Commons: Coercion and Dispossession in the Global Land Rush'. *World Development* 92: 1–12.
33 Narula, note 6.
34 Hepburn, Jonathan and Bellmann, Christophe. 2018. 'Overview'. In *Achieving Sustainable Development Goal 2: Which Policies for Trade and Markets?*, edited by Hepburn, Jonathan and Bellmann, Christophe. Geneva: ICTSD, 11.
35 Luchuo, Bain *et al.* 2013. 'Malnutrition in sub-Saharan Africa: Burden, Causes and Prospects'. *Pan African Medical Journal* 15(120): 2.
36 Mennen, Tiernan. 2015. 'Know Your SDGs: Land Matters for Sustainable Development'. <www.chemonics.com/blog/know-your-sdgs-land-matters-for-sustainable-development>.
37 Hanstad, Tim. 2017. 'Moving the Needle Forward on Land Rights in the Sustainable Development Goals'. 21 November <www.landesa.org/blog-moving-needle-forward-land-rights-sustainable-development-goals/>.

38 *Ibid.*
39 Ashukem, note 25, 1–29.
40 Mennen, note 36.
41 Roopanand, Mahadew. 2020. 'Land grabbing in Ethiopia and Madagascar: Balancing Respect for Human Rights of Victims with Development Needs through Land Investments'. LLD Thesis, University of the Western Cape, 21–22.
42 Oakland Institute. 2011. 'Understanding Land Investment Deals in Africa: Country Report: Ethiopia'. </www.oaklandinstitute.org/sites/oaklandinstitute.org/files/OI_Ethiopa_Land_Investment_report.pdf>.
43 Human Rights Watch. 'Waiting here for Death: Displacement and Villagization in Ethiopia's Gambella Region'. <www.hrw.org/sites/default/files/reports/ethiopia0112webwcover_0.pdf>.
44 Aust, Philip and Du Plessis, Anél. 2018. 'Good Urban Governance as a Global Aspiration: On the Potential and Limits of Sustainable Development Goal 11'. In *Sustainable Development Goals: Law, Theory and Implementation*, edited by French, Duncan and Kotzé, Louis. London: Edward Elgar Publishing, 201–202.
45 Enemark, Stig. 2017. 'Land Governance in Support of the 2030 Global Agenda'. Conference paper, UNECE Working Party on Land Administration, Palais de Nations, Geneva, Switzerland, 8–9 March <www.unece.org/fileadmin/DAM/hlm/wpla/session/10th_session/presentations_day1/keynote/2.Stig_Enemark.pdf>.
46 Mahadew, note 41, 65.
47 Mennen, note 36.
48 Campbell, Bruce. 2015. '4 Ways Countries are Successful in Fighting Hunger'. *World Economic Forum*, 4 September <www.weforum.org/agenda/2015/09/4-ways-countries-are-successfully-fighting-hunger/>.
49 *Ibid.*
50 Doss, Cherry; Summerfield, Gale and Tsikata, Dzodzi. 2014. 'Land, Gender and Food Security'. *Feminist Economics* 20(1): 1–23; Levien, Michael. 2017. 'Gender and Land Dispossession: A Comparative Analysis'. *The Journal of Peasant Studies* 44(6): 1111–1134.
51 Target 7.1.
52 Target 7.2.
53 Target 7.3.
54 Target 7.A.
55 Ashukem, note 3, 82.
56 *Ibid*, 66.

References

Abegaz, Tekle. 2020. 'Displacement from Land as a Limit to the Realisation of the Right to Development in Ethiopia'. *Journal of Juridical Science* 45(1): 51–71.

Ashukem, Jean-Claude. 2016. 'A Rights-Based Approach to Foreign Agro-Investment Governance in Cameroon, Uganda and South Africa'. LLD Thesis, North-West University.

Ashukem, Jean-Claude. 2019. 'A Human Rights-Based Approach to Foreign Agricultural Investment in Uganda'. *African Journal of International and Comparative Law* 27(2): 268–291.

Ashukem, Jean-Claude. 2019. 'A Rights-Based Approach to Foreign Agro-Investment in Cameroon: Enhancing the Protection of Local Communities' Rights'. *Journal of African Law* 63(2): 163–191.

Ashukem, Jean-Claude. 2020. 'Land Grabbing and Customary Land Rights in Uganda: A Critical Reflection of the Constitutional and Legislative Right to Land'. *International Journal on Minority and Group Rights* 27(1): 121–147.

Ashukem, Jean-Claude. 2020. 'The SDGs and the Bio-Economy: Fostering Land Grabbing in Africa'. *Review of Africa Political Economy* 47(164): 275–290.

Aust, Philip and Du Plessis, Anél. 2018. 'Good Urban Governance as a Global Aspiration: On the Potential and Limits of Sustainable Development Goal 11'. In *Sustainable Development Goals: Law, Theory and Implementation*, edited by French, Duncan and Kotzé, Louis. London: Edward Elgar Publishing.

Borass, Saturnino and Franco, Jennifer. 2012. 'Global Land Grabbing and Trajectories of Agrarian Change: A Preliminary Analysis'. *Journal of Agrarian Change* 12(1): 34–59.

Cotula, Lorenzo *et al.* 2009. *Land Grabor Development Opportunity? Agricultural Investment and International Land Deals in Africa*. Rome: Food and Agricultural Organisation.

Cotula, Lorenzo. 2014. 'Addressing the Human Rights Impacts of Land Grabbing'. *European Parliament—Policy Department*, 1–48.

Davis, Kyle; D'Odorico, Paulo and Rulli, Maria. 2014. 'Land Grabbing: A Preliminary Quantification of Economic Impacts on Rural Livelihoods'. *Population and Environment* 36: 180–192.

De Schutter, Olivier. 2011. 'The Green Rush: The Global Race for Farmland and the Rights of Land Users'. *Harvard International Law Journal* 52(2): 504–559.

Deininger, Klaus. 2011. 'Challenges Posed by the New Wave of Farmland Investment'. *Journal of Peasant Studies* 38(2): 217–247.

Dell'Angelo, Jampel *et al.* 2017. 'The Tragedy of the Grabbed Commons: Coercion and Dispossession in the Global Land Rush'. *World Development* 92: 1–12.

Dell'Angelo, Jampel; D'Odorico, Paulo and Rulli, Maria. 2007. 'Threats to Sustainable Development Posed by Land and Water Grabbing'. *Current Opinions in Environmental Sustainability* 26(27): 120–128.

Doss, Cherry; Summerfield, Gale and Tsikata, Dzodzi. 2014. 'Land, Gender and Food Security'. *Feminist Economics* 20(1): 1–23.

Hepburn, Jonathan and Bellmann, Christophe. 2018. 'Overview'. In *Achieving Sustainable Development Goal 2: Which Policies for Trade and Markets?*, edited by Hepburn, Jonathan and Bellmann, Christophe. Geneva: ICTSD.

Hufe, Paul and Heuermann, Daniel. 2017. 'The Local Impacts of Large-Scale Land Acquisitions: A Review of Case Study Evidence from Sub-Saharan Africa'. *Journal of Contemporary African Studies* 35(2): 168–189.

Jaiyesimi, Rotimi. 2016. 'The Challenge of Implementing the Sustainable Development Goals in Africa: The Way Forward'. *African Journal of Reproductive Health* 20(3): 13–18.

Janoušková, Svatava; Hák, Tomáš and Moldan, Bedřich. 2018. 'Global SDGs Assessments: Helping or Confusing Indicators?' *Sustainability* 10(5): 1–14.

Kotzé, Louis. 2018. 'The Sustainable Development Goals: An Existential Critique Alongside Three New-millennial Analytical Paradigms'. In *Sustainable Development Goals: Law, Theory and Implementation*, edited by French, Duncan and Kotzé, Louis. London: Edward Elgar Publishing, 41–65.

Levien, Michael. 2017. 'Gender and Land Dispossession: A Comparative Analysis'. *The Journal of Peasant Studies* 44(6): 1111–1134.

Luchuo, Bain *et al.* 2013. 'Malnutrition in sub-Saharan Africa: Burden, Causes and Prospects'. *Pan African Medical Journal* 15(120): 1–9.

Narula, Smita. 2013. 'The Global Land Rush: Markets, Rights, and the Politics of Food'. *Stanford Journal of International Law*, 101–175.

Roopanand Mahadew. 2020. 'Land Grabbing in Ethiopia and Madagascar: Balancing Respect for Human Rights of Victims with Development Needs through Land Investments'. LLD Thesis, University of the Western Cape.

Sengupta, Arjun. 2001. 'Right to Development as a Human Right'. *Economic and Political Weekly* 36(27): 2527–2536.

Wolford, Wendy *et al.* 2013. 'Governing Global Land Deals: The Role of the State in the Rush for Land'. *Development and Change* 44(2): 189–210.

Zoomers, Annelies. 2010. 'Globalisation and the Foreignisation of Space: Seven Processes Driving the Current Global Land Grab'. *Journal of Peasant Studies* 37(2): 429–447.

13 Freedom from want and the constitutional right to development in Ethiopia

Urban productive safety net programme—The case of Addis Ababa

Tsion T. Dechassa and Gedion G. Jalata

Introduction

Food security, as a concept, emerged in the mid-1970s in the context of several incidences that prompted discussions on food supply and sufficiency at the aggregate level both at the global and national levels. However, the issue of sufficiency at the household or individual level, as well as nutritional quality and environmental sustainability, were neglected.[1] Within such discourse, the World Food Summit in 1996 provided a widely accepted definition of food security: 'food security exists when all people at all times have access to sufficient, safe, nutritious food to maintain a healthy and active life'.

Globally, about 870 million people do not get sufficient food, which implies that one out of every eight persons is a victim of food insecurity. Africa's regional overview of food security and nutrition reported that the food security situation on the continent appeared to be worsening.[2] The 2018 report on the issue confirms this trend. The worsening situation in Africa is due to severe global economic conditions and, in many countries, conflict and adverse climate conditions and sometimes in a combination of these factors. Economic growth slowed in 2016 due to weak commodity prices in particular for oil and minerals.[3] Food insecurity has worsened in countries affected by conflicts, often exacerbated by drought or floods, and in southern and eastern Africa, many countries have been adversely affected by prolonged drought.[4] However, even in the face of these challenges, many countries have been able to achieve sustained progress in reducing food insecurity.

In sub-Sahara African countries, ensuring food security in urban areas is a challenge due to a confluence of factors, including irregular income, limited attention to urban agriculture, lack of employment opportunities, absence of legal frameworks for urban land tenure security, lack of potable water, sustainable energy supply, housing, and education as well as health facilities, among others.[5]

In this context, different countries globally developed and implemented specific food security programmes in urban areas. The most notable ones are the Famine Safety Net Programme in Kenya, the Bolsa Familia in Brazil, the

Progresa Aparchnitades in Mexico and the Zero Famine Strategy in Brazil. However, the actual results achieved differ depending on the specific context and commitments of individual countries.[6]

Achieving food security, however, has been an insurmountable challenge in Ethiopia. More than 30 percent of the Ethiopian population lives below the poverty line, while over 31 million people are undernourished. By using the threshold of 2,550 kilocalories (Kcals) per adult equivalent per day, 40 percent of Ethiopian households were identified to be food insecure and under nourished.[7]

Various regimes of the Ethiopian government had implemented different food security strategies to address issues of drought and food shortage that millions of Ethiopians have been faced with. Decades of food security programming have assisted many and undoubtedly saved lives. However, interventions fall short of the expectation to achieve the objective of improving household food security for the majority of poor and food-insecure households. In contrast, food insecurity remains a persistent problem in the country.

Aware of this fact, the Ethiopian government and other stakeholders have developed food security enhancement strategies in the country. The major components of the food security programme in Ethiopia include the Household Asset Building Programme, the Voluntary Resettlement Programme; the Complementary Community Investment Programme, the Productive Safety Net Programme (PSNP) and the Urban Productive Safety Net Programme.

The government of Ethiopia introduced the Urban Productive Safety Net Project (UPSNP), particularly intending to alleviate the challenges of chronic food security in urban areas, including Addis Ababa. This study focuses on Woreda 08 of Arada sub-city, Addis Ababa, mainly because the sub-city is one of the ten sub-cities of the capital Addis Ababa, having a total area of 825,424.3 m^2 and an overall population of 9,625 people. It has the largest poor population that is beneficiary of the UPSNP in Addis Ababa. The majority of the residents in Woreda generate their livelihood incomes from engagement in the informal sector, small micro-businesses such as shops, employment at public and private companies, as well as daily menial labour.

Implementation of the UPSNP helps the nation achieve the goals to 'end poverty in all its forms everywhere; end hunger, achieve food security and improved nutrition and promote sustainable agriculture; and ensure healthy lives and promote well-being for all at all ages' that are listed in the 2030 Sustainable Development Goals[8] as well as get closer to fulfilling aspiration 1 of Agenda 2063, which envisages: A prosperous Africa based on inclusive growth and sustainable development which is achieved by fulfilling goals, such as (1) a high standard of living, quality of life and well-being for all citizens and having healthy and well-nourished citizens.[9]

The government of Ethiopia has introduced a safety net to national coverage and designed the UPSNS framework, but its unique nature from the PSNP and also its broad-scale implementation demands close documentation, gap identification and prediction for its successful implementation. However, there is an apparent lack of information and documentation on practices of the UPSNP, including the documentation of best experiences and challenges specific to the practice in Addis Ababa. For this reason, it is timely and pertinent to assess the implementation of the UPSNP and identify significant challenges that the programme may be facing. Determinants and prospects of food insecurity experiences in the city of Addis Ababa, taking the case of Woreda 8 of Arada sub-city, would specifically be considered.

The main objective of the study was to assess the implementation, challenges and prospects and to identify the determinants of food insecurity experienced by the UPSNP beneficiaries in the region of Woreda 08, Arada sub-city in Addis Ababa, Ethiopia. The specific objectives include assessing the implementation and identifying the challenges and prospects of the UPSNP. The study also aims to investigate the food security status of the UPSNP beneficiaries in the study area and analyse the food insecurity experience of UPSNP beneficiaries, identifying the determinants of food insecurity status of beneficiaries in the study area as well as the way the UPSNP actualises the right to development in helping to achieve the goals listed in the 2030 sustainable development goals as well as Agenda 2063.

This research study answered the following questions: What are the implementation gaps, challenges and prospects of the UPSNP in the study area? What are the determinants of food insecurity experienced by UPSNP beneficiaries in the study area? How does the UPNSP actualise the right to development?

This study applied a cross-sectional data collection (one-time research) method. With regard to the nature of the inquiry, the study combined the descriptive and explanatory as well as the quantitative and qualitative approaches in the analysis to ensure an in-depth understanding of the issues under consideration.[10] Quantitative analysis was used to generate data in quantitative form while the qualitative types of data were applied to explain the subjective assessment of attitudes and opinions of respondents about the subject matter of the study. At the time of the study, there were about 1,415 beneficiaries of the urban PSNP, constituting the target population. Out of this, 609 are male, while 806 are female beneficiaries.

Households were used as sources for the quantitative data collection since beneficiaries of the UPSNP are household members. The household, therefore, was used as a unit of analysis. A two-stage sampling procedure was employed to draw a sample from the beneficiaries. First, Arada Woreda 08 from ten subcities of Addis Ababa was selected because it has one of the most significant beneficiary populations of the UPSNP in Addis Ababa. Secondly, out of 6 Ketenas beneficiaries, 4 of the UPSNP Ketena beneficiaries (18/1, 18/2, 18/3 and 18/4) were selected by using simple random sampling techniques. Thirdly, an updated list of the UPSNP benefices in the Arada 08 Woreda was taken by the researcher from Woreda 08 urban job creation and food security offices, and appropriate numbers of samples, representing the district were selected in proportion to the population size using Yamane (1967).[11] Accordingly, the sample size is then calculated as

$$n = \frac{N}{1+N(e)^2} = \frac{1415}{1+1415(0.1^2)} \sim 94$$

where n = the sample size; N = the target population size; e = the level of precision (10 percent).

The required sample size was calculated using a 95 percent confidence level, and the level of precision equal to 10 percent is recommended to obtain a sample size required, which represents a true population. Both primary and secondary data types were used. Primary data was collected from selected household heads who live in the sample Ketenas and other relevant government bodies. Secondary data was gathered from reports and statistical data of the Woreda and Kebeles, earlier empirical studies in related areas, published materials and internet sources.

In addition, semi-structured interview questions were conducted to get information from key individual informants. A total of 16 Key Informant Interviews (KIIs) were carried out from different organisations such as the World Bank group, Federal Urban Job Creation, and Food Security Agency, Woreda 08 urban job creation and food security office, Woreda 08, microfinance and small enterprise, Woreda workers 08 and social affairs office, Oxfam in Ethiopia and World Vision Ethiopia. Focus group discussions (FGD) were also held with beneficiaries, elders of the community, female representatives of Ketenas, chairperson of the Ketenas and the UPSNP field officers to generate more information about their collective awareness and perception on the subject.

The raw quantitative data collected was summarised, and the information displayed in a readily digestible form. Descriptive statistics frequencies, tables, percentages, mean techniques and standard deviations were used in the process of examining and analysing the socio-economic characteristics of sample UPSNP beneficiaries. The results were then presented through histograms and tables. The Food Insecurity Experience Scale (FIES) methods were used to measure access to food at the individual or household levels. It measures the severity of food insecurity based on people's responses to questions about constraints on their ability to obtain adequate food.

A Binary logistic regression model was employed to study whether there is a dichotomous relationship between the variables, that is, the probability of a UPSNP beneficiary to be either food secure or insecure. The Tobit model can be used in analysing both food insecurity and the level of food insecurity determinants. This model assumes that the same factors affect food insecurity and the intensity of food insecurity in the same direction.

It is believed that the results of the study would contribute to efforts to improve the implementation of the ongoing UPSNP in Ethiopia. The study, therefore, fills the gap in the existing knowledge concerning the UPSNP and thereby provides new information in the study of food security concerning PSNP in Ethiopia. This new information will also provide insight as to how the programme can further be adjusted to help meet the 2030 SDGs and aspirations of Agenda 2063.

The study focused solely on the UPSNP and brought out the challenges and prospects of food insecurity experience of it using a case in point, i.e., Woreda 08 of Addis Ababa. The focus of this study is on the beneficiaries of UPSNP in Woreda 08 of Addis Ababa, and there was no comparison of the control and treatment group. The study relied only on the cross-sectional data and the study area.

In the course of the research, several problems were encountered. First, the literature at hand was startlingly meagre and only just evolving. In this regard, there have been problems in getting up-to-date materials and official government documents, some of which are classified. Second, it has also been a challenge to get experts for an in-depth informant interview from government officials and partners. Third, space, time and resources allocated for carrying out the study were also challenging.

The study consists of four major sections. The first section is the introductory part, which deals with the background, statement of the problem, objectives, research questions, scope, significance, limitations and organisation. The second section unveils freedom from want and the constitutional right to development in Ethiopia. The third section deals with the benefits and challenges of the Urban Productive Safety Net Programme. The final section provides conclusions and recommendations.

Freedom from want and the constitutional right to development in Ethiopia

The concept of freedom from want first emerged when the then-president of the USA, Roosevelt, coined the term in 1941 during a speech known as the 'Four Freedoms Speech'. In his speech, Roosevelt outlined four freedoms: Freedom of speech, freedom of worship, freedom from want, and freedom from fear. Furthermore, he defines freedom from want as economic understandings which will secure to every nation a healthy peacetime life for its inhabitants.[12] These four freedoms then became the inspirations behind a mission that First lady Eleanor Roosevelt undertook called the United Nations Declaration of Human Rights.[13]

African governments attempted to address freedom from want by improving food security, nutrition and incomes in their prospective countries. Furthermore, Africa is addressing food insecurity, nutrition and incomes across the continent through the Comprehensive Africa Agriculture Development Programme (CAADP). Through CAADP, African leaders are playing a critical role in supporting science and innovation on which increased agricultural productivity depends.

The right to development was proclaimed in the Declaration on the Right to Development, adopted in 1986 by the United Nations General Assembly (UNGA) in its Resolution 41/128. This right is also recognised in the African

Charter on Human and Peoples' Rights and the Arab Charter on Human Rights and re-affirmed in several instruments including the 1992 Rio Declaration on Environment and Development, the 1993 Vienna Declaration and Programme of Action, the Millennium Declaration, the 2002 Monterrey Consensus, the 2005 World Summit Outcome Document and the 2007 Declaration on the Rights of Indigenous Peoples.[14]

This has, in turn, been enshrined in the Constitution of the Federal Democratic Republic of Ethiopia, which provides that 'The People of Ethiopian as a whole, and each Nation, Nationality, and People in Ethiopia, in particular, have the right to improved living standards and sustainable development'.[15] However, Tura[16] argues that the country has not yet adopted sufficient legislative and judicial measures to enforce the right to food and urges the government to introduce a framework law on the right to food to end hunger in the context of achieving national food security. In this regard, ensuring food security can be considered as a prerequisite to achieving improved living standards and sustainable development.

On the other hand, food security as a concept emerged in the mid-1970s in the context of several incidences, which prompted discussions on food supply and sufficiency at the aggregate level both at the global and national levels. The issue of sufficiency at the household or individual level were, however, neglected.[17] Within such discourse, the World Food Summit in 1996 provided a widely accepted definition of food security. It states that 'food security exists when all people at all times have access to sufficient, safe, nutritious food to maintain a healthy and active life'.[18]

The decade-long quest for a food security strategy in Ethiopia has culminated in Ethiopia's current Food Security Programme with the adoption of the Productive Safety Net Programme in 2005. The PSNP is a multi-year food security strategy that makes foreseeable cash or food transfer to food-insecure people.

Food security is a challenge in urban areas because of a confluence of factors including irregular income, limited attention to urban agriculture, lack of legal frameworks for urban land tenure security, lack of potable water, sustainable energy supply and housing, among others.[19] In this context, different countries globally developed specific food security programmes and implemented them to tackle food insecurity in urban areas. The most notable ones are the Famine Safety Net Programme in Kenya, the Bolsa Familia in Brazil, the Progresa Aparchnitades in Mexico and the Zero Famine Strategy in Brazil.

In Ethiopia, repeated shocks faced by the rural and urban poor as a result of extreme climatic variability compounded by severe environmental degradation means that food insecurity remains at the heart of Ethiopia's social protection challenge. Although this internationally financed humanitarian assistance was substantial and saved many lives, evaluations showed that it did little to protect livelihoods, generate community assets or preserve household assets.[20]

To tackle this problem, the government of Ethiopia, in collaboration with donor partners, launched the PSNP that aimed at developing an alternative to crisis response to support the needs of chronic, predictably food-insecure households as well as to develop long-term solutions to the problem of food insecurity. It replaced the emergency appeal system initially for around 5 million chronically food-insecure people, which was subsequently scaled up to cover 7.5 million people, and now with PSNP 4, it reaches 8.5 million people.[21] Encouraged on the successes of rural PSNP, the government of Ethiopia introduced the Urban Safety Net Strategy that seeks to guide the implementation of interventions that will alleviate the varying needs of the urban poor. The UPSNP is a flagship national programme and one of the few large Urban Safety Nets in Africa—and globally.[22]

The first phase of the programme runs from 2016 to 2021 and will focus on putting in place basic safety net building blocks, including productive and predictable transfers through public works, livelihood interventions and capacity building, as a response to the government's plan for immediate safety net support for the urban poor. The UPSNP has three major interrelated sub-programmes. These are unconditional direct support, the public work sub-programme and the livelihood sub-programme. The long-term objective of the programme is to reduce poverty and vulnerability among the urban poor by implementing productive and predictable urban safety nets and complementary livelihood interventions for 10 years and beyond.[23]

This first phase of the UPSNP is targeted to reach over 600 thousand people in 11 major cities (such as Adama, Addis Ababa, Asosa, Dessie, Dire Dawa, Gambela, Hawass, Harar, Jigjiga, Mekele and Semera).[24] As of April 2018, about 448,885 people are addressed, of which about 370,343 are selected for the public work sub-component, and the remaining 78,542 are considered for direct support. The conceptual framework of the UPSNP consists of three components: (i)Support (Safety Net Support), (ii) Livelihood, (iii) Institutional Strengthening, Project Management and Coordination.

The second component of the analytical framework is about UPSNP. This is indicated on the right side of the analytical framework. The third section of the analytical framework examines the impacts of the first section (including components of UPSN) and the second section, UPSNP as illustrated in Figure 13.1 with a two-directional relationship between the two sections, and UPSNP, and ultimately on the food security policy of the country.

The final section of the analytical framework investigates two kinds of relationships with the major actors involved in urban PSNP, including the government, the beneficiary urban poor and partners, as illustrated in the lower segment of Figure 13.1. The first relationship indicates their contribution to the programme, and the second relationship suggests the effect of the programme on the actors.

Benefits and challenges of the Urban Productive Safety Net Programme

In this section, the main findings of the study are presented and discussed. The first part presents the demographic and socio-economic characteristics of the

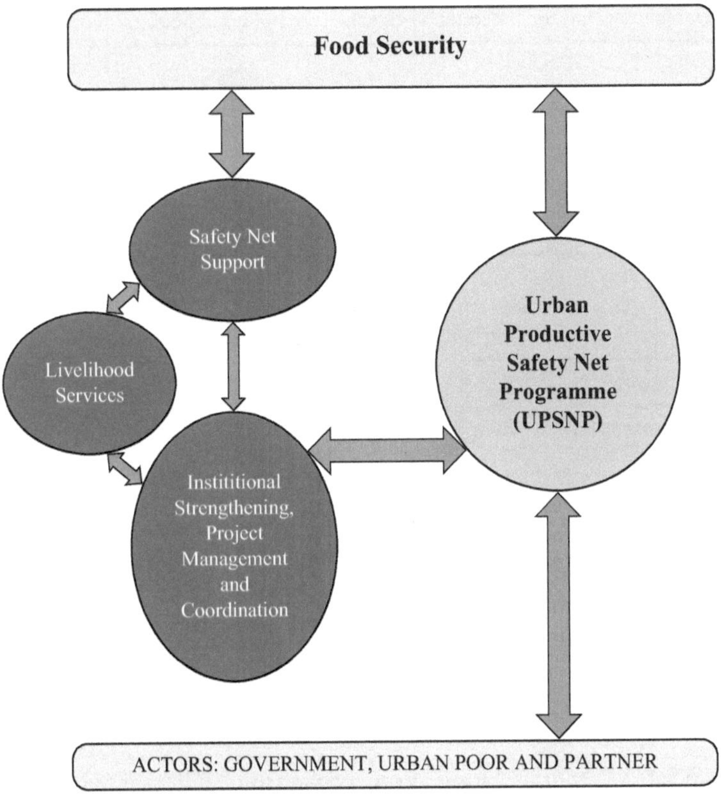

Figure 13.1 Analytical framework for Urban Productive Safety Net. *Source*: Analysed and illustrated by the researchers

UPSNP beneficiaries. Implementation of the UPSNP, the roles played by government institutions and partners in its promotion, challenges and prospects of the UPSNP are also presented. The second part discusses econometric results i.e., determinants of the UPSNP. The mean age of the household heads for the UPSNP beneficiaries was 42.75 years. Out of the 94 UPSNP beneficiaries interviewed, 50 percent of them were female. Of the total sample of the UPSNP beneficiaries, about 47.9 percent were married, 6.4 percent were single, 19.1 percent were divorced and 26.6 percent were widowed. The mean family size of sample households was found to be 4.6. The mean number of economically active members of the households was 2.36. The mean year of education for the UPSNP beneficiaries was found to be 4.7 years. The overall mean year of participating in the UPSNP was found to be 1.9 years as shown in Table 13.1.

Table 13.2 reflects income-poverty status and food security/access in Woreda 08, Arada sub-city as follows.

Table 13.1 Demographic and socio-economic characteristics of households

Demographic and socio-economic variables	*Mean*	*SD*
Age of the respondents	42.75	13.23
Number of economically active members of the households	2.36	1.96
Family size of the respondent of the household	4.6	1.46
Level of education of the respondent	4.7	4.07
Experience of UPSNP programme	1.96	0.42

Table 13.2 Livelihood activities of the sample households

Income-generating activities	*Frequency*	*Percent*
Involved only on UPSNP	20	21.3
Petty trade	11	11.7
Daily labour	18	19.1
Washing clothes	14	14.9
Fruits and vegetables retailing	3	3.2
Injera selling	4	4.3
Handicrafts (carpentry, masonry)	2	2.1
Other (pension, gift from family)	22	23.4
Other (commercial sex)	3	3.2
Total	94	100

House ownership is another characteristic of the sampled households taken into consideration where 61.7 percent of the sample households were found to be living in kebele houses, 24.5 percent in rented houses, 7.4 percent in self-owned houses and the remaining 6.4 percent in other types of housing arrangements.

Another characteristic of the households taken into account is asset protection and building. The UPSNP has been improving the household's capacity to improve children's school enrolment rates where 73.4 percent of the respondents confirmed that they enrolled their children in school as a result of benefiting from the UPSNP. Moreover, the UPSNP enabled beneficiaries to access health-care facilities. About 41.5 percent of households confirmed obtaining access to health-care facilities after becoming beneficiaries of the UPSNP.

As indicated by the increase in the UPSNP beneficiaries' access to health-care facilities, it can be said that the programme has helped immensely in improving the living standards forthe beneficiaries. This means that the programme is enhancing the capacity of the citizens for development, including access to quality healthcare, which is fundamental to development. Selling tangible home assets and interrupting savings are characteristic of food-insecure households. On the converse side, avoiding households selling off assets and continuous saving, acquiring new household assets, adapting to working culture, knowledge or skills after being a member of the UPSNP are considered as asset-building strategies and as protective of households. The survey results show that around 72.3 percent of the respondents avoided having to sell household assets to buy food after they joined the UPSNP.

The survey also indicates that around 72 percent avoided using their savings to buy food. In comparison, approximately 77.6 percent of the respondents responded that they acquired a new working culture, knowledge and skills after becoming participants of the UPSNP. However, still around 51 percent of the respondents responded that they did not keep their children in school for longer after the UPSNP; about 58 percent of the respondents did not access health-care facilities after the UPSNP and around 64 percent of the respondents also did not acquire any new household assets. The questionnaire to assess asset protection and build characteristics of the sample households is shown as follows in Table 13.3.

Poverty is one of the biggest hindrances to actualising human rights. The UPSNP, therefore, is a programme that was created to combat food insecurity issues in one of the poorest regions in Addis Ababa. By combating food insecurity, the programme improves the living standards of its beneficiaries and, in turn, allows the citizens to improve their lives and possibly rise out of poverty. This gives the citizens the chance to effectively exercise their constitutional right to development.

Frequently receiving enough access to service and support from the UPSNP, the government and partners have increased the beneficiaries' knowledge about business orientation, proper use of resources and market information. Out of the 94 UPSNP beneficiaries surveyed, around 88.5 percent confirmed

Table 13.3 Asset protection and building assessment response

Asset protection and building items	*Yes (%)*	*No (%)*
Have you enrolled your children in school?	73.4	26.6
Have you kept your children in school for longer after the UPSNP than before?	48.94	51.1
Have you used more health-care facilities after the UPSNP than before?	41.5	58.5
Have you consumed more food or diet food after the UPSNP than before?	38.3	61.7
Have you avoided having to sell household assets to buy food after participating in the UPSNP?	73.4	26.6
Have you avoided having to use your savings to buy food after the UPSNP?	72.3	27.66
Have you acquired any new household assets?	35.1	64.89
Have you acquired any new working culture, knowledge or skills after you became a member of the UPSNP?	77.65	20.2

access to different types of services or support from the UPSNP office. Out of this, approximately 23.2 percent got financial assistance through direct support, about 22.1 percent got financial assistance for medical care and about 3.2 percent got financial assistance for education. In comparison, about 2.1 percent of the respondents got another kind of support. Nonetheless, 11.6 percent of the respondents indicated that they did not get any form of support.

With regard to frequency of service and support from the UPSNP, about 20 percent got help daily, about 12.6 percent got support weekly, about 13.7 percent got support twice in a month, about 6.3 percent received support three times in a month, and about 47.4 percent got support monthly.

As shown in Table 13.4, beneficiaries receiving money from the UPSNP use it for different purposes. The survey results show that the beneficiaries used their stipends for a variety of purposes, such as for food consumption, school fees, house rental, health and other services and also for saving. Table 13.5 illustrates the frequency of payments and the purposes for which the money was used.

The food security status of the UPSNP beneficiaries was analysed from the FIES questionnaire items, and the results are shown in the following pie chart in Figure 13.2.

Based on the Household Food Insecurity Experience Scale, the households' food security status is categorised into three: mildly, moderately and severely food insecure. The survey results show that in Woreda 08, Arada sub-city

Table 13.4 Types of services extended to beneficiaries

Types of services	*Frequency*	*Percent*
Financial assistance through direct support	22	23.2
Financial assistance through conditional support	21	22.1
Counselling	28	29.5
Financial assistance for medical care	8	8.4
Financial assistance for education	3	3.2
Other	2	2.1
No support	11	11.6
Total	94	100

Table 13.5 UPSNP salary received and usage

Purpose of UPSNP Payments	*Frequency*	*Percent*
Food consumption only	68	70.2
Food consumption and school fees	3	3.2
Food consumption and renting a house	15	15.8
Health and other services	9	9.5
Food consumption and saving	1	1.1
Total	94	100

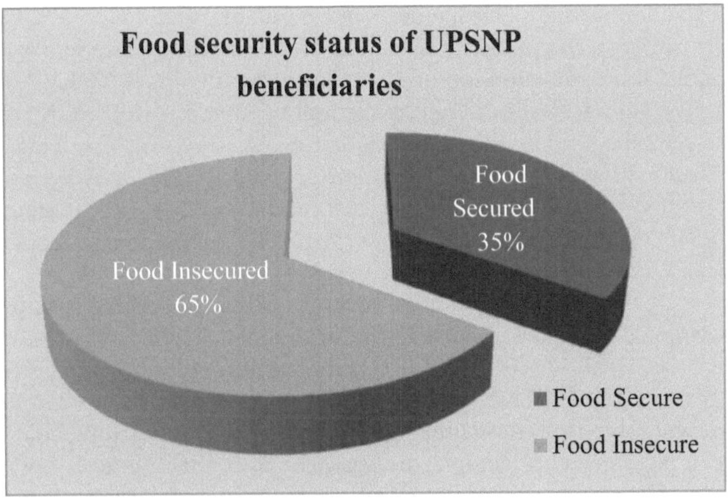

Figure 13.2 Food security status of UPSNP. *Source*: Analysed and illustrated by the researchers

about 35 percent of households were mildly food insecure, 34 percent were moderately food insecure and 31 percent were severely food insecure.

Abinary logit model was used to analyse the food security status of the households in Woreda 08, Arada sub-city. The dependent variable was the food security score, and the explanatory variables were factors that affect the food security status of the households reviewed from literature and experience. The results of the chi-squared statistic for the binary logit model for the food security status of households shows the goodness of fit of the model, that is, (LR chi2 (15) = 48.7; Prob > chi2 = 0.000) and rejects the null hypothesis that there is no relation between being mildly food insecure and explanatory variables.

Therefore, the following factors significantly influence the food security status of the UPSNP beneficiaries: Age, education, family size, number of economically active members of the household, number of rooms in the house, the use of savings to buy food, having a business, enrolled children at school, own or rented house, and free cash payment, among others. In order to see the contribution of each variable, we used the odds ratio to interpret the results of the binary logit model, where 2 represents no and 1 represents yes. The interpretation of the odds ratio does not follow the conventional interpretation.

According to the FIES model, household beneficiaries fall into three categories. These are mildly food insecure, moderately food insecure and severely food insecure. In order to analyse the food insecurity experience of the households, the food insecurity experience index was constructed using the three levels of food insecurity experiences, namely mildly food insecure, moderately food insecure and severely food insecure. Dependent variables, that is, food insecurity experience indices were analysed using the Tobit model, which accounts for the boundedness of the index value between 0 and 1.

For mildly food-insecure households, results of the chi-squared statistic for the Tobit model show the goodness of fit of the model, that is, (LR chi2 (7) = 30.13; Prob > chi2 =0.001) and rejects the null hypothesis that there is no relation between being mildly food insecure and explanatory variables. Thus, the model results show that being mildly food insecure is influenced by the socio-demographic variables of family size, education, number of economically active members of the household, own or rented house, number of rooms and finally, the UPSNP experience.

For moderately food-insecure households, results of the chi-squared statistic for the Tobit model show the goodness of fit of the model, that is, (LR chi2 (7) = 33.83; Prob > chi2 = 0.000) and rejects the null hypothesis that there is no relation between moderately food insecure and explanatory variables. Thus, the model result (Table 13.5) shows that moderately food-insecure households are influenced by the socio-demographic variables of sex, family size, own house, number of economically active members of the household, avoidance of selling household assets, how often households are getting service and support and finally, getting the house after the UPSNP.

The result of the chi-squared statistic for the Tobit model for severely food-insecure households shows the goodness of fit of the model, that is, (LR chi2

Table 13.6 The odds ratio of determinants of food security of UPNSP beneficiaries

Explanatory variables	*Odds ratio of food security*	
More food or diet	0.098**	(0.836)
Age	1.05*	(0.027)
Economically active	1.84*	(0.255)
Family size	0.50**	(0.256)
Avoided selling household assets	0.01**	(1.624)
Education	0.75**	(0.109)
Avoided using savings to buy food	7.44*	(1.064)
Currently running a business	0.22*	(0.839)
Enrolled children	3.47*	(0.567)
Own or rent house	0.34*	(0.521)
Number of rooms	5.17*	(0.665)
How often getting service and support	0.60*	(0.275)
Get service from the UPSNP	0.07*	(1.054)
Received free cash	1.67*	(0.298)
How much monthly rent	1.00*	(0.001)
N	94	
LR chi2(7)	48.7	
P < 0.0001	0.0000	
Log-likelihood	−36.54	

Standard errors in parentheses
*p< 0.05, **p< 0.01, ***p< 0.001

(6) = 28.00; Prob > chi2 = 0.000) and rejects the null hypothesis that there is no relation between severely food-insecure households and explanatory variables. Thus, the model result shows that severely food-insecure households are influenced by the demographic and asset variables of the number of economically active members of the household, number of households and lack of savings to buy food, among others.

Conclusion and recommendation

Food insecurity remains a significant challenge in Africa; this is mainly due to climate change, weather vagaries, soaring food prices, inadequate food production, political instability and conflicts. To ensure food security and mitigate these problems, African leaders need to protect the environment and mitigate the impact of human activities on the environment. It is also essential to take care of rural communities. Good practices must be identified, and examples of successful agricultural development practices should be publicised.

The results of the study vividly indicate that the government of Ethiopia played various roles in promoting the UPSNP by coordinating activities such as facilitating the allocation of resources from donors to beneficiaries and monitoring, evaluating and giving feedback on the performance of the project implementation to stakeholders. It is imperative that the government develops manuals, creates awareness, facilitates training, provides financial guidelines and assigns roles and responsibility to all governmental organisations to achieve positive outcomes.

Based on the FIES model, households' food security status has been categorised into three categories: Food secure, moderately food insecure and severely food insecure. The survey results showed that about 30.5 percent of households were found to be moderately food insecure, 47.4 percent were found to be severely food insecure, while 22.1 percent of households were found to be at accepted food security status. The binary logit models showed that the variables that have the greatest effect on the food security status of households are sex, age, marital status and education level in households.

The findings further show that more than half of the respondents had experienced the eight indicators of food insecurity on the FIESl to varying degrees. Based on the FIES, households' food security status has been categorised into the following three categories: Mild, moderate and severe. If the household at least responds positively to one of the eight items, the household was determined to be food insecure or otherwise food secure. Accordingly, about 65 percent of the households responded as food insecure, while 35 percent responded as a food secure.

The overall food insecurity experience was calculated by the average percent value of mild, moderate and severe food insecurity from analysis of the food insecurity experience scale. Thus, the results showed that about 35 percent of households were mildly food insecure, 34 percent were moderately food insecure and 31 percent were severely food insecure.

The determinants of food insecurity experience from the Tobit regression model show that female households are more likely to be moderately food insecure by 23.0 percent. Being educated increases the likelihood of being mildly food insecure, represented by 1.8 percent of the households. The household that owns a house is less likely to be moderately food insecure, represented by 32.5 percent. Households with a higher number of rooms are more likely to be mildly food insecure, represented by 24.1 percent, while households less likely to be severely food insecure are represented by 13.6 percent. Households that can afford to pay for rent are less likely to be severely food insecure than their counterparts. Having many economically active members of the household increases the likelihood of being mildly or moderately food insecure.

It is also indicated that the following challenges could hold back the proper implementation of the UPSNP. These are selection problems, of inclusion and exclusion, challenges in relation to the institutional arrangement, challenges of transparency and political intervention, lack of coordination among partners, lack of awareness about the project implementation manual, low and delayed financial payment problems and lack of support and service from partner organisations.

The following identified areas need to be developed. First, the government of Ethiopia should lay more emphasis on determinant factors that positively influence the food security status of households. The government should ensure that political willingness, appropriate policies and institutions are in place. A strategic partnership between the government and other entities is key to the successful implementation of the UPSNP.

When it comes to developmental partners, the gap in monitoring, evaluation and learning UPSNP objectives should adequately be dealt with. As such, monitoring, evaluation, accountability and learning (MEAL) mechanisms should be exercised. Partners can also help the support and service component of the UPSNP, mainly in creating awareness on the essence of the UPSNP for the beneficiaries and the community at large. Private companies and small and medium enterprises (SMEs) can contribute to the success of the UPSNP considering it as part of their social responsibility by providing financial support and donations when they pay their utility bills such as water and electricity. They can also provide apprenticeships and employment opportunities to beneficiaries of the UPSNP.

It is also established that the benefits of the UPSNP go beyond just improving the food security of its beneficiaries. Through this programme, its beneficiaries have had better access to health-care facilities and educational facilities for their children. These benefits have long-term effects, such as giving them better chances to improve their living standards, which constitutes the core goalof the right to development. Unfortunately, the benefits do not remain for long after the end of the programme. Adjusting these shortcomings could make a massive difference in the lives of beneficiaries of the programme and not just for the duration of the programme. Beneficiaries of the UPSNP are encouraged to understand the essence of the programme and make optimal use of it to enhance their livelihood status. This can be done through experience sharing and using the UPSNP to diversify their source of income through urban agriculture and direct linkages with rural producers and other activities.

Notes

1 Nagothu, Udaya and Flatin, Ingela. 2017. 'Food Security in the Context of Global Environmental and Economic Change'. In *Food Security and Development: Country Case Studies*, edited by Nagothu, Udaya. New York: Routledge; Kahsay, Shishay and Mulugeta, Messay. 2014. 'Determinants of Rural Household Food Insecurity in Laelay Maichew Woreda Tigray, Ethiopia'. *African Journal of Agriculture and Food Security* 2(1): 106–112.

2 Food and Agricaultural Organisation. 2017. 'The Future of Food and Agriculture—Trends and Challenges'. *Rome* <www.fao. org/3/a-i6583e.pdf>.

3 African Development Bank Group. 2018. 'African Economic Outlook'. *African Development Bank Group.*

4 Food and Agricaultural Organisation *et al.* 2017. 'The State of Food Security and Nutrition in the World in Brief: Building Resilience for Peace and for Food Security'. *Food and Agricultural Organisation.*

5 Federal Democratic Republic of Ethiopia—Ministry of Urban Development and Housing. 2016. 'Urban Food Security Strategy'. Addis Ababa: Ethiopia; The World Bank. 2015. *Project Appraisal Document on A Proposed Credit for FDRE for UPSNP.* Washington DC: The World Bank.

6 Tomazini, Carla. 2017. 'Bolsa Família and "Progresa/Oportunidades/Prospera": Consensual Reforms?' *International Policy Centre for Inclusive Growth One Pager 366.*

7 COMPACT. 2016. 'Ethiopia Ending Hunger and Under Nutrition: Challenges and Opportunities'. Scoping Report for Roundtable Discussion; Declaration on the Right to Development at 25. <www.un.org/en/events/righttodevelopment/>; UN Economic Commission for Africa and World Food Programme. 2010. 'The Cost of Hunger in Ethiopia Implications for the Growth and Transformation of Ethiopia'. Summary Report.

8 Transforming Our World: The 2030 Agenda for Sustainable Development Development. *Sustainable Development Knowledge Platform* <https://sustainabledevelopment.un.org/post2015/transformingourworld>.

9 AU Commission. 2015. 'Agenda 2063: The Africa We Want'. *African Union.*

10 Kothari, C.R. 2004. *Research Methodology: Methods and Techniques* (2nd rev edn). New Delhi: New Age International.

11 Yamane, Taro. 1967. *Statistics: An Introductory Analysis* (2nd edn). San Francisco: Harper and Row.

12 Roosevelt, Franklin. 1941. 'Excerpted from the State of the Union Address to the Congress'. 6 January.

13 White, Elwyn *et al.* 1942. *The United Nations Fight for the Four Freedoms.* Washington DC: National Government Publication.

14 DRTD at 25, note 7.

15 Constitution of the Federal Democratic Republic of Ethiopia Negarit Gazeta 1st Year, No 1, 1995, art 43(1).

16 Tura, Husen. 2019. 'Achieving Zero Hunger: Implementing a Human Rights Approach to Food Security in Ethiopia'. *Third Wolrd Quarterly* 40(9): 1613–1633.

17 Nagothu and Flatin, note 1.

18 United Nations. 2010. *Natural Hazards, Unnatural Disasters: The Economics of Effective Prevention.* Washington DC: World Bank Publication.

19 Federal Democractic Republic of Ethiopia, note 5; The World Bank, note 5.

20 Food and Agricultural Organisation *et al*, note 4.
21 Productive Safety Net Programme. 2014. 'Ethiopian Productive Safety Nets Project: Programme Implementation Manual'. <http://projects.worldbank.org/P146883?lang=en>.
22 The World Bank, note 5.
23 FUJCFSA. 2018. 'Urban Productive Safety Net Programme Joint Review and Supervision Aide Memoire'. Federal Democratic Republic of Ethiopia—Ministry of Urban Development and Housing Urban Job Creation and Food Security Agency, Addis Ababa.
24 *Ibid.*

References

African Development Bank Group. 2018. 'African Economic Outlook'. African Development Bank Group.

Federal Democratic Republic of Ethiopia—Ministry of Urban Development and Housing. 2016. 'Urban Food Security Strategy'. Addis Ababa: Ethiopia.

Food and Agricultural Organisation *et al.* 2017. 'The State of Food Security and Nutrition in the World in Brief: Building Resilience for Peace and for Food Security'. Food and Agricultural Organisation.

FUJCFSA. 2018. 'Urban Productive Safety Net Programme Joint Review and Supervision Aide Memoire'. Federal Democratic Republic of Ethiopia—Ministry of Urban Development and Housing Urban Job Creation and Food Security Agency, Addis Ababa.

Kahsay, Shishay and Mulugeta, Messay. 2014. 'Determinants of Rural Household Food Insecurity in Laelay Maichew Woreda Tigray, Ethiopia'. *African Journal of Agriculture and Food Security* 2(1): 106–112.

Kothari, C.R. 2004. *Research Methodology: Methods and Techniques* (2nd rev edn). New Delhi: New Age International.

Nagothu, Udaya and Flatin, Ingela. 2017. 'Food Security in the Context of Global Environmental and Economic Change'. In *Food Security and Development: Country Case Studies*, edited by Nagothu, Udaya. New York: Routledge.

The World Bank. 2015. *Project Appraisal Document on a Proposed Credit for FDRE for UPSNP.* Washington DC: The World Bank.

Tomazini, Carla. 2017. 'Bolsa Família and "Progresa/Oportunidades/Prospera": Consensual Reforms?' *International Policy Centre for Inclusive Growth One Pager 366.*

Tura, Husen. 2019. 'Achieving Zero Hunger: Implementing a Human Rights Approach to Food Security in Ethiopia'. *Third World Quarterly* 40(9): 1613–1633.

United Nations. 2010. *Natural Hazards, Unnatural Disasters: The Economics of Effective Prevention.* Washington DC: World Bank Publication.

Yamane, Taro. 1967. *Statistics: An Introductory Analysis* (2nd edn). San Francisco: Harper and Row.

14 Corporate social responsibility and the right to development in local communities

The case of the Democratic Republic of Congo

Muyamba Mangu and Serges Djoyou Kamga

Introduction

Originally, the human in rights discourse seeks to protect the human person against the might of states that are the main duty bearers of human rights. Yet in recent times, transnational companies (TNCs) have become extremely powerful, especially in the developing world where they operate. Because of the economic weight they pull around, sometimes more than most developing countries, these companies now pose a greater threat to human rights in general[1] and the right to development (RTD) in particular. To shield people from the threat posed by TNCs, the international community has developed numerous instruments including the Organisation for Economic Cooperation and Development (OECD) Guidelines for Multinational Enterprises (1976), the International Labour Organisation (ILO) Tripartite Declaration of Principles Concerning Multinational Enterprises and Social Policy Reform (1977), the United Nations (UN) Global Compact (2000), the UN Norms on the Responsibilities of Transnational Corporations and Other Business Enterprises with Regard to Human Rights (2003) as well as the UN 'Protect, Respect and Remedy' Framework (2011) (Ruggie Framework).

Not only are these principles highly controversial, with many arguing that they cannot shield people from businesses,[2] they remain in the sphere of soft law with a mere moral value, often disregarded by businesses seeking to make money. Meanwhile, these companies often rely on corporate social responsibility (CSR) to advance human rights. CSR can be defined as initiatives undertaken by corporate businesses intended to benefit the community. It comprises the extent to which companies promote human rights, democracy, community improvement and sustainable development objectives throughout the world.[3] Against this backdrop, this chapter examines the extent to which extractive companies realise the RTD through CSR in the Democratic Republic of Congo (DRC).

The term 'extractive corporates' is used to describe the 'processes that involve different activities that lead to the extraction of raw materials from the earth (such as oil, metals, mineral and aggregates), and their processing and utilisation by consumers'.[4] The focus is on extractive companies in the DRC, which is one of the wealthiest countries in the world in terms of mineral resources and as such attracts several TNCs. Secondly, the discussion focuses on the RTD because it is a composite right embodying civil and political rights as well as socio-economic rights. In addition, this right is binding in the African human rights system and enshrined in the Congolese Constitution. Ultimately, the chapter explores prospects for realising the RTD through CSR, necessitating a deconstruction of CSR.

Deconstructing corporate social responsibility

According to Blowfield and Murray, much has been written and is being written about CSR with intensifying attention being given to the extractive industry.[5] Besides, there is still a lack of consensus on the definition of the term CSR in the current global era where the issues relating to equality, human rights and sustainable development are used to measure and to safeguard global security and peace in terms of economic, social and environmental protection.[6] According to Hamann and Kapelus, there is an agreement in general that companies must contribute to society.[7] Nevertheless, the notion of CSR is especially contentious concerning the impact, implementation and measurement of the concept.

CSR refers to the business as well as initiatives that benefit the community. These initiatives fall into three categories habitually called the triple bottom line that comprises social, environmental and economic aspects, also known as the three pillars of sustainable development that play a major role in the construction of the concept of CSR, which largely contributes to the controversies associated with the concept. As defined by Blowfield and Murray,[8] CSR is a concept by which companies integrate social and environmental concerns in their activities and interaction with stakeholders on a voluntary basis. According to the World Business Council for Sustainable Development,[9] CSR is the ethical behaviour of a company towards society. It requires companies to acknowledge that they should publicly account not only for their financial performance but also for their social and environmental impact to show how it makes a positive contribution to society. It comprises the extent to which companies should promote human rights, democracy, community improvement and sustainable development objectives throughout the world.[10]

It is true that several companies do not take CSR earnestly. They pretend to do so when it comes to environmental aspects or what is referred to as greenwashing. Nevertheless, since people are becoming aware and more educated on social issues like global warming and unethical labour practices for instance, they are now asking about policies that benefit the entire community instead of just the business. Fundamentally, there is a public value and interest in promoting CSR. Progressively, companies are involved directly in production and service delivery that impacts directly on individual and community welfare; this includes employment, health and safety, water, housing, food, education and environmental well-beingand sustainability. The UN Sub-Commission on the Promotion and Protection of Human Rights acknowledges that companies and other businesses have the capacity to foster economic well-being and development and have a capacity to cause harmful impacts on human rights and the lives of individuals through their core business practices and operations including employment, environmental policies, relations between suppliers and consumers, interactions with governments and other activities.[11]

Recognising these impacts and spheres of activities and influence, particularly as they pertain to the realisation of fundamental human rights, there is a strong public interest in, and value associated with, the conduct of business and corporate affairs to influence positively not only relevant financial interests, but also on the realisation of the RTD in local communities.[12] Corporations have the potential to contribute to the advancement of human rights but have often rather violated these rights instead. The voluntary character of CSR cannot possibly advance the realisation of the RTD for local communities, which arguably is only possible if CSR, governments, non-governmental organisations and civil society play a key role.

Extractive corporates as agents of the RTD in the DRC

Before focusing on the prospects of achieving the RTD through CSR, it is important to examine the normative and policy landscape on the RTD in the DRC.

Normative and policy framework on the RTD in the DRC

It is important first and foremost to note that the DRC is party to the African Charter on Human Peoples' Rights, which provides for the RTD in article 22. To give domestic effect to this treaty obligation, the RTD has explicitly been enshrined in the Constitution of the DRC. Article 58 reads: 'All the Congolese have the right to enjoy national wealth' and the 'State has the duty to redistribute the wealth equitably and to safeguard the right to development'. Not only does the Constitution provide for the RTD, its realisation is clearly linked to the enjoyment of the national wealth by Congolese people. In addition, article 53 provides that all 'persons have the right to a healthy environment that is favourable to their development'. This also is key to the realisation of the RTD, especially in the context where mining by extractive companies is likely to pollute the environment and jeopardise the enjoyment of the RTD.

Recognising the importance of the mining sector in the country, the government of the DRC has taken major legislative measures in that sector to protect communities from the might of extractive companies. Among others, the government adopted Law No 007/2002 on 11 July 2002 called the Mining Code, which is supported by the Decree No 028/2003 adopted in March 2003. The Mining Code provides the legal framework for regulating the environmental practices of mining companies in the country. The Code requires companies to submit environmental impact assessments and management plans drawn with the participation of the community. Implementation of the Mining Code has, however, not been successful, especially because mining companies seemed to be more powerful than the government of the DRC, coupled with institutionalised corruption at all levels in the country and the fragility of the state.

As a result, the government proceeded to adopt a new Mining Code, which was promulgated by the President of the DRC on 09 March 2018, to amend Act No 007/2002 of 11 July 2002. The new Code underlines the CSR of business and especially, on taxes and administrative aspects. It makes provision for increments in royalties on mineral resources. In the same vein, article 258(b) on endowment for contribution to community development projects stipulates that the holder of a mining right or authorisation to exploit permanent quarries is obliged to constitute, free of the tax on profits, an endowment for contribution to community development projects whose minimal amount is equal to 0.3 percent of the turnover for the year in which it is incorporated. The article stipulates that with regard to compensation of the occupants of the land in the case of displacement of the population, the operator must first proceed to the compensation and resettlement of the populations concerned.

Mining sector and CSR obligations

This section explores the extent to which some key companies deliver CSR and examines the extent to which they contribute to the realisation of the RTD in the DRC, which remains very attractive to mining companies. Among others, *la Société Générale des Mines* (SGM), an American group with mines in the United States, South America and Asia was a copper cobalt company owned by the Ramu Nickel Foundation.[13] This company is important as its mining concessions cover around 1600 km^2, including 43 villages and two cities. As far as CSR per se is concerned, the social responsibility of mining companies is entrenched in domestic and international norms. The DRC Mining Code enshrines the requirement that any applicant for an operating permit must attach to their request the plan for the contribution in the project for the development of the neighbouring communities.[14] This makes it possible for the government and the community to assess the contribution of the mining project in the sustainable development of the community where the business operates.

There have been reports that the SGM invests only 0.3 percent of its revenues in community development, education and health care.[15] Nonetheless, on a positive note, during the establishment of the company, communities were consulted and compensated appropriately, including in cases of relocation.[16] From the same perspective, schools were built, and school buses offered to the communities. Moreover, in May 2009 the SGM signed an agreement to fight HIV and malaria and to foster mother–child health and to provide safe drinking water and health infrastructures. These are great initiatives to advance the RTD in the local communities.

Similar to the GSM, *la Générale des Carrières et des Mines* (GÉCAMINES) is another important extractive company that plays an active role through its CSR initiatives. Its investments are mostly in health care, education, family promotion and the environment. In terms of health care,[17] the company offers its workers a good health system. The GÉCAMINES health care system is an exceptional network of medical formations extending 300km on the Lubumbashi-Kolwezi axis. To date, the health care system of GÉCAMINES remains a resort of choice and a place of hope for patients in the province of Katanga. In terms of education, the main motivation for the creation of schools by GÉCAMINES is to stabilise their agents to make production profitable, by protecting them from the major concern around schooling and to guarantee the education of their children and the local communities.

In terms of housing, GÉCAMINES has subsequently built houses for workers who, in addition, received free water and sanitation services and electricity. As part of its social responsibility policy, GÉCAMINES sells to its workers the houses in which they have been accommodated in the cities of Lubumbashi, Likasi and Kolwezi as well as in the localities of Kipushi, Kambove, Kakanda, Fungurume and Luena. In terms of the environment, since the entry into force of the Mining Code of 2002 and its implementation measures contained in the 2003 Mining Regulations, GÉCAMINES, as well as other mining companies refer to these two texts for all aspects relating to the protection of the environment. These are indeed great initiatives for the realisation of the RTD.

The other example that comes to mind is the case of China Molybdenum Co Ltd (CMOC). This company is engaged in the mining and processing, smelting and deep processing of base and rare metals like molybdenum, tungsten, copper and niobium and possesses a complete industry chain which incorporates upstream and downstream procedures, including mining, processing, broiling and chemical products. The CSR of the CMOC is based on environmental social impact assessment, employees and community development. In terms of the environment, the CMOC develops its business operations in harmony with the environment by maintaining a mature environmental management system (EMS) and protecting the biodiversity, promoting the recycling of resources and minimising emission and waste generation.[18] In terms of employees, the CMOC gives more importance to the safety, health and career development of its employees. An active social investment policy covers many areas such as livelihoods, health care, education, agriculture infrastructure and resident housing. The company upholds the values of sustainable development in accordance with its environmental social impact assessment policy; and continues to develop the initiative in a way that promotes sustainable development for the project and neighbouring communities.

Notwithstanding the efforts put into CSR by the CMOC, community members have not been impressed, as they still do not feel satisfied with the level of investment by the company in the communities. This raises questions on the nature of its CSR. This notwithstanding, some TNCs like Glencore do very little in terms of CSR, yet is one of the biggest role players in the DRC where it owns two companies: Mutanda Mining (MUMI)—a copper and cobalt open pit mine—and Kamoto Copper Company (KCC), which is a group of copper and cobalt mines. Both companies are located near the city of Kolwezi, a region rich in copper and cobalt reserves.[19] The KCC and MUMI hold some of the largest copper and cobalt reserves in the world, making Glencore one of the world's largest producers of copper and cobalt.[20] Glencore's turnover according to the report published by three international non-governmental organisations, namely Bread for All and Fastenopfer with Rights and Accountability in Development (RAID) is more than 30 times the DRC's national budget, a scandalously poor country despite the wealth from its natural resources.[21]

According to Katanga Mining Limited Annual Information, it is recorded that the company owns two investments in Katanga Province;[22] 75.2 percent of Katanga Mining Limited (Yukon, Canada), which in turn owns 75 percent of Kamoto Copper Company. By virtue of a special agreement, Glencore buys 100 percent of KCC's cooper and cobalt production, 69 percent of the Mutanda group as a result of a merger between Mutanda Mining and Kansuki.

As far as CSR is concerned, Glencore's policies regarding CSR were few and underdeveloped as illustrated in the 2011 comparative survey indicating that Glencore CSR is still rudimentary and even inexistent but was positioned as the first player in the mining sector because of its reporting regarding environmental, social and governance matters.[23] In the same vein, Bread for All and Fastenopfer, quoted by Chantal Peyer *et al* note that 'Glencore corporate practice was difficult to ascertain whether it was genuinely committed to respecting the human rights and environmental standards set out in a range of international treaties and instruments'.[24] As an alternative, it seems to be understood that Glencore had described its own framework by using vague words and imprecise concepts. The reference to human rights in the report was narrowed down to restrictions in terms of forced labour, child labour and some labour rights, whereas socio-economic rights such as the right to water, the right to food and other concepts such as free, prior and informed consent were noticeably absent.[25]

Environmental issues and the right to food

Pain pour le Prochain and *Action de Carême* identify KCC and MUMI as agents that pollute the environment in the villages of Moloka, Kaindu and Tshamundenda.[26] These mining companies do so by diverting the residues from their exploitation towards the villages, which contributes to damaging the crops and the fields. The KCC and MUMI did not react to farmers' complaints and denied having caused pollution in the fields. The intentional or unintentional spills clearly have a negative impact on the right to food of the people who live around the KCC and MUMI mining sites.

The right to health

In the health sector, Glencore supports polio and measles immunisation campaigns for more than 30,000 children in local communities as well as the DRC's national HIV/AIDS programme. It also implemented a control programme to reduce the impact of malaria on its employees and in local communities. Accordingly, almost 20,000 homes were sprayed to protect more than 115,000 people.[27] However, these activities are qualified as trompe-l'oeil, given the gravity of the situation on the ground.[28] It has been affirmed that the KCC and MUMI are principal causes of health problems, especially because during the dry season, the inhabitants of Musoni and Luilu (cities around the mine sites) suffer from the dust that comes from the passage of their trucks and those of other companies as well as private vehicles.[29]

The right to clean water

For several years, different companies such as the KCC have stunned the river Luilu and this situation has been considered as a blockade to the realisation of the right to clean portable water. Despite the fact that the KCC supports the construction of substations to improve access to clean water, a large part of the inhabitants of the city of Luilu will not be able to benefit from the initiative.

The right to education

According to *Pain pour le Prochain* and *Action de Careme*, Glencore has contributed to the construction, rehabilitation and equipment of local schools. Looking at Glencore's resources, this is extremely insignificant. Even though Glencore is making efforts to spend more money on major infrastructures such as roads, bridges, renovation of the Kolwezi airport, the central problem of Glencore's community work is the absence of a rights-based approach. Similar to Glencore, Chemical of Africa (CHEMAF)—another giant company in the extractive sector in the DRC—does not comply with CSR regulations. In 2001, the Minister of the Environment expressed concerns over the environmental impact of CHEMAF's activities in the community and demanded to not allow the company to operate for reasons that its activities would be detrimental to the lives and health of local communities. In general, CHEMAF's activities violate human rights as the company does not integrate local concerns in its operations, which leads to recurring forced displacement of the local populations from their lands.

In the same register, while Rwashi Mining builds schools and other infrastructures in the community to tick its CSR box, populations are often forcibly removed from their lands to give space to the company's mining operations.[30] It is worth reiterating that mining creates pollution. Mining waste is dumped in the Lwashi Creeks and thus pollutes the fields of local farmers and handicaps the growth of their plants. The polluted Lwashi Creek joins the Lwano River which supplies water to several villages and this has an impact on the cultivation site that is located in the mining sites of Rwashi Mining.[31] This situation raises questions on the potential of CSR towards the socio-economic and cultural development of local communities. On 14 October 2004, there was insurrection in Kilwa. To protect its interests, Anvil Mining assisted the police with weapons and equipment, including aeroplanes to help crush the rebellion.

On behalf of victims, the Institute for Human Rights and Development in Africa and others, after exhausting local remedies without satisfaction, took the matter to the African Commission on Human and Peoples' Rights.[32] Not only did the Commission find the DRC to have violated numerous provisions of the African Charter including the RTD (article 22), it also condemned Anvil Mining and emphasised the need and legal imperative for entities engaged in extractive industries to undertake their activities with due regard to the rights of host communities. Although some companies commit to discharge their CSR obligations, it remains a voluntary mechanism, which begs the question about its potential to contribute to the realisation of the RTD.

Realisation of the RTD through CSR—some challenges

The benefits of CSR cannot be ignored. However, as indicated above, CSR is a voluntary mechanism and as such becomes undependable for its potential to contribute to the realisation of the RTD. It is therefore imperative to change the approach to CSR because it is known to be a mere voluntary commitment by companies,[33] which usually give back to communities in order to improve their reputation or to be altruistic without genuine commitment to human rights. While there is no universally accepted definition of the concept of CSR, it tends to be characterised as the continuing commitment of business to behave ethically and contribute to economic development while improving quality of life of workers and their families and of the local communities and society at large. This attitude according to Kamga and Ajoku[34] portrays CSR as an unnecessary expense with little-or-no financial or other benefits to companies.

In spite of this, spending on CSR is an investment for the sustainability of the enterprise. For example, investing in schools, health care facilities, roads and bridges and other infrastructures for the benefits of workers and the community goes a long way in developing the capacity and sustainability and promoting the brand name of the company. As a result, the company improves its credibility and attracts more investors, clients and workers. Considering its voluntary features, CSR needs to be transformed if it is to contribute to the achievement of the RTD.

CSR as atool for the realisation of the RTD

The transformation envisaged in this context entails turning CSR into binding human rights obligations, educating companies on the business attributes of CSR and holding states accountable for violation of human rights by companies.

Turning CSR into binding human rights obligations

This raises the question of how to compel TNCs to comply with their CSR obligations. As stated earlier, a starting point could be to look at the preamble of the 1948 Universal Declaration of Human Rights (UDHR), which obliges every organ of society to play its part in the promotion and protection of human rights.[35] Although not binding, this provision calls on individual and all entities of society or organs of society, including the private, public or TNCs operating within a state to ensure respect for human rights. The basic requirements of the UN Declaration on the Right to Development (UNDRTD) are summarised as follows: put people at the centre of development; ensure their free, active and meaningful participation; ensure non-discrimination; equitable distribution of the benefits of development; respect self-determination and sovereignty over natural resources and all this in a process that advances other civil, political, economic, social and cultural rights.[36]

In a country like South Africa, various laws and policies have been adopted to protect communities from abuse by extractive companies. The country recognises in the memoranda of association of extractive companies that they are bound by the Bill of Rights and are responsible for their realisation to the extent that they bear responsibility for them.[37] The 1996 Constitution of South Africa, the Mining Charter of 2018[38] which derives from the Mineral and Petroleum Resources Development Act of 2002 and the implementation Guidelines for the Mining Charter 2018 require the mining industry to carry out human resources development with 10 percent on essential skills development activities, and the development of mining community with 100 percent approved social and labour plan.

In the context of Australia, the Government's Corporations and Market Advisory Committee was asked to consider and report on several issues regarding corporate responsibility. In the same year, Federal Government's Parliament Joint Committee on Corporations and Financial Services, conducted an inquiry into corporate responsibility. These inquiries focused on the regulatory regime and the extent to which the Australian law (The Corporation Act 2001) requires corporate decision makers to have regard for the interests of stakeholders other than shareholders and the broader community.[39]

In its 2006 report, the Parliamentary Joint Committee on Corporations and Financial Services (PJCFS) agreed that most corporations' desire to escape regulation should voluntarily allow business to 'improve responsible corporate performance' and thus the current legal structure, which relies on businesses to adopt an educated self-interest approach to incorporate notions of human rights and corporate responsibility into their operations, is largely adequate.[40] In line with this approach, the PJCFS recommended the creation of a corporate responsibility network as a means for the industry to increase the level of corporate responsibility performance in the country. The increase of cooperation in the sector is a positive step but the voluntarism has its limits especially when the initiative remains in the business domain without any public transparency.

Corporate social responsibility in India has traditionally been viewed as a philanthropic activity.[41] It is regulated by clause 135 of the Companies Act of 2013,[42] passed by both Houses of Parliament, and consented to by the President in the same year. The Act introduces several new provisions which change the face of Indian corporate business. In this context, the concept of CSR rests on the ideology of give and take. By performing the task of CSR activities, the companies give something back to society. In this context, the results of the KPMG India CSR 2015 report on its corporate citizenship survey reveals that most companies have an annual report in accordance with company law.[43] Hence, several companies have begun understanding that it is reasonable to take up CSR activities and incorporate them into their business procedures.

The fundamental goal of CSR nowadays is to maximise a company's overall impact on society and stakeholders. Corporations are becoming increasingly aware of their role towards society. They are responsible bodies that feel a sense of duty towards the common welfare and the environment. India is the first country in the world to make CSR mandatory, after a change to the Company Act in April 2014. Businesses can invest their profits in areas such as education, poverty, gender equality and hunger.[44]

Educating companies on the business attributes of CSR

CSR is a business responsibility that is gaining popularity in the 21st century. It aims to put in place a self-regulatory mechanism that allows companies to control and guarantee effective compliance with the spirit of the law, international norms and ethical standards. It is a managerial commitment to make a move that secures and improves both the welfare of society and the organisation. CSR is then an ongoing commitment of business to act morally and contribute to economic development while improving the quality of life of workers and their families alongside the local communities and society at large.

Kaur defines CSR as a business approach that contributes to sustainable development by delivering economic, social and environmental benefits to stakeholders and local communities.[45] It means that CSR is a corporation's initiatives in assessing and taking responsibility of the company's effects on environmental and social well-being. Companies that value CSR stand the chance of having many advantages such as enhanced business reputation, boosted relationships,public trust is built, sustainability to remain in business for as long as possible, the deterence of government regulations, promotion of long-term profit for business, enhanced employee relationships and production innovation and it can also foster a strong relationship with communities through stakeholder engagement.

The above advantages were considered by Ksiezak who declared that CSR undeniably provides multiple benefits to a company.[46] These benefits are 'building a brand image, making profits, attracting investors, employee motivation, licence to operate, avoiding legal actions, risk mitigation, desire to do good, improve quality, customer loyalty, good relations with stakeholders, cost reduction and differentiation from other brands'.[47] The violation of human rights by TNCs often happens directly in their operations as private actors, in collusion with the state, a state entity or other public actors, or as a partaker in joint venture or complicit relations with another human rights violator.[48] The argument about TNCs' responsibilities to remain accountable for their actions regarding human rights does not in any manner dismiss the state's primary responsibility to protect and enforce human rights within their boundaries.

Safeguarding human rights is increasingly becoming a vital issue for the business operations of TNCs. Besides the self-evident obligation on every TNC to act in a philanthropic and ethical manner, there are other economic reasons for prioritising human rights. The reinforcement of political freedoms promotes economic development for example, by improving investment possibilities and free movement or training opportunities. On the contrary, the encroachment of human rights and the absence of the rule of law are immensely detrimental to economic activities and therefore hamper companies and the operations of TNCs. Hence, TNCs are morally and for economic reasons obligated to promote and respect human rights and thus, contribute to their recognition by applying then.

The Australian government through its Parliamentary Secretary to the Treasury asked the Governor of the Advisory Committee to revise the Corporation Act of 2001 to include CSR as a means to protect the best interests of stakeholders or the broader community in making decisions about CSR. Australian law require corporations to comply with human rights and to ensure that the standards formulated are in line with human rights. In addition to the human rights standards which Australian companies are legally obliged to comply with under domestic law, there is a broader range of human rights that are relevant to corporate activities.

Depending on the specific activities carried out by a corporation, these might include the right to liberty and security of the person, the right to an adequate standard of living (including adequate food, water, shelter and clothing), the right to education, the right to health, and other economic, social and cultural rights, the right to freedom of thought, conscience and religion, the right to freedom of expression and the rights of indigenous peoples. A number of Australian companies incorporate into their CSR policies the human rights standards they are obligated to comply with under Australian law. Some of the companies go beyond domestic legal requirements and participate in voluntary initiatives on CSR and human rights, including the UN Global Compact and the Global Reporting Initiative.

State accountability for human rights violation by business

International human rights law sets down obligations, which states are compelled to comply with. By becoming parties to international treaties, states take responsibility and obligations under international law to respect, to protect and to fulfil human rights. This means that states must abstain from meddling with or restraining the enjoyment of human rights. Through the ratification of these treaties, the DRC government undertakes to put domestic measures and other legislation in compatibility with its treaty obligations and duties. In the DRC, the exploration and development phases of extractive companies are often characterised by a lack of consultation with the local communities.

Consequently, extractive companies violate the right to participation for local communities. It is similar when it comes to the operation phase where extractive corporates violate the right to land ownership through expropriation, displacement and the disruption of livelihood, including the right to food as the land which constitutes the source of food production is often taken without compensation. Besides the right to land and the right to food, extractive corporations also violate other human rights such as the right to a safe environment, the right to health and the right to life. During mining, mineral wastes are thrown into the atmosphere and in rivers. The resultant pollution impacts on the environment, causing ecological disasters and a host of other negative consequences that compromise health standards and loss of livelihood for the local communities and thus, also create a human rights crisis.

The right to work is guaranteed in the Constitution of the DRC.[49] The labour market is another area in the extractives industry wherein it is observed that workers do not enjoy good working conditions. According to RAID, workers in mining companies in the DRC are exposed to the harmful dust of the minerals they handle, such as radioactive ores, and also, most of the workers are temporarily employed and not covered by any insurance policies.[50] Accordingly, states have the primary obligation to ensure that companies operating within their territorial borders conform to universal human rights standards. The Constitution of the DRC provides in article 16 that the human person is sacred and, thus, imposes an obligation on the state to respect and protect them. The right to property is equally recognised as a sacred entitlement and the state is obligated to guarantee to every individual or group to acquire property in conformity with the law or custom.[51]

States are the prime bearers of rights and duties under international law and are at the same time the primary actors in international relations, which allows them the central function, for example, of proposing and disposing of agreements. Most significantly, perhaps, multinationals themselves owe their influential position to the (non) regulatory approach that states adopt. Currently, states are neither obliged under international human rights law to regulate the extraterritorial activities of businesses domiciled in their territories or jurisdictions, nor are forbidden from doing so, which however, must follow the jurisdictional basis. Nevertheless, within these parameters some human rights treaty bodies recommend that home states take steps to prevent abuse abroad by business enterprises within their jurisdiction.

There are strong policy reasons for home states to set out clearly the expectations for businesses to respect human rights abroad, especially where the state itself is involved in support. The primary obligation of the state to protect the human rights of its people is strengthened by various human rights instruments. General Comment 31 of the Human Rights Committee stipulates that

> the positive obligations on State Parties to ensure that provisions of the Covenant will only be discharged if individuals are protected by the State, not against violations of covenants rights by its agents, but also against acts committed by private persons or entities.[52]

The Commission is clear in affirming that states could be considered to have breached their commitments if they neglect to take suitable measures or to practice due diligence to prevent, punish, investigate and redress the harm caused by the actions of private persons or entities.

The duty of promoting the rule of law and human rights in a given country lies with the host government and this duty can only be discharged where stable structures and strong political will are a reality. In the case of the DRC where the government faces difficulties in controlling some areas of its territory because of the war, the country can only expect support from the international community to hold the TNCs operating in the country accountable. The international community should consolidate the existing capacity of the DRC toregulate and adjudicate harmful actions by TNCs. Majinge argues that at the international level, states are competing to get mineral resources from weak countries, which constitutes a significant challenge that undermines the political will and credibility to address human rights violations.[53] The only credible alternative is an internationally binding legal framework that will impose rights and duties on TNCs under international law. An internationally binding treaty will provide a genuine and effective oversight role and considerably constrain TNCs and compel them to start adopting comprehensive strategies to address human rights violations and support rule of law initiatives in the regions where they operate.

Conclusion

The chapter has explored the extent to which CSR by extractive companies could provide a mechanism for the realisation of the RTD in the DRC. While CSR has a great potential to achieve the RTD, its voluntary nature is a matter of concern in the sense that companies only engage in social responsibility as a moral obligation and are therefore generally not obliged or compelled to do so. It is argued that to remedy this problem, it is imperative to transform the operational modalities of CSR in a manner that companies can be able to incur human rights obligations. It entails also educating companies on the value addition and benefits of CSR to their business operations and holding states accountable for the violation of human rights committed by companies when they later fail to comply with human rights standards in their operations.

Notes

1 Hazemberg, Jilles. 2016. 'Transnational Corporations and Human Rights Duties: Perfect and Imperfect'. *Human Rights Review* 17(4): 479.
2 Suraya, Deva. 2013. 'Treating Human Rights Lightly: A Critique of the Guiding Principles Complicity in Undermining the Human Rights Obligations of Companies'. In *Human Rights Obligations of Business: Beyond the Corporate Responsibility to Respect?,* edited by Suraya, Deva and Bilchitz, David. Cambridge: Cambridge University Press, 78–104; López, Carlos. 2013. 'The Ruggie Process: From Legal Obligations to Corporate Social Responsibility'. In *Human Rights Obligations of Business: Beyond the Corporate Responsibility to Respect?,* edited by Suraya, Deva and Bilchitz, David. Cambridge: Cambridge University Press.
3 Blowfield, Michael and Murray, Alain. 2014. *Corporate Responsibility*. Oxford University Press, 8.
4 Sigam, Claudine and Garcia, Leonard. 2012. *Extractive Industries: Optimizing Value Retention in Host Countries*. New York/Geneva: UNCTAD.
5 Blowfield and Murray, note 3.
6 Cronjé, Freek; Reyneke, Suzanne and Chenga, Charity. 2017. 'Corporate Social Responsibility in the Zambian Mining Sector: An Overview of Three Distinctive Operational Eras'. *Koers-Bulletin for Christian Scholarship* 82(1): 2.
7 Hamann, Ralph and Kapelus, Paul. 2004. 'Corporate Social Responsibility in Mining in Southern Africa: Fair Accountability or Just Green Wash?' *Society for International Development* 47(3): 85.
8 Blowfield and Murray, note 3.
9 Moir, Lance. 2001. 'What do we Mean by Corporate Social Responsibility?' *Corporate Governance International Journal of Business in Society* 1(2): 16–22.
10 Blowfield and Murray, note 3.
11 UN Sub-Commission on the Promotion and Protection of Human Rights. 2005. 'Report of the United Nations High Commissioner on Human Rights on the Responsibilities of Transnational Corporations and Related Business Enterprises with Regard to Human Rights'. [24] UN Doc E/CN.4/2005/91.
12 Lynch, Philip. 2006. 'Human Rights and Corporate Social Responsibility: Submission to the Corporations and Markets Advisory Committee Inquiry into Corporate Social Responsibility'. *Human Rights Law Resource Centre.*
13 Mària, Josep and Devuyst, Emmanuelle. 2011. 'CSR and Development: A Mining Company in Africa'. *Journal of Management Development* 30(10): 955–967.
14 Action against Impunity for Human Right. 2010. 'Chinese Private and Public Investments in the Mining Sector in Katanga: Good Governance and Human Rights'. Report on the Violation of the Rights of Workers in the Chinese Companies in Katanga and the Imbalance in the DRC-China Consortium Agreement: Companies in View: Jiaxing Mining, Congo Dong Fang International Mining and Lida Mining, 23.
15 Mària and Devuyst, note 13.
16 *Ibid.*
17 Gécamines. 'Medical Services'. <www.gecamines.cd/medical.html>.
18 China Molybdenum. 2017. 'Environmental, Social and Governance Report'. *Environmental, Social and Governance Report* <www.chinamoly.com/06invest/DOC/2019/E 03993 E036.pdf>.
19 Peyer, Chantal and Mercier, Francois. 2012. 'Glencore in the Democratic Republic of Congo: Profit before Human Rights and the Environment'. *Bread for All and Swiss Catholic Lenten Fund* 6.
20 *Ibid*, 7.
21 *Ibid.*
22 Katanga Mining Limited. 2013. 'Annual Information Form for Year Ended'. 7.
23 Peyer and Mercier, note 19, 24.

24 *Ibid.*
25 *Ibid.*
26 *Pain Pour le Prochain and Action de Carême.* 2018. '*Glencore en RD Congo: Une Diligence Raisonnable Incomplète*'. *Rapport de Pain Pour le Prochain et Action de Carême, Ecrit en Collaboration avec les Organisations Partenaires en RDC, l'Observance Africaine des Resources Naturelles (AFREWATCH) et le Centre d'Aide Juridico-Judiciaire (CAJJ)* 3–5.
27 *Ibid*, 25.
28 Interview with a Judge in Lubumbashi 22 December 2019.
29 Pain pour le prochain and Action de Carême, note 26, 5.
30 The Carter Center. 2017. 'Affaire d'Etat: Privatisation du Secteur du Cuivre en République Démocratique du Congo'. <www.cartercenter.org/resources/pdfs/news/peace_publications/democracy/congo-report-carter-Nov-2017-french.pdf>.
31 Scheele, Fleur; de Han, Emmanuel and Kiezebrink, Vincent. 2016. 'Cobalt Bleus: Environmental Pollution and Human Rights Violations in Katanga's Copper and Cobalt Mines'. *Centre for Research on Multinational Corporations* (SOMO), 17.
32 *Institute for Human Rights and Development in Africa and Others v the DRC* Comm 393/10.
33 Aguirre, Daniel. 2005. 'Corporate Social Responsibility and Human Rights Law in Africa'. *African Human Rights Law Journal* 5(2): 239–265.
34 Kamga, Serges and Ajoku, Ogechuwu. 2014. 'Reflections on How to Address the Violations of Human Rights by Extractive Industries in Africa: A Comparative Analysis of Nigeria and South Africa'. *Potchefstroom Electronic Law Journal* 17(1): 453–612.
35 *Ibid*, 493.
36 Declaration on the Rights to Development 1986.
37 Bilchitz, David. 2008. 'Corporate Law and Constitution: Towards Binding Human Rights Responsibilities for Corporation'. *South African Law Journal* 124(4): 781.
38 Implementation Guidelines for the Broad-Based Socio-Economic Empowerment Charter for the Mining and Mineral Industry, 2018.
39 Kinley, David; Nolan, Justine and Zerial, Natalie. 2007. 'The Politics of Corporate Social Responsibility: Reflections on the United Nations Human Rights Norms for Corporations'. *Company and Security Law Journal* 25(1): 30–42.
40 *Ibid.*
41 Dhruv, Sabharwal and Sumit, Narula. 2015. 'Corporate Social Responsibility in India: Introspection'. *Journal of Mass Communication and Journalism* 5(7): 1–3.
42 India Company Act of 2013.
43 Kaur, Simran and Tandon, Nidhi. 2017. 'The Role of Corporate Social Responsibility in India'. *Research Journal of Commerce and Behavioural Science* 6(3): 29–34
44 Ministry of Corporate Affairs Notification. 2014. Government of India Press, New Delhi, 27 February <www.mca.gov.in/Ministry/pdf/CompaniesActNotification3_2014.pdf>.
45 Kaur, Mandeep. 2014. 'Capitalism and Corporate Social Responsibility'. *Cardia Multidisciplinary Research Journal* 3(1): 1–45
46 Ksiezak, Paulina. 2016. 'The Benefits from CSR for a Company and Society'. *Journal of Corporate Responsibility and Leadership* 3(4): 53–65.
47 *Ibid.*
48 Paust, Jordan. 2002. 'Human Rights Responsibilities of Private Corporations'. *Vanderbilt Journal of Transnational Law* 35(1): 801–825.
49 Constitution of the Democratic Republic of Congo (2006), art 36.
50 RAID. 2009. '*Les Entreprises Minières Chinoises au Katanga République Démocratique du Congo*'. *Résumé Exécutif* 3–4.
51 DRC Constitution, note 49, art 34.
52 Human Rights Committee. 2004. General comment No 31(80): The Nature of the General Legal Obligation imposed on State Parties to the Covenant, UN Doc. CCPR/C/21/Rev.1/Add.13 adopted 29 March.

53 Majinge, Charles. 2011. 'Can Multinational Corporations Help Secure Human Rights and the Rule of Law? The Case of Sudan'. *Law and Politics in Africa, Asia, and Latin America* 44(1): 7–31.

References

Aguirre, Daniel. 2005. 'Corporate Social Responsibility and Human Rights Law in Africa'. *African Human Rights Law Journal* 5(2): 239–265.

Bilchitz, David. 2008. 'Corporate Law and Constitution: Towards Binding Human Rights Responsibilities for Corporation'. *South African Law Journal* 124(4): 754–789.

Blowfield, Michael and Murray, Alain. 2014. *Corporate Responsibility*. Oxford University Press.

Cronjé, Freek; Reyneke, Suzanne and Chenga, Charity. 2017. 'Corporate Social Responsibility in the Zambian Mining Sector: An Overview of Three Distinctive Operational Eras'. *Koers-Bulletin for Christian Scholarship* 82(1): 1–18.

Dhruv, Sabharwal and Sumit, Narula. 2015. 'Corporate Social Responsibility in India: Introspection'. *Journal of Mass Communication and Journalism* 5(7): 1–3.

Hamann, Ralph and Kapelus, Paul. 2004. 'Corporate Social Responsibility in Mining in Southern Africa: Fair Accountability or Just Green Wash?' *Society for International Development* 47(3): 85–92.

Hazemberg, Jilles. 2016. 'Transnational Corporations and Human Rights Duties: Perfect and Imperfect'. *Human Rights Review* 17(4): 479–500.

Kamga, Serges and Ajoku, Ogechuwu. 2014. 'Reflections on How to Address the Violations of Human Rights by Extractive Industries in Africa: A Comparative Analysis of Nigeria and South Africa'. *Potchefstroom Electronic Law Journal* 17(1): 453–612.

Kaur, Mandeep. 2014. 'Capitalism and Corporate Social Responsibility'. *Cardia Multidisciplinary Research Journal* 3(1): 1–45.

Kaur, Simran and Tandon, Nidhi. 2017. 'The Role of Corporate Social Responsibility in India'. *Research Journal of Commerce and Behavioural Science* 6(3): 29–34.

Kinley, David; Nolan, Justine and Zerial, Natalie. 2007. 'The Politics of Corporate Social Responsibility: Reflections on the United Nations Human Rights Norms for Corporations'. *Company and Security Law Journal* 25(1): 30–42.

Ksiezak, Paulina. 2016. 'The Benefits from CSR for a Company and Society'. *Journal of Corporate Responsibility and Leadership* 3(4): 53–65.

López, Carlos. 2013. 'The Ruggie Process: From Legal Obligations to Corporate Social Responsibility'. In *Human Rights Obligations of Business: Beyond the Corporate Responsibility to Respect?*, edited by Suraya, Deva and Bilchitz, David. Cambridge: Cambridge University Press, 58–77.

Lynch, Philip. 2006. 'Human Rights and Corporate Social Responsibility: Submission to the Corporations and Markets Advisory Committee Inquiry into Corporate Social Responsibility'. *Human Rights Law Resource Centre.*

Majinge, Charles. 2011. 'Can Multinational Corporation Help Secure Human Rights and the Rule of Law? The Case of Sudan'. *Law and Politics in Africa, Asia, and Latin America* 44(1): 7–31.

Mària, Josep and Devuyst, Emmanuelle. 2011. 'CSR and Development: A Mining Company in Africa'. *Journal of Management Development* 30(10): 955–967.

Moir, Lance. 2001. 'What do we Mean by Corporate Social Responsibility?' *Corporate Governance International Journal of Business in Society* 1(2): 16–22.

Paust, Jordan. 2002. 'Human Rights Responsibilities of Private Corporations'. *Vanderbilt Journal of Transnational Law* 35(1): 801–825.

Peyer, Chantal and Mercier, Francois. 2012. 'Glencore in the Democratic Republic of Congo: Profit before Human Rights and the Environment'. *Bread for All and Swiss Catholic Lenten Fund.*

RAID. 2009. 'Les Entreprises Minières Chinoises au Katanga République Démocratique du Congo'. *Résumé Exécutif* 3–4.

Scheele, Fleur; de Han, Emmanuel and Kiezebrink, Vincent. 2016. 'Cobalt Bleus: Environmental Pollution and Human Rights Violations in Katanga's Copper and Cobalt Mines'. Centre for Research on Multinational Corporations.

Sigam, Claudine and Garcia, Leonard. 2012. *Extractive Industries: Optimizing Value Retention in Host Countries*. New York/Geneva: UNCTAD.

Suraya, Deva. 2013. 'Treating Human Rights Lightly: A Critique of the Guiding Principles Complicity in Undermining the Human Rights Obligations of Companies'. In *Human Rights Obligations of Business: Beyond the Corporate Responsibility to Respect?*, edited by Suraya, Deva and Bilchitz, David. Cambridge: Cambridge University Press, 78–104.

15 Laws on natural resource ownership in Ghana

A perceived hindrance to community development

Bridget Kafui Anthonio-Apedzi

Introduction

Economic development remains a primary objective in the development agenda for most countries in the Global South, some of which factor in on the provision of social amenities like good roads, water, electricity, schools, hospitals and jobs. These are achieved through exploration of natural resources, and it is the duty of every government to ensure that every citizen is able to enjoy these basic needs of life. One issue that poses as a hindrance to the enjoyment of citizens' rights over their natural resources is the domestic laws on governance of natural resources. It is the responsibility of the State to safeguard citizens' rights through the formulation of appropriate laws, policies and strategies that enhance the general well-being of citizens, empower them to participate in decisions that concern them and ensure the equitable distribution of the benefits resulting from those resources.[1] The right to development cannot be isolated from other rights such as economic, social, cultural and political rights, because all these together make up for self-determination.

The right to development is an inalienable right which everyone is entitled to and states have the mandate to ensure that everyone is empowered to enjoy the right to full sovereignty over natural wealth and resources associated thereto.[2] According to Sengupta, the human person is the focus of the right to development and therefore, must be the centre of all policies and laws relating to the full sovereignty over natural resources.[3] The sovereignty a state has over its natural resources is explicitly stated in most international legal instruments. For example, the right of nations to exercise permanent sovereignty over their natural resources is espoused in General Assembly Resolution 1803 (XVII) of 14 December 1962. Other international instruments such as the Convention on Biodiversity (CBD)[4] and the Rio Declaration[5] also affirm the sovereign rights of states to exploit their own natural resources in accordance with their own developmental policies. According to Wenar, because international legal instruments of this nature are ratified by many nations, their legal obligations are to integrate both state and private ownership.[6] Wenar's assertion seems to suggest that private ownership of natural resources is possible but that for private ownership to be achievable, domestic laws must be enacted to render ownership of natural resources transferable from public to private ownership.[7]

In most cases, states do not explore the natural resources themselves but rather enter into exploration agreements with body corporates where the corporate in turn pays royalties to the state. Such is the case in Ghana where most of its mines were sold as part of the economic recovery programme requirements.[8] Before the structural adjustment and economic recovery programmes, introduced by The World Bank and the International Monetary Fund (IMF), the majority of the mines in Ghana like the Dunkwa Goldmines and Ghana Consolidated Diamonds among others were state-owned. However, with the economic pressure from the World Bank and the IMF, most of the state-owned enterprises were divested and thus, became dependent more on the royalties paid by the private companies for development. How the royalties are dispensed of for the benefit of the people is a question of interest that this chapter explores in the sense that the inability of citizens to enjoy the proceeds of their natural resources is against the spirit and letter of most legal instruments that promote peoples' right to self-determination and the right to development.

This view point is articulated in the following manner. The section 'Practical realities on natural resource ownership in Ghana' that follows, examines the practical realities on natural resource ownership in Ghana. The law on ownership of natural resources in Ghana is then further explored in the section 'Law on ownership of natural resources in Ghana'. The section 'Royalties' looks at the issue of royalties that accrue from the exploitation of mineral resources. Empirical findings from the case study on the management of mineral resources in Kanyasi No 2 is presented in the section 'Management of mineral Resources—case study of Kenyasi no 2'. The chapter concludes with some recommendations.

Practical realities on natural resource ownership in Ghana

The African Charter on Human and Peoples' Rights, which Ghana has ratified on 24 January 1989, guarantees the right of peoples to dispose of their wealth and natural resources as they may deem necessary to the satisfaction of their exclusive interests.[9] A purposive reading of the article 21 provision of the Charter implies that ownership of natural resources is attributed to peoples while the state, as a guarantor of rights, has the duty to ensure the realisation of that right. As a state party to the African Charter, Ghana is thus, obligated to take legislative and other measures to ensure that the right to natural resource ownership guaranteed by the Charter to all the peoples of Africa is implemented at the domestic level. This has, however, been a matter of contention in some instances, particularly with regard to claims of indigeneity by some communities and ownership of the natural resources found within their territories.

In areas where natural resources are in the geographical location of indigenous peoples, there is usually the tension between the benefits derived by the state for the general good of the citizens at large, as against the right of ownership of the indigenous groups. Cases like the Ogonis of Nigeria[10] and the Endorois of Kenya present examples of this conflict.[11] Another problem is the identification of the indigenous groups. In countries with foreign settler communities that displace original settlers, such as the Aboriginal people of Australia or the Indian, Inuit and Métis peoples of Canada,[12] the identification of indigenous groups are not in contention. Comparatively, most African countries have indigenous groups that are difficult to identify using the UN criteria[13] because the majority of the people are inherently indigenous. In Ghana, the term indigenous is used loosely to refer to communities that live in the rural areas and are largely agrarian or depend mostly on the land and its natural resources to make a living. Hence, the manner in which these resources are managed is of great importance because bad management would reduce the people to impoverished lifestyles.

Unlike for mineral resources and the acquisition of land, the law in Ghana is not explicit on ownership and management of natural resources. Land is either owned communally by a 'Stool', a 'Skin', individually, or by families. The Stool is the sacred traditional emblem in most Akan communities of the southern part of Ghana. The supremacy of the Stool is acknowledged by the community and the community is bound to serve and respect it in accordance with the traditional laws and customs.[14] According to Ollenu and Woodman, the Stool occupier, the Chief, represents the sovereignty of his people so long as he is occupying the Stool and is the custodian of the land. In some Northern parts of Ghana, it is the Skin instead of the Stool that is the sacred traditional emblem. Ollenu and Woodman explain that in some parts of the Northern, Upper East and West Regions of Ghana where the Skin applies, the traditional priest, known as the Tindana retains a strong control over the land.[15]

In places where land is owned by the Skin or the Stool, the people who owe allegiance to the Stool or Skin are the rightful owners of the land and the occupier of the Stool or the Skin only holds the land in trust for the people. Mineral resources are found either on land or underneath, and therefore, issues of ownership either of the mineral resources or the land cannot be divorced from each other. The law does not differentiate any geographical region or community in stating the ownership of mineral resources, and thus, all mineral resources irrespective of where they are located are held by the state in trust for the people of Ghana as a whole. The trusteeship created by these provisions obliges the government to manage the country's mineral resources to the benefit of all the peoples of the Republic of Ghana. For the purposes of this chapter the researcher focuses on the mineral resources in the gold mining community of Kenyasi No 2 where land is communally owned by the Stool.

In Ghana, the state is mandated to ensure optimisation of economic development through policy and strategies in order that every citizen is guaranteed adequate means of livelihood and appropriate customary and cultural values are upheld.[16] With emerging problems of conflict between members of communities on the one hand, and traditional leaders, state institutions and exploration companies on the other hand, over the management of natural resources for community development, the researcher investigated the challenges at the primary source in Kenyasi No 2.

The research aimed to demonstrate, by use of the empirical legal method, the extent to which the law ensures the right to development to local communities through the use of proceeds from the mineral resources found on their land. Empirical legal research is noted to be 'valuable in revealing and explaining the practices and procedures of legal, regulatory and redress',[17] as it is a method concerned with the development of substantive law, the practice and impact of law on the life of citizens.[18] Questionnaires and focus group discussions were used for data collection. Structured questionnaires were administered to garner the views of community members on the laws relating to the governance or management of natural resources, ownership of mineral resources and the benefits accruing therefrom. Interviews were conducted with the aid of research assistants. In all, 60 participants were interviewed, including opinion leaders. The data collected was transcribed and analysed using qualitative research methods.

Law on ownership of natural resources in Ghana

The law is not explicit on the ownership of natural resources in Ghana. However, the 1992 Constitution states that no one, including the government, can grant any right for the exploitation of mineral, water and *other natural resources* of Ghana except by resolution of Parliament supported by the votes of not less than two-thirds of all the members.[19] The Constitution also provides that Parliament must provide for the establishment of Commissions to manage natural resources within the country.[20] Management here means the formulation of policies and regulations that relate to exploitation of the natural resources. Although there are no explicit provisions relating to ownership of natural resources, exploitation of natural resources cannot be a preserve of any single person. No one has the authority to exploit them. It can be inferred then from the constitutional provisions that, what constitutes natural resources must be determined by a state institution and once a resource has been identified as a natural resource, it will be managed as other regulated natural resources such as oil, gas, minerals and water.

It is also stated in the Constitution that every mineral in Ghana in its natural state; in, under or upon any land, rivers, streams or water courses, is the property of the Republic of Ghana and shall be vested in the President on behalf of, and in trust for the people of Ghana.[21] This constitutional provision is replicated in the Mining and Mineral Act 703 of Ghana, 2006. In effect, mineral resources are given constitutional protection as belonging to the people of Ghana, not only for the people in the geographical area where the minerals are located. In accordance with the trusteeship created by law, which vests legal ownership and property rights to the state, for the benefit of the citizens of Ghana. As the principle of the law of trust goes, it is the duty of the state to place the interest of beneficiaries, in this case the citizens, above all things.

The state being the trustee must manage the mineral resources in order for it to be beneficial to the citizens for growth and development. This is usually not the case, especially in countries ruled by authoritarian governments, where proceeds from natural resources are appropriated by these governments without recourse to citizens' legal rights as is the case in Equatorial Guinea,[22] or in countries ruled by corrupt governments in pursuit of self-seeking interest against national interest, as portrayed in Nigeria.[23] Invariably, citizens are condemned to live in abject poverty due to lower rates of economic development and incessant civil conflicts.

The issue of land ownership and mineral resources are intrinsically connected since mineral deposits are found on, in or underneath the land. Ownership of land in Ghana is in three categories: public, customary and private. Other classifications present two categories: public and private, while defining customary ownership as part of private ownership.[24] To provide a better understanding of land ownership in Ghana, the three classifications of public, customary and private ownership are considered. These three categories are capable of holding the allodial title. Public lands are those owned by the government and are vested in the president for public use. The government acquired these lands through purchase from customary landowners or private individuals or through compulsory acquisition in accordance with the State Lands Act of 1962.[25] Customary lands are communal lands owned by the Stool, Skin, families or clan; in other words, communally owned by members of the community who pay allegiance to the Stool or the Skin, members of the family or the clan. The law defines 'Stool land' to include an interest in, or right over a land controlled by a 'Stool or Skin, the head of a particular community or the captain of a company, for the benefit of the subjects of that Stool or the members of that community or company'.[26]

Land in this category is usually held in trust by the Chief, head of family, clan, or fetish priests, depending on the traditional area in question, for the benefit of the members of that group.[27] Private ownership of land (that is the allodial title vested in a private individual) can be acquired by way of a grant, sale, gift or marriage.[28] This type of ownership is now rare owing to constitutional restrictions.[29] However, it is possible that a private person who acquired an allodial title before the coming into force of the Constitution is capable of transferring same.

When minerals are found under the land, the government need not acquire the land, because the owners still maintain their surface rights and are allowed to continue their activities such as grazing of livestock and farming on the land, provided these activities do not interfere with mining operations.[30] However, the owners of the land with surface rights cannot erect any building or any structure on the land for which mineral lease has been granted even without first seeking the consent of the lease holder.[31] Technically, the land owner is still somehow constrained to exercise the right. Where minerals are found on the land, it must be acquired in its entirety and compensation paid. Exploitation of the mineral resources found on or under public lands poses no complications. It is customary lands and lands owned by individuals that pose complexities, since minerals found in or under these lands do not belong to the owners of the land. On minerals found on or underneath such lands, the president has the mandate to acquire the land (compulsorily) as stated by the law,[32] for development of the minerals found thereon, subject to the payment of fair, prompt and adequate compensation.[33] The law also stipulates that land must be made available for mineral rights, which indicates that the state can exercise the right of eminent domain at any time without any restraint.[34] Every mining activity requires the granting of a permit for mining rights regardless of the title owners of the land, and no individual can apply for a mining right but only body corporates and partnerships.[35] The implication of these provisions are that though you may have a title to the land, you cannot extract any minerals found on the land, or even prospect for minerals as an individual.[36]

Royalties

The government of Ghana receives royalties from mining explorations and exploitation of other natural resources, which are then distributed according to law. The law sets royalties to be paid to the Republic at between 3 percent and 6 percent of the total revenue obtained from mineral exploration.[37] The government established by law, an institution called the Office of the Administrator of the Stool (ASL) for the disbursement of royalties.[38] It entrusts the ASL to set up accounts into which royalties and other revenues are paid; and to collect and disburse the royalties and other revenues accordingly. These royalties are the percentage the government pays to the customary owners of the land, the majority of which are Stool lands, hence the name ASL. This could be seen as a benevolent gesture from the government, an effort towards a fair and even distribution of royalties. Although the minerals are the property of the entire state, part of the royalties is given to the customary owners of the land. This could also be seen as a proper management of mineral resources, so that areas where such resources are found, the indigenes are guaranteed to enjoy more of the benefits.

The money paid to the Stool land's account is not only royalties but other revenues accruing to the Stool. In disbursement of such revenues, the Constitution states that 10 percent of all revenue shall be paid to the ASL to cover administrative expenses and out of the rest of the 90 percent, 55 percent is paid to the District Assembly within the area of authority in which the Stool lands are situated; 25 percent is given to the Stool through the traditional authority for the maintenance of the Stool in keeping with its status and another 20 percent is allocated to the traditional authority.[39] To ensure proper management of the revenue from mineral resources, the government passed the Mineral Development Fund Act (Act 912) in 2016. The purpose of the Fund is to ensure that moneys collected as royalties are used for the right purpose and also to the benefit of the community.

Management of mineral Resources—Case study of Kenyasi No 2

Kenyasi No 2 is a community located near a gold mine site operated by the Newmont Ghana Gold Ltd (NGGL). The community is in the Brong Ahafo Region of Ghana and is one of the ten communities in the region where Newmont Mines operates. Kenyasi No 2 was chosen because the town has an institutionalised Traditional Council and a Youth Group. The study examined the Environmental and Social Assessment Impact reports and also went to the community to gather information on community members' experience of the impact of the operation of the mines on their livelihood, and environment. Questions addressed the following thematic areas:

- Ownership of mineral resources;
- Involvement of the community in the decision-making process;
- Ease with which information about the mining project, documents relating to environmental and social impacts and royalties are accessed;
- Impact of mines on their socio-economic and cultural lives;
- Perceived impact of the mines on their environment.

Ownership and management of mineral resources

Whether communities, clan, family or an individual who own the land should be the owners of the mineral resources found thereon or underneath was a question for consideration during the research. With regards to land ownership in Ghana, land is communally owned in the area where the research was conducted, and hence ownership resides in the Stool, which is the unification symbol of the community. The occupant of the Stool, that is, the traditional leader or the Chief is the custodian of the land. All revenues accruing to the Stool land is paid to the Chief who must use it for the benefit of the entire community.[40] Thus, another form of trusteeship is created which vests Stool revenues in the Chief for the beneficial use of the community, aside from the trusteeship which vests legal ownership and property rights to the state, for the benefit of the citizens of Ghana. However, the people often do not trust their traditional leaders to manage revenues from gold mining. There were complaints about the royalties not being used by the traditional authorities for the development of the area. This presupposes that the community does not benefit from such royalties. It has been reported that most royalties are enjoyed only by the Chief, that is, the Stool occupant or custodian of the Stool and his family. The only remedy the law gives is in relation to the disbursement of Stool revenue.[41] The assertion that royalties received are not used for the benefit of the community could, however, not be established. The scheduled meeting with the Chief and the

Traditional Council could not be held as the Chief reiterated that the research team must first seek permission from the Newmont Mining Company. This in itself raised some ethical questions with regard to the possibility of a collusion of some sort between the Chief or the traditional authorities and the mining company. Why would the Chief, as custodian of the land not be willing to talk to the research team unless with the permission from a foreign company that has no business in the traditional affairs of the community?

It was suggested to the participants of the research that the law on ownership could be reviewed to allow the communities to assert direct ownership of the mineral resources found on their lands. The suggestion was, however, met with reservations. Some argued that the law should be left as it is because transferring ownership to the communities would mean giving the Chief the utmost and unfettered authority over the resources. It was the view that although the Chiefs are mandated to hold the land in custody for their subjects, some Chiefs arrogate power to themselves as though the land belongs to them in their personal capacity and thus, are not obligated to use the royalties to the benefit of the community.

This, they said, is already evident in the way royalties are enjoyed by only the Chiefs and their families. They cited the issue of 'galamsey' (a term commonly used for illegal gold mining) and its antecedent problems as a classical example of the inept attitude of Chiefs in the management of natural resources. There is, therefore, an overwhelming agreement that the Chiefs are unable to do a good job in managing the proceeds from the mineral resources. They argue that royalties that are paid do not get to them or are not used for the benefit of the community and therefore, there would be no guarantee that the situation would get better if the ownership of the gold resided with the Stool. Participants are of the view that they would be better served if government also intervenes in the management of the natural resources and places some form of obligation and sanctions on the Chiefs to compel them to do things right.

Impact of mining on socio-economic and cultural life in the community

The study aimed to ascertain how mining exploration has impacted on the livelihood and general well-being of the Kenyasi No 2 community. The study used the following specific indicators in determining whether the mining activities in the community have

(a) Led to an increase in economic activities;
(b) Increased means of livelihood;
(c) Brought in more jobs;
(d) Led to the destruction of existing habitats from activities such as dumping of waste and vegetation clearance;
(e) Led to changes in social, cultural and economic patterns within the community;
(f) Increased the tourism potential of the area.

Participants were asked on a scale of 1 to 5, where 1 is 'strongly agree' and 5 is 'strongly disagree' with the above statements. It was observed that the scores were higher with the indicators that the participants strongly disagreed with, which centred on increase in economic activities, jobs and means of livelihood among others, demonstrating that members of the community have not derived any economic benefits from the mines.

The consensus above is corroborated by a follow-up question on the participants' overall perception of the mines on their livelihood generally. Out of the 60 participants interviewed, 56 responded in the negative. For this category of people, the mining operations that take place in the community are of no significance to them. Only four of the participants said the mines have any positive impact on their lives; however, they could not explain the kind of positive impact they have experienced.

The questions per se however, do not address the issue of development in terms of statistics; they were only to gather the perceived views on the impact of the mining activities on their lives regarding social and economic activities. Nonetheless, it was assumed that members of the community would be able to point out some on-going development projects that have taken place or some vibrant economic activities, including job creation, food security, primary health care and education among other minimum indicators of the right to development,[42] as a result of the mining operations, to depict some community development and general well-being. However, they mentioned an office building allegedly constructed by NGGL for the traditional authorities; a durbar arena and a minibus, which they maintained are of no use to them. They expressed as a main concern the fact that Newmont Mining Company should have provided them with jobs and that their workers should have stayed in the community to boost economic activities instead of just providing a building.

In all, comments that were common or had the highest percentage were comments on the failure of the mining companies to fulfil their promises on job creation and on patronising local goods. It was observed that these complaints of unfulfilled promises and the failure to provide development projects in the community are not the preserve of NGGL. At best, they might be regarded as a Corporate Social Responsibility (CSR). The comments in the diagram in Figure 15.1 portray the general dissatisfaction of the community:

Questions on pollution of rivers; destruction of farmlands to make way for gold mining; adequate compensation and prohibitions from farming in the area without an alternative source of work were not properly answered, indicating participants' lack of interest in these matters.

Conclusion and recommendations

It is concluded that vesting of mineral resources in the state for the benefit of the citizens has no setbacks on community development, as the research findings and the reaction of participants towards the amendment of laws on

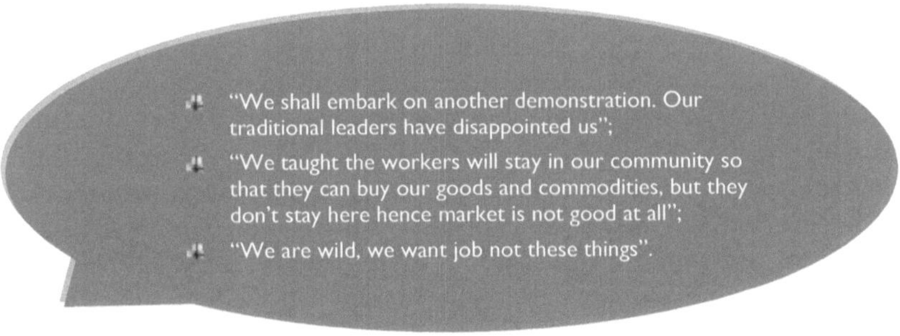

Figure 15.1 Community feelings. *Source:* Researcher

mineral resources indicate. It can therefore be concluded that the law on natural resources per se poses no hindrance to the community's right to development either. It is rather the system that needs to be tightened and a mechanism of holding traditional authorities accountable that is relevant and needs to be put in place so that community members may be empowered to enforce their right to development.[43]

The enactment of the Minerals Development Fund Act 912 of 2019 sets a purposeful approach towards the management of proceeds from mineral resources. The law establishes the Minerals Development Fund (MDF) into which shall be paid not only royalties but also other sources of income such as grants, donations, gifts and voluntary contributions.[44] The MDF is to be used among other things to redress the harmful effects of mining on affected communities and persons;[45] and promote local economic development and alternate livelihood projects in the communities affected by mining.[46] These provisions are practicable and are consistent with the main object of the Fund, which is the facilitation of socio-economic development in the mining community.[47] However, the law in its entirety is not comprehensive.

A legislative instrument could be passed to flesh out more details to the law. Modalities need to be designed in order to ensure effective implementation of the law. The disbursement of funds to the Local Management Committee, for instance, which according to the law, seems to be the main actors of the object of the Fund within the mining communities have not been stated. Again, there should be some description regarding what constitutes a 'gift' as a source of income for the Fund. This will ensure that 'gifts' meant for the Fund will not end up in the wrong hands. It is stated that part of the fund may be used to redress harmful effects of mining within the community. To what extent does this provision absolve mining companies from their duty to ensure the healthy and safe environment of the communities within which they operate? This may breed a situation where the company donates money to the Fund, aside from royalties, and yet destroys the environment.

The research also unveils a misperception on CSR of the company to the community and the lawful duty of the traditional authority and the local government towards community development. It all comes down to the issue of education or lack of it. Participants' complaints were mostly about unfulfilled promises on the part of NGGL to provide jobs and better social amenities, issues which could be considered as a CSR or corporate citizenship of the mining company. So, the community clearly needs orientation to appreciate the difference between the principles of CRS, which is a philanthropic gesture by a company to boost its image as well as to give back to the society, and the legal duty of the authorities to develop the community. Periodic education of the community members should be carried out to the effect that companies are not obligated to embark on community development projects. However, mining companies may do more than just pay royalties by ensuring that the royalties they pay are used well. They may also disclose openly the amount of money paid each year and make records on royalties paid and the community development projects undertaken to the benefit of the entire community. This is debatable because mining companies must not be seen to be too involved in the domestic affairs of a sovereign state; nevertheless, it will go a long way to maintain a peaceful coexistence between the company and the community. In this regard, the company's community relations officer may be made a member of the Local Management Committee of the Mineral Development Fund scheme.

Ghana's economy thrives on the exploitation of natural resources. The benefits that the people are interested in flow from the development boom associated with the activities of the mining industry. The laws relating to natural resources should be reviewed to give more rights to community members to enable them, where necessary, to seek redress in a court of law or alternative dispute resolution institutions in relation to mismanagement or misappropriation of royalties and other mineral resources. In this regard, any person or groups that are aggrieved that their right to development is being abused may have the capacity to sue, including Chiefs and traditional authorities. In this way, the community or individuals are empowered to enforce their right to development, since the Declaration on the Rights to Development cannot be enforced directly in any legal system. Convicting a Chief has serious implications in Ghana, as it can lead to the Chief's 'destoolment'. The cultural practice reiterated by case law in the case of *Gyamfi and Another v Owusu and Another*,[48] which states that Chiefs cannot be chastised openly and are not accountable to their subjects should be reviewed because such a rule can encourage Chiefs to embezzle with impunity. To ensure transparency, governments may publish the list of companies and the amounts of monies paid as royalties each year as well as the amount of monies individual traditional authorities receive periodically for community development.

Notes

1 Declaration on the Right to Development Resolution A/RES/41/128 adopted by the UN General Assembly on 4 December 1986, art 2(3).
2 *Ibid*, art 1.
3 Sengupta, Arjun. 2000. 'Realizing the Right to Development'. *Development and Change* 31(3): 553, 568.
4 Convention on Biological Diversity adopted by the UN General Assembly in 1992 principle 3.
5 Rio Declaration on Environment and Development adopted at the United Nations Conference on Environment and Development at Rio de Janeiro, Brazil, 3–14 June 1992, principle 2.
6 Wenar, Leif. 2008. 'Property Rights and the Resource Curse'. *Philosophy and Public Affairs* 36(1): 11.
7 *Ibid.*
8 Ghana launched an economic recovery program (ERP) in 1983 under the guidance of the World Bank and the IMF to reduce Ghana's debts and to improve its trading position in the global economy. The government under the ERP divested itself of many assets through privatization. <www.imf.org/external/np/pfp/ghana/ghana0.htm>.
9 African Charter on Human and Peoples' Rights adopted by the Organisation of African Unity in Nairobi Kenya on 27 June 1981, OAU Doc CAB/LEG/67/3 rev. 5, art 21.
10 Coomans, Fon. 2003. 'The Ogoni Case before the African Commission on Human and Peoples' Rights'. *The International and Comparative Law Quarterly* 52(3): 749–760.
11 *Centre for Minority Rights Development (Kenya) and Minority Rights Group International on behalf of Endorois Welfare Council v Kenya* Comm 276/2003 (2009) AHRLR 75 (ACHPR 2009).
12 The Constitution Act 1982 sect 35(2).
13 UN advocates for self-identification of indigenous people by historical continuity using criteria such as retained social, cultural, economic and political characteristics that are distinct from those of the dominant societies in which they live. UN Department of Economic and Social Affairs Indigenous Peoples. <www.un.org/development/desa/indigenouspeoples/about-us.html>.
14 Ollennu, Amaa and Woodman, Gordon. 1985. *Ollennu's Principles of Customary Land Law in Ghana*. Birmingham: CAL Press.
15 *Ibid*, 8.
16 The Constitution of Ghana, 1992, Chap. 6.
17 Genn, Hazel; Partington, Martin and Wheeler, Sally. 2006. *Law in the Real World: Improving Our Understanding of How Law Works*. London: Nuffield Foundation, 1.
18 *Ibid*, paras 4 and 5.
19 Constitution of Ghana, note 16, art 268.
20 *Ibid*, art 269.
21 *Ibid*, art 257 (6).
22 Wenar, note 6, 3.
23 Ogbeidi, Michael. 2011. 'Political Leadership and Corruption in Nigeria since 1960: A Socio-Economic Analysis'. *Journal of Nigeria Studies* 1(2): 3.
24 Sittie, Rebecca. 2006. 'Land Title Registration: The Ghanaian Experience—Shaping the Change'. Paper presented at XXIII FIG Congress Munich, Germany, 3.
25 State Lands Act 1962.
26 Office of the Administrator of Stool Lands Act 1994 sect 17.
27 Constitution of Ghana, note 16, art 267(1).
28 Sittie, note 24, 3.
29 Constitution of Ghana, note 16, art 267(5).
30 Mining and Minerals Act of Ghana 2006, sect 72 (3).
31 *Ibid*, sect 72(4).

32 *Ibid*, sect 2.
33 Constitution of Ghana, note 16, art 20.
34 Mining and Minerals Act, note 30, sect 3.
35 *Ibid*, sect 10.
36 *Ibid*, sect 9.
37 *Ibid*, sect 25.
38 Office of the Administrator of Stool Lands Act, note 26.
39 *Ibid*, sect 7(1); Constitution of Ghana, note 16, art 267(6).
40 *Ibid*, sect 7(1); *Ibid*, art 267(6).
41 Sect 7(2) of Act 481 states that 'A beneficiary of Stool land revenue aggrieved by the apportionment of the Stool land revenue may submit a complaint to the Commission of Human Rights and Administrative Justice'.
42 Sengupta, note 3, 553.
43 African Charter, note 9, art 22(2).
44 Minerals Development Fund Act 2016, sect 3.
45 *Ibid*, sect 5(a).
46 *Ibid*, sect 5(b).
47 *Ibid*, sect 17.
48 *Gyamfi and Another v Owusu and Another* [1981] GLR 612–633.

References

African Charter on Human and Peoples' Rights adopted by the Organisation of African Unity in Nairobi Kenya on 27 June 1981, OAU Doc CAB/LEG/67/3 rev. 5.

Coomans, Fon. 2003. 'The Ogoni Case before the African Commission on Human and Peoples' Rights'. *The International and Comparative Law Quarterly* 52(3): 749–760.

Declaration on the Right to Development Resolution A/RES/41/128 adopted by the UN General Assembly on 4 December 1986.

Genn, Hazel; Partington, Martin and Wheeler, Sally. 2006. *Law in the Real World: Improving our Understanding of How Law Works*. London: Nuffield Foundation.

Gyamfi and Another v Owusu and Another [1981] GLR 612–633.

Ogbeidi, Michael. 2011. 'Political Leadership and Corruption in Nigeria since 1960: A Socio-Economic Analysis'. *Journal of Nigeria Studies* 1(2): 1–25.

Ollennu, Amaa and Woodman, Gordon. 1985. *Ollennu's Principles of Customary Land Lawin Ghana*. Birmingham: CAL Press.

Sengupta, Arjun. 2000. 'Realizing the Right to Development'. *Development and Change* 31(3): 553–578.

Sittie, Rebecca. 2006. 'Land Title Registration: The Ghanaian Experience—Shaping the Change'. Paper presented at XXIII FIG Congress Munich, Germany.

Wenar, Leif. 2008. 'Property Rights and the Resource Curse'. *Philosophy and Public Affairs* 36(1): 2–32.

16 Policy implications for the right to development through the South African Mining Charter

Mofihli Teleki

Introduction

The South African Mining Charter adopted in 2018 serves as a regulatory instrument that provides guidelines for 'effective symbiotic growth and mutual transformation'.[1] The guidelines as spelt out in the Mining Charter partly seek to cater for the so-called 'host communities' with reference to certain aspects of the right to development, including housing, adequate living conditions, health care and education, amongst other things. These aspects are found in article 8 of the United Nations Declaration on the Right to Development (UNDRTD) which states that state parties should undertake measures to provide access to education, health services, food, housing, employment and the fair distribution of income.[2] As the Mining Charter aims to give effect to aspects of the right to development, it is important to note that in recent times, matters pertaining to housing, education, decent remuneration and health care have been subject to scrutiny in host communities where mining activities are prevalent.

The Mining Charter needs to, therefore, be interrogated for its prospects to inform policy creation over and above the guidelines referred to in the Charter. For clarity purposes, it is important to highlight that the Mining Charter is not in itself a policy document; it only provides guidelines for regulating mining activities and to ensure best mining practices. Thus, besides the Charter, a policy framework needs to be put in place to regulate mining activities and ensure best practices within the mining sector. In this instance, the Charter is portrayed merely as a regulatory document that provides the South African government the basis for devising a policy framework on social responsibility for mining companies in as far as community development is concerned.

This chapter taps into the transposition of the guidelines of the Mining Charter into a policy framework that can effectively regulate mining companies and the actions of government in relation to provisions on the right to development. The chapter is structured as follows: The section 'Approach and context' explores the approach and the context of mining in South Africa with reference to problems and challenges. The section 'Polemics between the state and mining companies' explains the actual polemical issues of the mining sector and makes reference to why principles of the right to development are needed in the implementation of the Mining Charter. The section 'The mining charter and the right to development' consists of a debate about the Mining Charter and its connection with the right to development. The sections that follow present arguments on envisioned policy directives.

Approach and context

Framing of historical factors

The current state of affairs in the mining industry in South Africa may be explained by revisiting the past. These historical factors explain the connection between mining companies and their attitude towards human rights and the rights of the communities that host mining activities. Colonialism and apartheid with its policy of separate development that was responded to by widespread social strife portrayed South Africa as a non-respecter of human rights prior to the advent of democracy in 1994. The post-liberation period ushered in an era of human rights, which were non-existent during the colonial and apartheid periods. This chapter explains the historical socio-economic factors that warrant a higher appreciation of the right to development in the context of mining. A reflection of the polemics of mining amid the challenges faced by mining communities reminds of the issues that need to be redressed to ensure that the negative legacy in the relationship between mining companies and host communities are redefined through the right to development.

Commonalities between the Mining Charter and the right to development

Although I point out a great deal of reference to elements of the right to development within the Mining Charter, it does not by default imply the realisation of the right to development in the communities that host mining activities. The Mining Charter puts an emphasis on the provision of housing, healthcare and education, amongst other things. Even though these aspects are contained in the Mining Charter, their advancement cannot be limited to the context of the regulation, especially because a Charter is by nature not a policy per se, but rather a regulatory instrument that governs the relationship between host mining communities and mining companies.

The Mining Charter in the context of this chapter should be viewed as an aspiration for the development of policy that aims to protect the rights of host mining communities in as far as the right to development is concerned. The connection between the right to development and the elements of it that are found in the Mining Charter are explained in the section 'The mining charter and the right to development'. The nexus created under that section of this chapter seeks to draw aspirations from the right to development policy for host mining communities and mining companies. In other words, the ambition of drafting a policy on the right to development can be easily drawn from the existing text of the Mining Charter. In this way, particular elements of the right to development as contained in the Mining Charter could make it easy to infer the policy directives that are inherently borrowed from the UNDRTD.

Key considerations in policy formulation

Public policy is considered an intent of the state to advance or solve particular problems within society. The policies intended for a specific purpose may exist according to provisions of the law, regulatory frameworks. Ultimately, public policy as an applied discipline functions in determining what societies' inspirations or aspirations or even wishes are regarding a particular problem. These aspects of law, policy or regulations need to be coherent with solving problems of a country. In essence, in order to solve particular problems of a country or society it is important that the solutions accompanied by laws, policies and regulations could be coherent enough to have consistency. This is important in that all the solutions either in law or policy should have ingredients that are aimed at solving a common problem. This specific context is very much linked to the approach of this chapter.

Be that as it may, it should be borne in mind that envisioning policy should take into account all of the factors that necessitate the formulation of such a policy. It is the reason why I explore the form and type of policies needed to solve the particular problems associated with mining activities as well as the host mining communities. Inspiration herein is drawn from the theme of the Mining Charter, which is referenced as aiming to achieve 'symbiotic and mutual transformation'. In assessing the form and type of policy that can be used to infer the right to development, a paradigm is needed. Hence, I draw inspiration from the theme of the Mining Charter to determine whether the narrative constructed therein aligns with or reflects South Africa's national agenda for transformation in relation to racialised labour, mining ownership and beneficiation.

This caution as I explain in this chapter is important because some of the policy prescriptions within the mining sector are transactional in nature. It is therefore important to take careful consideration of the fact that the right to development as a concept is not only about provisions and transactions, but also about the maintenance and sustainability of development. A major part of my criticism of the Mining Charter is in the fact that the Charter encourages a transactional approach rather than encouraging the protection of the right to development for sustainability. The discourse on rights or on rights-based development or even the right to development is based on entitlements, which cannot shift from one point to another depending on the nature of the transaction.

Polemics between the state and mining companies

South Africa's mining industry spans over a century of extraction of minerals in various shapes and forms.[3] From time immemorial, mining activities have largely centred on extraction of natural resources such as coal, diamonds, platinum and gold, amongst others.[4] One of the main drivers of mining in South Africa was the discovery of minerals, which resulted in the so-called 'mining boom' at the turn of the previous century.[5]

The history of mining in South Africa cannot be dissociated from the exploitation of migrant labour;[6] a practice wherein mining companies gratuitously opt fora cheap labour force to work in the mines as a means to maximise profits from their mining operations.[7] In essence, the labour force was constrained to take a hazardous task for pittance. Over decades, mining companies were looked upon to build schools, and clinics, amongst other social amenities.[8] These are elements contained in article 8 of the UNDRTD as well as in articles 16 and 17 of the African Charter on Human and Peoples' Rights. This notwithstanding, it is important to point out that decent work and decent pay have been amongst the key demands that mining unions and host mining communities have agitated about for decades[9] owing to the fact that the rights of workers have often been subject to transgression by mining companies.[10]

The extraction of minerals in the 1940s led to the establishment of new towns and cities whose survival depended on the activities of the mining companies.[11] At some point, South Africa was deemed to be a high mining energy complex economy as a result of the wide range of mining activities across the country.[12] In the early 1990s the economy suffered as a result of mining strikes that took place in the late 1980s.[13] In other words, the mining sector went through intensive labour strikes around the late 1980s and 1990s[14] provoked by grievances resulting from low wages that created disagreements between workers and trade unions and the mining companies.[15] In the greater part of the 1990s, South Africa was plagued with several cases of silicosis and tuberculosis, which emanated from adverse mining practices that gravely affected mining communities.[16] Tuberculosis became so widespread within mining communities to the extent that it appeared to be normal.[17] By the early 2000s, several mining towns by default transformed into ghost towns as a result of mining companies relocating overseas.[18] Some of these companies include Anglo-America, and Harmony Gold, to name a few.

Around 2010 and beyond, the issue surrounding decent work and decent pay intensified to the point where it climaxed into the Marikana massacre in 2012 where 64 mine workers were killed for legitimately demanding a fair increase in their wages.[19] Marikana remains a symbol of what would typically happen when basic amenities, decent pay for decent work and social conditions of mining communities are not taken care of through a policy framework that directly advances the right to development. In the case of the Marikana massacre, workers on the platinum belt of South Africa protested and went on strike in demand for a minimum wage of R12 500 (USD950).[20]

The settlement reached for the minimum salary for mine workers of the platinum belt was R8500 (USD610).[21] Within the global context, a mine worker earns about USD150 000 annually,[22] which is far higher than the average of South Africa which is equated to only about USD9 000 (annually).[23]

An assessment of this scenario indicates that decent work for decent pay is one of the aspirations of the right to development.[24] This is in the same way that decent shelter and education constitutes elements of the right to development.[25] The issues of housing and healthcare were also in the peripheral demands of the workers who went on strike in Marikana.[26] In essence, the principles of the right to development indirectly led to the massacre of miners who were gunned down by the South African police in the course of their protests. Apart from the socio-economic factors that had become highly contested in the mining communities, there was also a socio-cultural element to the plight of the miners. The mining communities are generally populated by bread winners who came from the rural areas of South Africa, often far away from their families to supply the labour force for the mining companies.[27]

The polemics of mining in South Africa and the poor social conditions in the mining communities are an antithesis of the spirit of the preamble to the African Charter, which seeks to 'preserve the harmonious development of the family and to work for the cohesion and respect of the family; to respect his parents at all times, to maintain them in case of need'.[28] It became common at some point for the mine workers to establish a new family in the mining communities.[29] This had a great impact on the spiritual connection between the mine workers and their family roots. In most cases, some of the mine workers would only return to their rural homesteads to be buried after suffering from silicosis, TB and HIV related diseases.[30] In essence, loss of life in the mining communities could directly be linked to the use of cheap labour, and conditions that adversely impact on human lives.

Fatal diseases associated with mining activities and deaths became common in the mining towns in the mid-2000s.[31] Several cases of claimants seeking compensation from the mining companies also became common.[32] Several non-governmental organisations encouraged litigation of TB and silicosis class actions, seeking adequate compensation for the suffering and death of mine workers.[33] By implication, key aspects of the right to development, including matters relating to health care and conditions of work became the subject of class actions before the courts.

Migration to the mines for labour purposes also brought about the degeneration of family structures in rural areas[34] leading in many instances to child-headed households as a result of parents dying from HIV/AIDs.[35] It is therefore safe to assume that the ripple effect of mining to the families of mine workers affected the next generation of offspring. This again implicates the right to development in the sense that the government is obligated to guarantee the right to socio-economic and cultural development for future generations.[36]

Even though at some point in the mining history of South Africa, provisions of the right to development were met through corporate social responsibilities of mining companies, on the most part, all of the aspects of the right to development seem to have been compromised if one reflects on the 100-year period of mining in South Africa.

The Mining Charter and the right to development

As indicated earlier, the South African Mining Charter serves as a regulatory instrument that provides guidelines to ensure 'effective symbiotic growth and mutual transformation' within the country's mining sector.[37] In this light, the Charter presents as an effective tool that may be used to ensure there is effective growth both within the mining sector and mining communities, and thus, it necessitates transformation amongst the companies that operate within the sector. The latter, which relates to the object of transformation, is given reference through the government's Broad-Based Black Economic Empowerment (BBEE) policy.[38]

The government's BBEE policy has always been used as a tool to ensure that there is racialised transformation in the context of industry, but with reference to ownership of companies and direct beneficiation of the outcomes of businesses.[39] This reference to BBEE is arguably indicative of the transformation agenda that the government has been advancing over the decades.[40] The symbiosis that the Charter makes reference to is linked to the relationship that exists between mining companies and the host communities where mining activities take place. The main policy theme in this regard is aimed at regulating the future relationship between mining communities and mining companies as well as the transformation that is envisaged to take place within the mining sector. One would therefore deduce that the government of South Africa has made good strides to regulate the interface between the mining communities and the companies that enter into these communities for the purposes of mining over a period of time.

The UNDRTD stipulates that development should be a meaningful participation of people in the development process in relation to the fair distribution of benefits resulting therefrom.[41] The provision ties in with the aspirations contained in the Mining Charter, which locates individual members of the mining communities as an entity that is actively involved in the development process of mining. The UNDRTD further highlights the recognition that 'the human person is the central subject of the development process and that development policies should therefore make the human being the main participant and beneficiary of development'.[42] In line with this provision, the Mining Charter provides assurance that mining communities should be able to directly benefit from the mining operations located within those communities. This is made possible by a section of the Mining Charter, which states the position of mining communities and the protection they ought to have in relation to possible abuse by mining companies that have been prevalent in the past and recent scenarios of the social polemics of mining in South Africa.[43]

The UNDRTD provides in articles 2(3) and 10 that steps should be taken to formulate policies on the right to development.[44] In article 15 of the African Charter, it is stated that all individuals are entitled toenjoy the right to work under fair conditions.[45] Further to that, all peoples should have favourable conditions to the aspects of development.[46] The preamble to the African Charter partly states that people should enjoy their rights as guaranteed by state parties.[47] The Mining Charter stipulates that there should be direct benefit from the mining rights and that the development needs of the communities should be identified so that funds arising from proceeds of mining directly benefit the people.[48]

In this case, the Charter as a document intended to promote good mining practices implies that the communities will be part of the decision-making processes regarding the sharing of development benefits and that the communities would in accordance, realise their development objectives. The principles of living conditions as per the Mining Charter are proper healthcare services, affordable and equitable health care systems and a balanced nutrition. A guarantee is provided that these things are positioned as the rights of mining communities.[49] It is worth reiterating that the Mining Charter is not a piece of legislation and the approach to which it utilises the notion of the right to development is likely to be problematic in the future. Reference to income generation is mentioned (in the Mining Charter) alongside the issues of beneficiation of the mining communities. The Charter mentions that this income should then be distributed by authorities.[50]

It seems that the relationship between the host communities is more transactional than evolutional. The issue here is not just about the transfer of profits but the commitment and outcomes of that commitment to development as a fundamental human right for the mining communities. So, the frame of reference is important in this instance. The Mining Charter directly addresses that particular aspiration by making it possible for part of the income generated from mining activities to be shared with the host communities. Without acknowledgement of the right to development in the process of wealth generation, the mining companies might not be held accountable for their failure to comply with the right to development. Mining activities are located at the centre of economic processes that generate enormous wealth. In this instance, without adequate regulation, mining companies may escape accountability if they fail to ensure that host communities benefit from the wealth they generate from their mining operations.

Envisioned policy provisions on the right to development as per the Mining Charter

Policy regime on the right to development in South Africa

Several scholars have made the point that the right to development seems evasive in the Constitution of South Africa and subsequent policy instruments.[51] In the first instance, the right to development does not exist in the Constitution of South Africa. It should be mentioned that the Constitution failed to integrate certain aspects that are indicative of South Africa's history and values.[52] Even though the Constitution of South Africa is revered the world over as a human rights-based constitution expounding the notions of equality, the Constitution is still imperfect in light of the right to development.[53] Some scholars lament the fact that the Constitution needs to have been infused with certain values that are unique to the South African environment.[54] One would argue that the reason why the South African Constitution embodies universal principles as opposed to issues of particularity is due to the fact that it was not a piece of document that was written only by South Africans but included international partners who gave a lending hand in the drafting process.[55]

On the one hand, this could be beneficial to South Africa in the sense that the country could easily respond to global issues borrowing from its universal appeal and on the other hand, it should be noted that the involvement of international partners could be to the detriment of policy creation.[56] The reason for my assertion is related to the fact that the West has always held a controversial view in light of the right to development as a fundamental human right.[57] The differing views in this regard between countries of the North and those of the South may have an impact on the policy regime of South Africa on the right to development.[58]

It is indeed the case that South Africa's Constitution would eventually not have any reference to the right to development in the final document that was adopted in 1996. The reason for this is related to the fact that the missing text on the right to development leaves policy makers to their own devices in relation to drafting policy on the right to development. The missing text on the right to development may create a missing link that could have integrated the socio-economic as well as civil and political rights into one form. It has been argued consistently that this missing link also creates a problem when these rights ought to be claimed by society in general.[59] It is highly plausible that if there is such a link missing in framing the right to development in the Constitution, the policy that follows will not have reference to the right to development. Even though that is the case, scholars agree that although the right to development is not framed in the Constitution, its spirit exists in the South African constitutional order, and thus, the right to development can legitimately be claimed by citizens.[60] The reason why there is much lamentation on the state of the Constitution and the missing text on the right to development is that the Constitution has not given clear aspirations on the subject matter. This is indicative of how Article 10 of the UNDRTD stipulates that states have a responsibility to develop national policy on the subject matter.[61]

Form and shape of the policy integrating the right to development and mining in South Africa

As mentioned earlier, there are polemical issues associated with the long history of mining communities in South Africa. These polemics can easily and at best be described by parts of the preamble of the UNDRTD, which makes provision for the protection and advancement of peoples' rights and freedoms.[62] The applied racialised labour force in South Africa as an object of the policy of separate development charted by the apartheid state further complicated matters for mining communities.[63] The UNDRTD mentions human rights throughout its content, implying that it was drafted with principles of human rights in mind. It is obvious on the other hand that South Africa never had a human rights culture until the dawn of democracy in the mid-1990s.[64]

Over several decades, mining operations were not only an object of internal economic activity but there have been several multinational companies that have had an interest in mining exploration in South Africa.[65] It is perhaps for this reason that the Mining Charter places foreign investors and foreign mining companies in their role and responsibilities in relation to the mining communities. The polemical issues of the past and present status quo should actually be used to shape a paradigm for policy creation in relation to mining communities and mining companies. Below, I try to justify why it is important for such a paradigm to be drawn out into the policy-making processes.

Policy paradigm on the right to development for mining communities

A policy paradigm refers to a system of standards and ideas which specify goals of the intended policy instruments that are used to attain such goals as well as the problems that they ought to address.[66] As explained above, the overarching matter is that the Mining Charter seeks to imbue a transactional approach in its aspirations rather than a social basis for rights-based development. This is where social policy is often subordinated by neoliberal economics which permits minimal space for the intervention of the state. In other words, the shift from an economic policy agenda to a social investment agenda is needed in order to tackle problems of poverty and inequality.[67] It is therefore in this context that South Africa should make considerations of its paradigm, in view of the fact that the previous era or century of mining was based on economic policies that warranted a transactional mode for development rather than a social one; hence the need for the shift in policy thinking.

Policy is bound to shift for many reasons. One of those reasons that necessitates a shift is the decision to move from one form of a developmental model to another. For instance, the South African policy regime has in the past 25 years made reference to transformation, affirmative action and problems with human rights.[68] This move was necessitated by the highly racialised labour force from the time of apartheid, separate development and a non-human rights state.[69] South Africa's recent past in as far as apartheid was concerned necessitated a shift in national policy in order to cater for indigents and address stark inequalities that were created on purpose by the apartheid government. This is not peculiar to South Africa in that there are several traces of national policy shifts to address particular socio-economic problems. For example, a country such as Australia has gone through similar shifts in national policy, moving away from the neoliberal and macroeconomic policy thinking to a social investment policy shift.[70]

Changes made in policies that are being challenged create a general shift in the policy directive.[71] In this instance, a case is made for a national policy framework for host mining communities. This shift should be made possible because of how mine workers have continued to challenge government on the rights of workers and protection from polemics which in the past resulted in class action courts cases, low wages and health risks associated with mining.Policy scholars have often observed that policy paradigms are influenced by sociological reasons but never on scientific grounds.[72] The examination of policy paradigms in relation to its decline or rise is important because of the relations between people and government. When there are unexpected policy outcomes, there tend to be disputes. These disputes arise from the networks of policy makers and unions, inclusive of government and non-governmental actors. The networks may be stable and long lasting over a period of time as was the case in how labour movements existed for over decades to challenge for the rights of mining communities and the protection of mine workers from diseases that emanate from mining activities.[73]

Policy scholars argue that paradigm policy change is necessary for application towards the de-marginalisation of a specific population.[74] In this instance, the de-marginalisation of mining communities would need to be achieved given the polemics I have explained above. In shaping a paradigm, this allows for ideas about dealing with social problems and how these problems can be solved as distinct entities.[75] Based on the issues and reasons advanced above, it seems that policy may be dictated to by the need for a change in paradigm for mining entities in South Africa. The envisioned policy aspects for mining communities should therefore cater for the right to development in the following manner:

Transformation

The transformation that the government refers to in the Mining Charter should be of such a nature that it does not start and only end with BBEE as it is stated in the implementation document. In order to sustain transformation, there would be a need for government to ensure that new role players enter the market of mining with the intention to build sustainability of those businesses. The policy aspect should take into account the type of paradigm where new businesses are able to take part in the mining process in order to ensure sustainability.

The issue of sustained transformation is important to ensure that mining communities do not become ghost towns when mining happens.In other words, it is important that there is sustainability in the host mining towns to avoid the 'ghost mining' towns phenomenon. Therefore, the type of transformation that ought to happen in the case of communities that host mining activities should happen in a manner that ensures longevity in business activities even if those activities surpass the mining industry. The importance of this rests in the idea that mining companies are beginning to explore the possibility of mechanising the mines. In this case, the notion of decent work as an object of the right to development should be foreseen and anticipated as a future prerogative of policy directive. It may, therefore, be important that an envisioned policy for advancing the right to development for host mining communities locates the problem of joblessness at the centre of human development. The type of transformation that policy should drive is where mining activities should be supported by an ancillary industry that isable to continue to exist long after the mining activities within a community have ceased.

Mutual beneficiation

Mutual beneficiation as per the Mining Charter exists in a transactional perspective where the host communities may yield parts of the profits. The transactional arrangement of this kind is not adequate because mutual beneficiation should exist beyond mining profits in order to ensure that the long-term needs of the communities are taken care of. Mutual beneficiation in relation to profits must further be given a descriptive stance with a view to formulating policy that guides communities to use the benefits derived from the proceeds of mining by putting them into a fund that can be ploughed back into the host mining communities.

Advancement and protection of the social and cultural life in host communities

In the earlier sections of this chapter, I have explained at length the polemical issues of a total breakdown of community life in towns that host mining activities. This is illustrated by how migration and other forms of life systems in the mining communities tend to be broken by social ills that become prevalent when the state fails to monitor the advancement of social problems. In this context, policy needs to be derived from the need to ensure the protection of social and cultural life of each mining community. The descriptive nature of policy in this regard would therefore make it possible for accountability to occur. When role players are identified in policy action plans, it becomes easier to detect the responsibility of those who may be held accountable for any failure of policy implementation.

Conclusion

The aspirations contained in the South African Mining Charter as they relate to the objective of mutual symbiotic transformation makes it a powerful document of relevance to communities that host mining activities. The three aspects above, which consist of transformation, mutual beneficiation and advancement of social and cultural life should inform the policy paradigm to be created in the advancement of the right to development for mining communities. The reason for such a paradigm is that policy intervention on the right to development in the context of host communities should prevail over the future and the current polemics that need to be corrected. This paradigm should explain the form and shape of transformation in a manner that is sustainable, but not in the current form where transformation seems to be associated with transactional activities of shareholding in mining activities. Transformation should be about the evolution of people's lives over a period of time. The lives I refer to are those of the people who reside in mining communities.

The polemics associated with mining companies and host communities in the past has had an impact on the social and cultural life of people. The transactional nature of the Mining Charter should not influence the form and shape of policy that could be made in the future because this could further create social and cultural problems for host communities. It should be recalled that the UNDRTD as well as the African Charter expressly mention the importance of development as an economic, social and cultural process rather than merely focusing on one aspect. It can therefore be argued that in the past century, development in relation to mining was viewed with the lens of economic development, excluding other aspects of development that are provided for in the UNDRTD and the African Charter.

Notes

1 Republic of South Africa. 2018. 'Broad Based Socio Economic Empowerment Charter for the Mining and Minerals Industry 2018'. *Government Gazette*, 27 September.

2 UN Declaration on the Right to Development Resolution A/RES/41/128 adopted by the UN General Assembly on 4 December 1986.

3 Simson, Howard. 1974. 'The Myth of the White Working Class in South Africa'. *African Review* 4(2): 189.

4 Kubicek, Robert. 1990. 'Mining: Patterns of Dependence and Development 1870–1930'. In *Studies in the Economic History of Southern Africa: Volume 1*, edited by Zbigniew, Konczacki; Parpart, Jane and Shaw, Timothy. London: Frank Cass, 64.

5 Townsend, Robert and Van Zyl, Johan. 1998. 'Stages of Growth in the South African Economy: The Role of Agriculture'. *South African Journal of Economic and Management Sciences* 1(2): 204.

6 Wolpe, Harold. 1972. 'Capitalism and Cheap Labour-Power in South Africa: From Segregation to Apartheid'. *Economy and Society* 1(4): 425.

7 O'Meara, Dan. 1975. 'The 1946 African Mine Workers' Strike and the Political Economy of South Africa'. *Journal of Commonwealth and Comparative Politics* 13(2): 146; Mabin, Allan. 2003. *Dispossession, Exploitation and Struggle: An Historical Overview of South African Urbanization: The Apartheid City and Beyond*. New York: Routledge.

8 Demissie, Fassil. 1998. 'In the Shadow of the Gold Mines: Migrancy and Mine Housing in South Africa'. *Housing Studies* 13(4): 445.

9 Ncube, Mkhululi. 2008. 'The Labour Market and Decent Work Agenda in Southern Africa: Evidence and Challenges'. *International Labour Organisation—Issue Paper No 31* 18; Wilson, Francis. 2001. 'Minerals and Migrants: How the Mining Industry Has Shaped South Africa'. *Daedalus* 130(1): 99.

10 National Union of Mine Workers. 'About Us'. *Num.org.za* <https://num.org.za/About-Us/History>.

11 Winde, Frank and Stoch, E.J. (Leslie). 2010. 'Threats and Opportunities for Post-Closure Development in Dolomitic Gold Mining Areas of the West Rand and Far West Rand (South Africa)—A Hydraulic View Part 1: Mining Legacy and Future Threats'. *Water South Africa* 36(1): 75.

12 Fine, Ben. 2010. 'Can South Africa be a Developmental State?' In *Constructing a Democratic Developmental State in South Africa: Potentials and Challenges*, edited by Omano, Edigheji. Pretoria: Human Sciences Research Council Press, 169.

13 Marais, Lochner and Cloete, Jan. 2013. 'Labour Migration, Settlement and Mine Closure in South Africa'. *Geography* 98(2): 77.
14 Moodie, Dumbar. 2009. 'Managing the 1987 Mine Workers' Strike'. *Journal of Southern African Studies* 35(1): 45.
15 Schultz, Paul and Germano, Mwabu. 1998. 'Labor Unions and the Distribution of Wages and Employment in South Africa'. *ILR Review* 51(4): 680.
16 Hnizdo, Eva and Murray, Jill. 1998. 'Risk of Pulmonary Tuberculosis Relative to Silicosis and Exposure to Silica Dust in South African Gold Miners'. *Occupational and Environmental Medicine* 56(3): 496.
17 Kleinschmidt, Immo and Churchyard, Gavin. 1997. 'Variation in Incidences of Tuberculosis in Subgroups of South African Gold Miners'. *Occupational and Environmental Medicine* 54(9): 636.
18 Winde and Stoch, note 10, 69.
19 Mazibuko, Sibonginkosi. 2013. 'Exposure of State Coloniality and Labour Aristocracy'. *The Thinker* 30.
20 Botiveau, Raphael. 2014. 'Briefing: The Politics of Marikana and South Africa's Changing Labour Relations'. *African Affairs* 113(450): 128.
21 SA History.org.za. 2014. '2014 South African Platinum Strike: Longest Wage Strike South Africa'. *South African History Online*, 24 October <www.sahistory.org.za/article/2014-south-african-platinum-strike-longest-wage-strike-south-africa>.
22 Mining-Technology.com. 2014. 'Fortune Hunters: The Top Ten Best Paid Mining Jobs'. *Mining-Technology*, 16 April <www.mining-technology.com/features/featurefortune-hunters-the-top-ten-best-paid-mining-jobs-4214939/>.
23 ENCA.com. 2014. 'Minimum of R8,500 Wage Wanted for Mine Workers'. *Enca .com*<www.enca.com/south-africa/minimum-r8500-wage-wanted-mineworkers>.
24 MacNaughton, Gillian and Frey, Diane. 2010. 'Decent Work for All: A Holistic Human Rights Approach'. *American University International Law Review* 26(2): 441.
25 Marais, Lochner and Venter, Anita. 2006. 'Hating the Compound, But […] Mineworker Housing Needs in Post-Apartheid South Africa'. *Africa Insight* 36(1): 53.
26 Ngcebetsha, Tembeka and Jamela, Nomazizi. 2015. 'Revisiting Ubuntu in the Midst of a Violent Conflict: Reflections on the Marikana Tragedy in South Africa'. *International Relations and Diplomacy* 3(8): 520.
27 Kubicek, note 3, 64.
28 African Charter on Human and Peoples' Rights. The Organisation of African Unity(OAU), 27 June 1981, CAB/LEG/67/3 rev. 5, 21 I.L.M. 58.
29 Campbell, Catherine. 1997. 'Migrancy, Masculine Identities and AIDS: The Psychosocial Context of HIV Transmission on the South African Gold Mines'. *Social Science and Medicine* 45(2): 273.
30 Rees, David. 2010. 'Oscillating Migration and the Epidemics of Silicosis, Tuberculosis, and HIV Infection in South African Gold Miners'. *American Journal of Industrial Medicine* 53(4): 398.
31 Crush, Jonathan. 2005. 'Migration and HIV/AIDS in South Africa'. *Development Southern Africa* 22(3): 293.
32 McCulloch, Jock. 2009. 'Counting the Cost: Gold Mining and Occupational Disease in Contemporary South Africa'. *African Affairs* 108(431): 221.
33 Niselow, Tehillah. 2019. 'Everything you Need to Know about the R5bn Silicosis Settlement as Mining Companies Prepare to Payout'. *Fin24.com*, 3 August.
34 Murray, Colin. 1980. 'Migrant Labour and Changing Family Structure in the Rural Periphery of Southern Africa'. *Journal of Southern African Studies* 6(2): 139.
35 Pharaoh, Robin and Schonteich, Martin. 2003. 'AIDS, Security and Governance in Southern Africa: Exploring the Impact'. *Institute for Security Studies Papers* 14.
36 Declaration on the Right to Development Resolution A/RES/41/128 adopted by the UN General Assembly on 4 December 1986.

37 Republic of South Africa. 2018. 'Broad Based Socio-Economic Empowerment Charter for the Mining and Minerals Industry 2018'. *Government Gazette*, 27 September.
38 Republic of South Africa. 2019. 'Implementation Guidelines for the Broad-Based Socio-Economic Empowerment Charter for the Mining and Minerals Industry, 2018'. *Government Gazette*, 19 December.
39 *Ibid.*
40 Krüger, Louis. 2011. 'The Impact of Black Economic Empowerment (BEE) on South African Businesses: Focusing on Ten Dimensions of Business Performance'. *Southern African Business Review* 15(3): 207.
41 DRTD, note 35.
42 *Ibid.*
43 *Ibid.*
44 *Ibid.*
45 African Charter, note 28.
46 *Ibid.*
Ibid.
47 *Ibid.*
48 Implementation Guidelines, note 38.
49 *Ibid.*
50 *Ibid.*
51 Gutto, Shadrack. 2006. 'The Right to Development: An Implied Right in South Africa's Constitutional Order'. In *Reflections on Democracy and Human Rights*, edited by SAHRC. Cape Town: South African Human Rights Commission, 109–118.
52 Moosa, Ebrahim. 2000. 'Tensions in Legal and Religious Values in the 1996 South African Constitution'. In *Beyond Rights Talk* and *Culture Talk: Essays on the Politics of Rights and* Culture, edited by Mamdani, Mahmoud. New York: St Martin's Press, 100.
53 Ngang, Carol. 2018. 'Towards a Right-to-Development Governance in Africa'. *Journal of Human Rights* 7(1): 107.
54 Moosa, note 52, 100.
55 Corder, Hugh. 1994. 'Towards a South African Constitution'. *The Modern Law Review* 57(4): 491.
56 Widner, Jennifer and Contiades, Xenophon. 2013. 'Constitution-Writing Processes'. In *Routledge Handbook of Constitutional Law*, edited by Tushnet, Mark; Fleiner, Thomas and Saunders, Cheryl. New York: Routledge, 81.
57 Sengupta, Arjun. 2001. 'Right to Development as a Human Right'. *Economic and Political Weekly* 36(27): 2527.
58 Boyle, Kevin. 1993. 'Stock-Taking on Human Rights: The World Conference on Human Rights, Vienna 1993'. *Political Studies* 43(1): 79.
59 Teleki, Mofihli. 2017. 'Framing of the Right to Development Policy in the Context of South Africa'. *Africa Insight* 47(3): 78.
60 Shai, Isaac. 2019. 'The Right to Development, Transformative Constitutionalism and Radical Transformation in South Africa: Post-Colonial and De-Colonial Reflections'. *African Human Rights Law Journal* 19(1): 494; Ngang, Carol. 2019. 'Radical Transformation and a Reading of the Right to Development in the South African Constitutional Order'. *South African Journal on Human Rights* 35(1): 25.
61 DRTD, note 36.
62 *Ibid.*
63 Chari, Sharad. 2008. 'The Antinomies of Political Evidence in Post-Apartheid Durban, South Africa'. *Journal of the Royal Anthropological Institute* 14(1): S61.
64 Slye, Ronald. 2001. 'International Law, Human Rights Beneficiaries, and South Africa: Some Thoughts on the Utility of International Human Rights Law'. *Chi J Int'l L* 2(1): 59.
65 Fauconnier, Audrey and Mathur-Helm, Babita. 2008. 'Black Economic Empowerment in the South African Mining Industry: A Case Study of Exxaro Limited'. *South African Journal of Business Management* 39(4): 1.

66 Hall, Peter. 1993. 'Policy Paradigms, Social Learning, and the State: The Case of Economic Policymaking in Britain'. *Comparative Politics* 25(3): 275.
67 Deeming, Christopher and Smyth, Paul. 2015. 'Social Investment after Neoliberalism: Policy Paradigms and Political Platforms'. *Journal of Social Policy* 44(2): 297.
68 Roux, Nico. 2002. 'Public Policy-Making and Policy Analysis in South Africa amidst Transformation, Change and Globalisation: Views on Participants and Role Players in the Policy Analytic Procedure'. *Journal of Public Administration* 37(4): 418.
69 Satgar, Vishwas. 2008. 'Neoliberalized South Africa: Labour and the Roots of Passive Revolution'. *Labour, Capital and Society/Travail, Capital et Société* 41(2): 38.
70 Deeming and Smyth, note 67, 297.
71 Hall, note 66, 275.
72 *Ibid.*
73 *Ibid.*
74 Howlett, Michael. 2014. 'From the "Old" to the "New" Policy Design: Design Thinking Beyond Markets and Collaborative Governance'. *Policy Sciences* 47(3): 187–207.
75 *Ibid*, 187.

References

Botiveau, Raphael. 2014. 'Briefing: The Politics of Marikana and South Africa's Changing Labour Relations'. *African Affairs* 113(450): 128–137.

Boyle, Kevin. 1993. 'Stock-Taking on Human Rights: The World Conference on Human Rights, Vienna 1993'. *Political Studies* 43(1): 79–95.

Campbell, Catherine. 1997. 'Migrancy, Masculine Identities and AIDS: The Psychosocial Context of HIV Transmission on the South African Gold Mines'. *Social Science and Medicine* 45(2): 273–281.

Chari, Sharad. 2008. 'The Antinomies of Political Evidence in Post-Apartheid Durban, South Africa'. *Journal of the Royal Anthropological Institute* 14(1): S61–S76.

Corder, Hugh. 1994. 'Towards a South African Constitution'. *The Modern Law Review* 57(4): 491–533.

Crush, Jonathan. 2005. 'Migration and HIV/AIDS in South Africa'. *Development Southern Africa* 22(3): 293–318.

Deeming, Christopher and Smyth, Paul. 2015. 'Social Investment after Neoliberalism: Policy Paradigms and Political Platforms'. *Journal of Social Policy* 44(2): 297–318.

Demissie, Fassil. 1998. 'In the Shadow of the Gold Mines: Migrancy and Mine Housing in South Africa'. *Housing Studies* 13(4): 445–469.

Fauconnier, Audrey and Mathur-Helm, Babita. 2008. 'Black Economic Empowerment in the South African Mining Industry: A Case Study of Exxaro Limited'. *South African Journal of Business Management* 39(4): 1–14.

Fine, Ben. 2010. 'Can South Africa be a Developmental State?' In *Constructing a Democratic Developmental State in South Africa: Potentials and Challenges*, edited by Omano, Edigheji. Pretoria: Human Sciences Research Council Press.

Gutto, Shadrack. 2006. 'The Right to Development: An Implied Right in South Africa's Constitutional Order'. In *Reflections on Democracy and Human Rights*, edited by SAHRC. Cape Town: South African Human Rights Commission, 109–118.

Hall, Peter. 1993. 'Policy Paradigms, Social Learning, and the State: The Case of Economic Policymaking in Britain'. *Comparative Politics* 25(3): 275–296.

Hnizdo, Eva and Murray, Jill. 1998. 'Risk of Pulmonary Tuberculosis Relative to Silicosis and Exposure to Silica Dust in South African Gold Miners'. *Occupational and Environmental Medicine* 56(3): 496–502.

Howlett, Michael. 2014. 'From the "Old" to the "New" Policy Design: Design Thinking Beyond Markets and Collaborative Governance'. *Policy Sciences* 47(3): 187–207.

Kleinschmidt, Immo and Churchyard, Gavin. 1997. 'Variation in Incidences of Tuberculosis in Subgroups of South African Gold Miners'. *Occupational and Environmental Medicine* 54(9): 636–641.

Krüger, Louis. 2011. 'The Impact of Black Economic Empowerment (BEE) on South African Businesses: Focusing on Ten Dimensions of Business Performance'. *Southern African Business Review* 15(3): 207–233.

Kubicek, Robert. 1990. 'Mining: Patterns of Dependence and Development 1870–1930'. In *Studies in the Economic History of Southern Africa: Volume 1*, edited by Zbigniew, Konczacki; Parpart, Jane and Shaw, Timothy. London: Frank Cass.

Mabin, Allan. 2003. *Dispossession, Exploitation and Struggle: An Historical Overview of South African Urbanization: The Apartheid City and Beyond*. New York: Routledge.

MacNaughton, Gillian and Frey, Diane. 2010. 'Decent Work for All: A Holistic Human Rights Approach'. *American University International Law Review* 26(2): 441–483.

Marais, Lochner and Cloete, Jan. 2013. 'Labour Migration, Settlement and Mine Closure in South Africa'. *Geography* 98(2): 77–84.

Marais, Lochner and Venter, Anita. 2006. 'Hating the Compound, But […] Mineworker Housing Needs in Post-Apartheid South Africa'. *Africa Insight* 36(1): 53–62.

Mazibuko, Sibonginkosi. 2013. 'Exposure of State Coloniality and Labour Aristocracy'. *The Thinker*.

McCulloch, Jock. 2009. 'Counting the Cost: Gold Mining and Occupational Disease in Contemporary South Africa'. *African Affairs* 108(431): 221–240.

Moodie, Dumbar. 2009. 'Managing the 1987 Mine Workers' Strike'. *Journal of Southern African Studies* 35(1): 45–64.

Moosa, Ebrahim. 2000. 'Tensions in Legal and Religious Values in the 1996 South African Constitution'. In *Beyond RightsTalk and CultureTalk: Essays on the Politics of Rights and Culture*, edited by Mamdani, Mahmoud. New York: St Martin's Press.

Murray, Colin. 1980. 'Migrant Labour and Changing Family Structure in the Rural Periphery of Southern Africa'. *Journal of Southern African Studies* 6(2): 139–156.

Ncube, Mkhululi. 2008. 'The Labour Market and Decent Work Agenda in Southern Africa: Evidence and Challenges'. *International Labour Organisation—IssuePaper No 31*.

Ngang, Carol. 2018. 'Towards a Right-to-Development Governance in Africa'. *Journal of Human Rights* 7(1): 107–122.

Ngang, Carol. 2019. 'Radical Transformation and a Reading of the Right to Development in the South African Constitutional Order'. *South African Journal on Human Rights* 35(1): 25–49.

Ngcebetsha, Tembeka and Jamela, Nomazizi. 2015. 'Revisiting Ubuntu in the Midst of a Violent Conflict: Reflections on the Marikana Tragedy in South Africa'. *International Relations and Diplomacy* 3(8): 520–546.

O'Meara, Dan. 1975. 'The 1946 African Mine Workers' Strike and the Political Economy of South Africa'. *Journal of Commonwealth and Comparative Politics* 13(2): 146–173.

Pharaoh, Robin and Schonteich, Martin. 2003. 'AIDS, Security and Governance in Southern Africa: Exploring the Impact'. *Institute for Security Studies Papers*.

Rees, David. 2010. 'Oscillating Migration and the Epidemics of Silicosis, Tuberculosis, and HIV Infection in South African Gold Miners'. *American Journal of Industrial Medicine* 53(4): 398–404.

Roux, Nico. 2002. 'Public Policy-Making and Policy Analysis in South Africa amidst Transformation, Change and Globalisation: Views on Participants and Role Players in the Policy Analytic Procedure'. *Journal of Public Administration* 37(4): 418–437.

Satgar, Vishwas. 2008. 'Neoliberalized South Africa: Labour and the Roots of Passive Revolution'. *Labour, Capital and Society/Travail, Capital et Société* 41(2): 38–69.
Schultz, Paul and Germano, Mwabu. 1998. 'Labor Unions and the Distribution of Wages and Employment in South Africa'. *ILR Review* 51(4): 680–703.
Sengupta, Arjun. 2001. 'Right to Development as a Human Right'. *Economic and Political Weekly* 36(27): 2527–2536.
Shai, Isaac. 2019. 'The Right to Development, Transformative Constitutionalism and Radical Transformation in South Africa: Post-Colonial and De-Colonial Reflections'. *African Human Rights Law Journal* 19(1): 494–509.
Simson, Howard. 1974. 'The Myth of the White Working Class in South Africa'. *African Review* 4(2): 189–203.
Slye, Ronald. 2001. 'International Law, Human Rights Beneficiaries, and South Africa: Some Thoughts on the Utility of International Human Rights Law'. *Chi Journal of International Law* 2(1): 59–79.
Teleki, Mofihli. 2017. 'Framing of the Right to Development Policy in the Context of South Africa'. *Africa Insight* 47(3): 78–94.
Townsend, Robert and Van Zyl, Johan. 1998. 'Stages of Growth in the South African Economy: The Role of Agriculture'. *South African Journal of Economic and Management Sciences* 1(2): 204–218.
Widner, Jennifer and Contiades, Xenophon. 2013. 'Constitution-Writing Processes'. In *Routledge Handbook of Constitutional Law*, edited by Tushnet, Mark; Fleiner, Thomas and Saunders, Cheryl. New York: Routledge.
Wilson, Francis. 2001. 'Minerals and Migrants: How the Mining Industry Has Shaped South Africa'. *Daedalus* 130(1): 99–121.
Winde, Frank and Stoch, E.J. (Leslie). 2010. 'Threats and Opportunities for Post-Closure Development in Dolomitic Gold Mining Areas of the West Rand and Far West Rand (South Africa)—A Hydraulic View Part 1: Mining Legacy and Future Threats'. *Water South Africa* 36(1): 75–82.
Wolpe, Harold. 1972. 'Capitalism and Cheap Labour-Power in South Africa: From Segregation to Apartheid'. *Economy and Society* 1(4): 425–456.

17 Impact of artisanal and small-scale gold mining on livelihoods in Mutasa District, Zimbabwe

Cephas Mandirahwe

Introduction

The African Mining Vision recognises small-scale mining in promoting sustainable livelihoods and in reducing poverty in resource-rich countries. In Zimbabwe, informal gold mining is one of the livelihood strategies since the turn of the new millennium, which has been marked by a decline of the agricultural sector. This is confirmed by the Chamber of Mines of Zimbabwe,[1] which notes that the country has shifted from an agro-based to a mining country. In this context, the contribution of informal mining is appreciated as the sector contributes 60 percent of foreign currency earnings.[2] An understanding of artisanal and small-scale mining remains contested among scholars since there is no agreed definition of the concept. Be that as it may, some scholars associate the concept of artisanal and small-scale mining with low levels of production, poorly skilled labour, subsistence and illegality. Salient to note is the fact that artisanal and small-scale mining blurs the boundary between informal and registered small-scale mining, both of which are generally considered in the definition. Some scholars link the growth of artisanal and small-scale mining with poverty.[3] The above definitions are apt for the study, as they allow an understanding of the growth of artisanal and small-scale gold mining in Mutasa District in particular and the whole of Zimbabwe in general.

The national blueprint for development code-named the Zimbabwe Agenda for Socio-Economic Transformation (ZIMASSET) (2013–2018) identifies small-scale mining as the key driver of economic growth, which has triggered a debate on the impact of artisanal and small-scale gold mining on the livelihoods of local communities. The role of artisanal and small-scale gold mining in achieving sustainable development has experienced mixed reaction where some scholars see the availability of natural resources as a curse rather than a blessing wherein the presence of natural resources seems to retard development through corruption and conflicts. Additionally, due to the unsustainable activities of artisanal and small-scale gold mining, more often than not, it has culminated in disasters that have claimed large numbers of lives. From time immemorial, the community has survived through agriculture owing to the presence of fertile soils and the availability of water bodies for irrigation of crops. Hence, agricultural production has the potential to improve the livelihoods of the communities. However, this is compromised with the activities of artisanal and small-scale gold mining in the area.

The study seeks to answer the following major research questions: What are the factors leading to the growth and expansion of artisanal and small-scale gold mining? To what extent does artisanal and small-scale gold mining contribute to the livelihoods of local communities? To what extent do the activities of artisanal and small-scale mining contribute to the realisation of the right to development and finally, to what extent has Zimbabwe implemented the right to development in as much as artisanal and small-scale gold mining are concerned? This researcher argues that despite the potential of the artisanal and small-scale gold mining in boosting economic growth, it has rather compromised the livelihoods of communities and thus, violates the right to development for the people of Mutasa District in Manicaland Province of Zimbabwe. Artisanal and small-scale gold mining is indispensable in realising the right to development for communities endowed with natural resources like gold.

Interestingly, the right to development is explicitly articulated in section 13 and further expanded in sections 9 and 11 of the Zimbabwean 2013 Constitution. However, in as much as artisanal and small-scale gold mining is concerned, the implementation of the right to development has remained a chimera for the government of Zimbabwe. Despite the contribution of artisanal and small-scale gold mining to the exploration and production of gold in Zimbabwe, it has remained an informal activity, thus perpetuating corruption in the form of gold smuggling. The activities of artisanal and small-scale mining also pose a threat to the environment, which is exposed to pollution of water bodies as a result of the use of mercury in gold processing. The livelihoods of the local population is therefore compromised and this results in the violation the right to development. The chapter therefore explores the impact of artisanal and small-scale gold mining on the livelihoods of the local communities in Mutasa District of Manicaland Province of Zimbabwe.

Background to artisanal and small-scale mining in Zimbabwe

Artisanal and small-scale mining are some of the emerging activities providing alternative livelihoods in communities endowed with mineral resources. It is imperative to note that the artisanal and small-scale mining phenomenon is not unique to Zimbabwe but rather should be viewed from a global perspective. According to Maponga and Meck, there are many different nomenclatures associated with artisanal and small-scale mining.[4] In Zimbabwe, artisanal and small-scale mining operations are called *Makorokoza*, in Ghana they are called *Galamseys* and in Brazil they are called *Garimpeiros*. The massive involvement of communities in artisanal and small-scale gold mining operations is a typical gold rush stereotype.[5] The gold rush is a period of feverish migration of workers to an area where there is a sudden discovery of gold deposits. Major gold rush periods were recorded in the 19th century in Australia, Brazil, South Africa and the United States while small gold rush periods were recorded in other countries around the globe. In Africa in general and Zimbabwe in particular, small-scale mining contributes to both direct and indirect employment and hence, a source of livelihood. Mawowa notes that the artisanal and small-scale mining industry provides temporary survival strategies against poverty and unemployment.[6] For Hentschel, despite some level of recognition, implementation of legislation among mining communities remains problematic, causing many miners to lose faith in the ability or commitment of their governments to provide assistance and as a result miners tend to depend on middlemen, thus further exacerbating the situation in the mining operations and for this reason artisanal and small-scale mining has remained informal.[7]

In Zimbabwe, mining started in ancient times as an artisanal activity. During the pre-colonial era, the early and late Stone Age societies began mining of iron, which became an important social and economic activity. The availability of mineral resources gave rise to the pre-colonial states of Great Zimbabwe, Mutapa and Rozvi. The rulers of these states became powerful due to the control they had over mineral resources, which also had a significant role on the livelihoods of the peoples. During the colonial period, gold mining became popular and was administered by the British South African Company (BSAC). Resultantly, gold mines were established, including the Sabi Gold Mine in 1890, the Jena Gold Mine in 1898 and the Blanket Gold Mine in 1898. The decolonisation of Zimbabwe left some mine shafts underutilised, which attracted the local communities into informal mining as a way of living.

At the turn of the new millennium, mining was given priority over the agricultural sector. Scoones points out that the growth of artisanal or small-scale mining is one of the features of the post-2000 era in Zimbabwe.[8] This is mainly attributed to the decline in the agricultural sector, which was a source of livelihood for households in the rural areas. Moyo reiterates the estimates from the Ministry of Mines and Development suggesting that small-scale miners are responsible for 30 percent of the 11.79 tonnes of gold mined in 2012.[9] Hence, small-scale mining is widely acknowledged as the key driver of the economy as expressed by the then-President of Zimbabwe:

> Mining, in 2015 exhibits stronger growth of well above 3.5 percent, despite the impact of declining international prices. This growth is primarily driven by substantial output increases of gold, platinum, chrome, coal, among other minerals. In order for the country to fully benefit from this sector, government will maintain the policy thrust on beneficiation and value addition that is supported by other reforms which enhance transparency and accountability in the sector.[10]

The above sentiment clearly shows that mining was given precedence over agriculture, hence perceived as the key driver of the economy. Since 2015, artisanal and small-scale gold producers have accounted for a bigger proportion of Zimbabwe's annual gold deliveries to Fidelity Printers and Refiners, the country's sole gold buyer. Gold was Zimbabwe's largest foreign currency earner, contributing USD977 million in 2017, which translates to 39 percent of total merchandise exports. This implies that of the USD977 million earned by the gold sector, small-scale miners contributed more than 50 percent.[11] Artisanal and small-scale mining contributes immensely to the national economy as compared to the primary producers. Primary producers in this context refer to large-scale miners such as commercial companies. Fidelity Printers, which is a gold buying agent has from the period between January and June 2019, recorded higher gold purchases from artisanal and small-scale miners as compared to primary producers.[12] It is therefore worth noting that small-scale producers, despite facing a plethora of challenges, continue to dominate the country's gold mining sector, accounting for more than 60 percent[13] of total deliveries. Primary producers contribute the remainder proportion of the deliveries as shown in the Fidelity Printers illustration in Figure 17.1.

The youth who constitute a large proportion of the informal mining sector thus find some form of employment and livelihood. Areas such as Mutasa District, Manicaland Province in Zimbabwe, Shurugwi in Midlands Province and Esigodini in Matebeland South are widely known for small-scale mining where the miners are popularly known as *Makorokoza* in the local language. However, the growth of the artisanal and small-scale mining sector has not happened without an adverse impact on the livelihoods of the local communities. More often than not, small-scale mining activities have resulted in disasters that culminate in the loss of their livelihood for the entire community. A case in point is the Battlefields disaster where over 20 miners were reported dead and between 50 and 100 others were trapped in the mines.[14]

Also, the use of chemicals such as mercury further accelerates environmental changes resulting in land and water pollution. While artisanal and small-scale mining has been touted by policymakers in Zimbabwe as having the potential of improving economic growth, it has compromised the livelihoods of local communities through unsustainable mining activities. Livelihood strategies

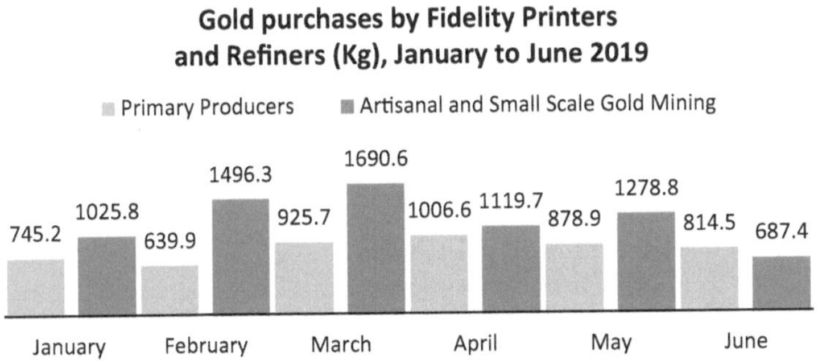

Figure 17.1 Gold purchases. *Source*: Fidelity Printers and Refiners (2019)

such as fishing and farming have been seriously affected in areas endowed with mineral resources. This analysis is set in the context of the discourse on natural resources and the right to development. Zimbabwe has ratified the African Charter on the Human and people's rights. Interestingly, article 22 of the Charter articulates the right to development, which is further divided into social, economic and political development. Even more reassuring is the fact that the right to development has been domesticated in the municipal law of Zimbabwe, precisely in Section 13 of the 2013 Constitution on national development.

The right to development is intertwined with the United Nation Sustainable Development Goals as well as well as the Africa Agenda 2063. Amartya Sen views the right to development as an umbrella right.[15] To this end, the right to development can be disaggregated into two elements for its realisation, that is, fair distribution of the benefits of development and popular participation. In Africa in general and Zimbabwe in particular, natural resources such as gold have benefitted the elite thus, making the communities with natural resources poorer. In this context the poor are always at the receiving end of the benefits accruing from the exploitation of their natural resources, which in many instances, amounts to violation of article 22 of the African Charter. By right to development standards, states have a duty to first and foremost, adopt measures to protect against damages caused by an unjust distribution of development benefits. The researcher argues that the government of Zimbabwe leaves a lot to be desired inasmuch as implementation of the right to development is concerned. The state has adopted some measures to encourage artisanal and small-scale mining activities. However, there is inadequate legislation in this regard to effectively regulate operations within the artisanal and small-scale mining sector. Paradoxically, the state acknowledges the contribution of artisanal miners on gold output, including for example, the report of the Trans Stabilisation Policy (TSP), which states that gold output in 2017 was 24.8 tonnes up from 21.4 tonnes in 2016.[16]

The increase is explained by a boost in production from artisanal and small-scale miners who continue to benefit from gold facilities made available by Fidelity Printers under the Gold Development initiative launched in 2016.[17] The 2013 Constitution of Zimbabwe provides fundamental objectives inasmuch as natural resources are concerned, including for instance the objective on ensuring that local communities benefit from the resources in their areas[18] and the objective on the empowerment of persons, groups and communities.[19] The government of Zimbabwe has, however, not been forthcoming in formalising artisanal and small-scale mining activities amidst allegations that senior governmental officials benefit from the informalisation of the sector through buying cheap gold from the informal gold sector.[20]

The researcher argues that formalisation of artisanal and small-scale mining through adopting an adequate legislative framework would go a long way in the realisation of the right to development. The availability of natural resources has not contributed to transforming the livelihoods for the communities endowed with those resources due to corruption, and the conflicts associated with extraction of mineral resources such as gold, which accordingly, impacts on the realisation of the right to development. More often than not, the government of Zimbabwe has politicised the extraction of such resources as gold and diamond. Mawowa makes the argument that artisanal and small-scale mining should be conceptualised as part of development, sustenance and reproduction of a patronage system controlled by those with state power and/or party positions.[21] The Centre for Natural Resource Governance (CNRG) notes that resource-rich communities in Zimbabwe are poorer than those communities that are not endowed with natural resources.[22]

Huber articulated the view that mineral resources have the potential to stimulate development if properly managed.[23] Scholars have diverged on defining development. The Brandt Report sees development as desirable social and economic progress[24] although what is desirable for one society may not be desirable for another. Rodney emphasises the inclusion of peoples in the development process.[25] According to him, development should be peopleoriented. This also resonates well with Africa's most respected leaders, for example, Julius Nyerere[26] reiterated the fact that the purpose of development is to meet the needs of the peoples, which in terms of article 21 of the African Charter allows the disposal of natural resources only to their exclusive interests. The situation in Zimbabwe on the contrary, demonstrates that the availability and exploitation of natural resources is rather associated with corruption and violence that impoverishes the livelihoods of local communities rather than contributing to the realisation of their right to development.

The researcher employed a sustainable livelihood approach to explain the impact of artisanal and small-scale gold mining on the livelihoods of local communities. Bebington defines livelihood as consisting of the capabilities, assets and materials like mining equipment and activities for sustaining a living.[27] In its simplest sense, livelihood is a means of gaining a living through adequate stocks and flows of food and cash to meet the basic needs of people.[28] Hence, the sustainable livelihood approach is apt for this study as it has an orientation towards the empowerment of people to make the most out of their own natural resources.

In the context of gold mining, artisanal and small-scale activities form a central aspect as a source of sustainable livelihood since the activity is reliant on a limited natural resource-base. In line with artisanal and small-scale gold mining in Zimbabwe, this refers to the ability of gold panning operations to maintain production yield levels without destroying the natural environment, destroying market forces or breaking the law.

It has remained veiled in obscurity among scholars whether gold panning maintains, enhances, depletes or degrades the local natural resources-base, especially looking at how it causes desertification, deforestation, soil erosion, declining water table and siltation of rivers. Maponga and Ngorima note that gold panning becomes a sustainable livelihood activity if it results in impro ved human health, raises income and improves the living conditions of the poor majority.[29] In this context, it enhances the realisation of the right to development. Maponga further articulates the view that gold panning becomes a sustainable livelihood activity when it creates gainful employment for the unemployed.[30] Hoadley and Limptlaw argue that small-scale gold mining has a long role to play in improving sustainable livelihoods among rural communities with the government's demonstrated commitment and the political will to support small-scale miners.[31]

While most rural development scholars agree that rural communities are predominantly sustained by subsistence farming, new trends are emerging in the rural parts of Zimbabwe following the occurrence of successive droughts. In this regard, gold panning is now seen as an alternative livelihood strategy absorbing unemployed youths. On the contrary, Mawere argues that gold panning is causing more harm than good to humans and the natural environment.[32]

Methodological approach

The research was conducted through a qualitative method. A qualitative research method enables a researcher to obtain data necessary to meet the study objectives. The researcher then employed a mixture of qualitative techniques through purposive sampling, snowballing as well as convenience sampling. Due to the illegality and informal nature of artisanal and small-scale mining activities, the researcher applied a snowballing sampling technique to identify artisanal small-scale miners on the basis of their willingness to participate in the study. To this extent, a total number of 60 artisanal small-scale gold miners were sampled using snowballing out of a population of 600 gold panners. Since the researcher sought to assess the impact of artisanal small-scale miners on the livelihoods of local communities, convenience sampling was used to select 30 households, including 15 male-headed households and 15 female-headed households. Purposive sampling was used to select ten key informants, including two from the Environmental Management Agency, two from Mutasa Rural District Council, two from Zimbabwe National Water Supply Authority and two from the District Administrator's Office.

The researcher further employed focused group discussion to complement other data collection tools. One group was established comprising 18 discussants, nine of whom were females and nine males. The researcher then used an observation guide to solicit data on the impact of artisanal small-scale miners on the livelihoods of local communities. This was achieved through four weeks of extensive transact walks[33] as the researcher sought to check the validity of the data collected using other data collection techniques.

Findings and discussions

Various factors have been identified by research participants to explain the expansion of artisanal and small-scale gold mining in Mutasa District. Among these, unemployment has been mentioned as an overriding factor that drives people to engage in gold mining as a livelihood strategy, especially as a result of the economic hardship that the people of Zimbabwe, particularly the youths suffer. To hedge against economic hardship, the youths have embarked on gold mining. From the turn of the new millennium, economic hardships have created a problem of unemployment despite some piecemeal interventions that have aimed at stabilising the economy through the indigenisation policy, for example. Many youths engage in artisanal and small-scale gold mining as an immediate response to the problem of unemployment.

Many unemployed youths from different areas of Manicaland Province have embarked on informal gold mining. Small-scale gold miners are said to be from different areas. In Mutasa District the majority are mainly from neighbouring villages like Chikanga and Grange, neighbouring cities and towns like Mutare and suburbs such as Tsvingwe, Penhalonga, Manica Bridge as well as distant areas like Masvingo and Kwekwe. In estimating their origin, a village head observes that 'the small-scale miners are composed of different groups with different backgrounds. The majority come from adjacent areas of Tsvingwe, Manicabridge, Watsomba and Grange in Mutasa District of Zimbabwe'.[34]

Since the beginning of this millennium, the level of unemployment among the youth has continuously increased due to the economic meltdown aggravated by a decline in agricultural production. Efforts earmarked on revitalising the economy in 2008 were short term as the economic hardships were experienced with the end of the Government of National Unity in 2013. With the coming of the new dispensation, unemployment levels seem to have experienced a surge, pushing many youths into informal mining as a survival strategy in the face of the economic meltdown.

Findings from this study reveal that inflation associated with the Real Time Gross Settlement System (RTGS) Zimbabwean Dollar has given rise to the demand for the United States Dollar (USD). As a result, professionals such as teachers, nurses and health practitioners from different areas have embarked on artisanal and small-scale gold mining with the intention of selling gold to the informal market where they can earn United States Dollar, which has more value than the Zimbabwean RTGS Dollar. Key informants interviewed remarked:

> We have opted for the parallel market where gold prize per gram is lucrative, that is USD30. Formal market which is responsible for purchasing gold such as Fidelity Printers in Mutare is buying in RTGS dollar, which has a low buying power as compared to the USD.

The researcher comes to the conclusion that some communities in Penhalonga area have improved their financial security through engaging in god mining.[35] It is salient to note the changes in land ownership resulting in expansion of the artisanal and small-scale mining sector. Mabena explains that changes in land ownership in Zimbabwe due to the land reform programme have increased the off farming activities, particularly focussing on artisanal and small scale.[36]

The African Charter stipulates that states shall have the duty, individually or collectively, to ensure the exercise of the right to development.[37] Indeed, economic policies such as the land reform programme and the indigenisation and black empowerment policies were crafted in pursuit of the obligation imposed on Zimbabwe to ensure the realisation of the right to development.

Contribution of artisanal gold mining to community livelihood

Small-scale miners have some comparative advantages that they score for themselves and the communities they serve. In pursuit of national development, Zimbabwe has adopted policies aimed at empowering the people of Zimbabwe as a means of realising the right to development as articulated in article 22 of the African Charter and article 13 of the national Constitution. However, the Reserve Bank of Zimbabwe position on artisanal and small-scale mining, particularly gold mining has been far from consistence. Controversially, it banned the activities of artisanal and small-scale mining while at the same time buying gold from the informal sector without verifying its source. Despite the criminalisation of such activities, the bank proceeded in 2004 to issue special licences to designated agents to buy gold from gold rush sites as part of its decentralisation of gold buying. Mawowa notes that most of the licensed gold buyers were in fact, Zimbabwe African National Union—Patriotic Front (ZANU-PF) officials.[38] In this context, there is no transparency and accountability inasmuch as controlling artisanal and small-scale gold mining activities is concerned and hence, impairing the realisation of the right to development of the rest of the peoples in the communities where the mining operations take place.

This notwithstanding, with regard to the high unemployment level in the country, the artisanal and small-scale mining sector has stood firm as a source of employment through which miners eke a living. Most of the income they generate is used within the communities and thus helps to boost the local economy. While most of the income is said to be spent on beer and other luxury commodities, at the end of the day it contributes to the gross domestic product (GDP) of the country. The term *Makorokoza* as artisanal and small-scale mining is known locally implies that the area has disposable income. In the areas where artisanal and small-scale mining operations take place, there are makeshift tuck shops that sell products that the miners need most, which means the retail sector is also benefiting immensely from the artisanal and small-scale mining sector.

Concerning the effects on community livelihoods, the respondents revealed that, although it is difficult to control how some of the artisanal and small-scale miners spend their income unprofitably, some of them are said to have invested in property. A good number are also reported to have purchased vehicles, stands for housing and commercial businesses, and mining equipment, while others have moved from artisanal to commercial mining using machinery and partnering with companies in the mining sector.[39] However, sexually transmitted diseases such as HIV and AIDS are also reported to be on the rise in areas where artisanal miners operate. A National AIDS Council report on HIV and AIDS testing and analysis reveals that health institutions in areas where the majority of small-scale miners reside and also operate are in the top ten clinics with the highest number of persons who have tested positive for HIV. These health centres are the catchment areas for *Makorokoza*. Old Mutare, St Augustine's and Premier are in the operational areas of *Makorokoza*.

It is unfortunate that the intended community development that small-scale miners were envisaged to bring has still a long way to come. There is need to educate miners to utilise the revenues from their mining operations sensibly and profitably. Promiscuity is noted to be increasing in the area. Community leaders said some marriages are breaking up as wives get attracted to small-scale miners whose spending spree is 'too good to resist'. In contrast, while small-scale mining is said to be crucial for advancing economic growth in the country, it is affecting livelihoods in local communities adversely.

A community leader indicated during interviews that all the area that is demarcated for land redistribution is being mined by small-scale miners. This has the long-term impact of reducing agricultural production on the plots in question, which means the right to food for the local communities is compromised. Findings from this study also reveal that in January 2017, 14 cattle died as a result of drinking cyanide contaminated water. This was confirmed by a key informant who stated:

> In January 2017, probably due to incisive rains, or substandard structures of the cyanide ponds or tanks or both, there was over spillage though direct precipitation or excessive leakage or rather leaching of the chemical through the soil into nearby streams and water bodies that feed into Mutare River. To this end, 14 (fourteen) of his cattle died instantly due to the cattle possibly grazing contaminated grass or drinking cyanide contaminated water from the spillages and leakages/leaching of the chemical from Armstrong Custom Mill cyanide tanks. This was picked after specialists did the post-mortem of the carcases of the cattle.[40]

The 2013 Constitution of Zimbabwe contains many provisions that enshrine human rights entitlements. Section 73(1) (a) of the Constitution provides that every person has the right

> to an environment that is not harmful to their health or well-being as well as the right "to have the environment protected for the benefit of present and future generations, through reasonable legislative and other measures that secure ecologically sustainable development and use of natural resources while promoting economic and social development".

What it means is that activities of small-scale gold miners in Penhalonga area in Mutasa District violate the rights of communities through the reckless disposal of cyanide in the environment.

Impact of artisanal gold mining on the environment

All alluvial gold miners operate on river banks and they process their ore manually along the rivers, which has a detrimental effect as the dust from the mineral contaminates the rivers. Besides aquatic animals losing their habitat, the communities that live downstream in areas where small-scale miners operate face difficulties in accessing safe water for domestic use. In fact, the district Environmental Management Agency officer highlighted that they have embarked on a rehabilitation project worth about USD8000. The Agency has been battling with illegal mining activities along the Mutare River since 2014.

The availability of gold deposits in the Penhalonga–Olda Mutare area has contributed significantly to the growth of artisanal and small-scale mining in the area. There was a recent gold rush at Premier Estate, along the Mutare River where an estimated 500 illegal gold miners have been engaged in mining on the site. Water bodies in the area have severely been affected by activities of artisanal small-scale miners, resulting from the use of mercury in gold processing. From the transact walks carried out by the researcher, siltation of river bodies was noticed, in areas where small-scale gold mining is carried out. The Environmental Management Agency rehabilitation project in the Mutare River (Premier estate) revealed environmental concerns over land degradation, river siltation and water turbidity. It was noted that at one time, the Mutare River water was in the yellow class, implying that the water was not environmentally friendly or fit for domestic or commercial use.

The use of mercury by small-scale gold miners has also contaminated water bodies in the Penhalonga area leading to the destruction of aquatic life. In the case of the Mutare River contamination by the mercury used, the levels may be difficult to determine owing to the fact that mercury has long-term effects as it has the properties of bio-accumulation and bio-magnification over time. The Mutare River pours its waters into the Odzi River, and then into the Save river. The unfortunate situation with regard to the use of mercury in the mining operations in the area is that it has long-term effects that may be difficult to assess in the immediate instance since signs and symptoms of exposure take a long time to manifest.

Conclusion and recommendations

It is critical to note that artisanal and small-scale mining is fundamental to the development of communities endowed with natural resources. Community development is possible with a supportive legal framework, which is the duty of member states to establish. Zimbabwe has constitutionalised the right to development, which is a positive step in acknowledging the relevance of a human rights-based approach to development. However, implementation of the right to development in Zimbabwe leaves a lot to be desired. While natural resources are envisaged to play a catalytic role in the realisation of the right to development, with regard to artisanal and small-scale mining, there is still a lot of inconsistency in regularising the sector in spite of its enormous contribution to the

Zimbabwean economy. Consequently, the impact of artisanal and small-scale mining on the environment and livelihoods is both diverse and complicated. It is diverse in the sense that it has both positive and negative implications which outweigh the positive ones. The complication arises from the fact that despite the adverse implications of small-scale mining, the miners involved in the operations are noted to contribute to national development but yet, little is being done to formalise their operations to become more productive to the local communities.

Unfortunately, the local communities, unlike the state, are not benefiting significantly from the small-scale mining ventures. In Mutasa District, the envisaged positive impact on livelihoods is noted to be rather negative instead of positive as the environment, vegetation and water sources are being destroyed in the process of small-scale gold mining. The open pits that the miners create are not reclaimed. This has had a detrimental impact on the health of both animals and human beings. Livestock, a major source of livelihood for the rural folk get trapped in the open pits. The small-scale miners are also doing little to bring concrete socio-economic development to the Mutasa District. They pay little attention to pertinent issues like building modern dwellings for themselves or even renovating old homes within the community. The majority of the miners seem to want to maintain the status quo and leave the communities underdeveloped.

Currently in Zimbabwe, there are no clear rules and regulations governing artisanal mining operations. At the time of finalising this study, it was established that the Mines and Minerals Bill (MMAB, 2015), which was introduced to address the challenges bedevilling the mining sector, fails to fully recognise artisanal and small-scale miners but rather only describes them as holders of small mining locations as compared to large-scale miners. No attempt has been made to comprehensively define artisanal mining. The bill is, however, yet to be adopted. The researcher recommends that the government of Zimbabwe, through the Mines and Mining Development Ministry speeds up the formulation of the mining policy, which would formalise the operations of artisanal and small-scale mining. The formulation process should be participatory and include stakeholders such as miners, mining communities, government and civil society.

Notes

1 Chamber of Mines. 2015. 'State of the Mining Industry Survey: Report on the Findings 2015'. <www.chamberofminesofzimbabwe.com/wp-content/uploads/2016/05/State-of-Mining-Industry-Survey-Report-2015>.

2 Reserve Bank of Zimbabwe. 2019. 'Mid Term Monetary Policy Statement Transition to Normalcy'. <www.rbz.co.zw/documents/mps/MPS-September-2019-.pdf>.

3 Hentschel, Thomas; Hruschka, Felix and Priester, Michael. 2002. *Global Report on Artisanal and Small-Scale Mining*. England: International Institute for Environment and Development and World Business Council for Sustainable Development.

4 Maponga, Oliver and Meck, Maideyi. 2003. 'Illegal Artisanal Gold Panning in Zimbabwe'. *Journal of Rural Studies* 23(3):349.

5 International Labour Organisation. 2012. 'Facts on Small-scale Mining'. *International Labour Organisation*. <www.ilo.org/communication>.

6 Mawowa, Showers. 2013. 'The Political Economy of Artisanal and Small-Scale Gold Mining in Central Zimbabwe'. *Journal of Southern African Studies* 39(4): 921–936.
7 Hentschel, Hruschka, and Priester, note 3.
8 Scoones, Ian. 2014. 'Zimbabwe's Gold Rush: Livelihoods for the Poor or a Patronage Economy or Both?' *Zimbabweland*, 27 January <https://zimbabweland.word press.com/2014/01/27/zimbabwes-gold-rush-livelihoods-for-the-poor-or-a-patronage-economy-or-both/>.
9 Moyo, Roman. 2013. 'Zimbabwe: Small-Scale Gold Miners Get $20m Cash Boost'. *AllAfrica*, 19 September <http://allafrica.com/storiesw/201309200278.html>.
10 Stanley, S. 2013. 'Contribution of the Mining Sector to Sustainable Economic Development of the Zimbabwean Economy'. *Institute of Administration and Commerce* <www.iac.co.zw/syllabus/ChamberofMines.pdf>.
11 Nyoni, Mthandazo. 2018. 'Government Urges Small-Scale Miners to Diversify Commodity Portfolio'. *Newsday*, 21 August <www.newsday.co.zw/2018/08/govt-urges-small-scale-miners-to-diversify-commodity-portfolio/>.
12 Fidelity Printers and Refiners Report 2019.
13 Reserve Bank of Zimbabwe, note 2.
14 Chidakwa, Blessing. 2019. 'Just In: Battlefields Mine Disaster Death Toll Rises to 26'. *The Herald*, 28 February <www.herald.co.zw/just-in-battlefields-mine-disaster-death-toll-rises-to-26/>.
15 Sen, Amartya. 2010. 'Human Rights and Development'. In *Development as a Human Right: Legal, Political and Economic Dimensions*, edited by Bård, Andreassen and Marks, Stephen. Cambridge: Intersentia, 11.
16 'No Questions Asked' Policy on Gold Buying.
17 Transitional Stabilisation Programme, 5 October 2018, para 1030.
18 Constitution of Zimbabwe 2013, sect 13(4).
19 *Ibid*, sect 14.
20 Mawowa, note 6.
21 *Ibid*.
22 Centre for Natural Resource Governance. 2015. 'Communities, Companies and Conflict in Zimbabwe'.
23 Huber, Maria. 2014. 'Extractivism and Transition in Africa: Opportunities and Challenges'. MPhil Thesis, University of Stellenbosch, South Africa.
24 Brandt, Willy. 1980. *North-South: A Programme for Survival—Report of the Independent Commission on International Development Issues*. London: Pan Books.
25 Rodney, Walter. 1972. *How Europe Underdeveloped Africa*. London: Bogle-L'Overture, 9.
26 Nyerere, Julius. 1973. *Freedom and Development/Uhuru na Maendeleo: A Selection for Writings and Speeches 1968–1973*. Dar es Salaam/Oxford: Oxford University Press, 59.
27 Bebbington, Anthony. 1999. 'Capitals and Capability. A Framework for Analyzing Present Viability, Rural Livelihoods and Poverty in World Development'. *World Development* 27(12): 2021–2044.
28 Bernstein, Henry. 1992. 'Agrarian Structures and Change: Latin America'. In *Rural Livelihoods: Crises and Response*, edited by Bernstein, Henry; Crow, Ben and Johnson, Hazel, 27–50. Oxford: Oxford University Press and The Open University, 36.
29 Maponga, Oliver and Ngorima, Clay. 2003. 'Overcoming Problems in Gold Panning Section through Legislation and Education: The Zimbabwean Experience'. *Journal of Cleaner Production* 11(2): 147–157.
30 Maponga, Oliver. 1999. 'Socio-Economic and Environmental Impacts of Alluvial Gold panning in Zimbabwe'. *University of Zimbabwe—A Progress Report*.
31 Hoardley, Marie and Limpitlaw, Daniel. 2004. 'The Artisanal and Small-Scale Mining Sector and Sustainable Livelihoods'. Paper presented at the Mintek Small Scale Mining Conference, Book Proceedings, Johannesburg.
32 Mawere, Munyaradzi. 2011. 'Gold Panning in Central Mozambique: A Critical Investigation of the Effects of Gold Panning in Manica'. *The International Journal of Politics and Good Governance* 2(4): 1–8.

33 Transact walks refers to the observatory walks carried out by the researcher.
34 Interview with an anonymous miner on 12 April 2019.
35 Interview with an anonymous miner on 14 April 2019.
36 Mabhena, Clifford. 2012. 'Mining with a "Vuvuzela": Reconfiguring Artisanal Mining in Southern Zimbabwe and Its Implications to Rural Livelihoods'. *Journal of Contemporary African Studies* 30(2): 219–233.
37 African Charter on Human and Peoples' Rights, adopted in Nairobi, Kenya, on 27 June 1981. Organisation of African Unity (OAU) Doc CAB/LEG/67/3 Rev.5(1981), art 22(2).
38 Mawowa, note 6.
39 Interview with an anonymous miner on 5 May 2019.
40 Interview with an anonymous miner on 10 February 2019.

References

Bebbington, Anthony. 1999. 'Capitals and Capability. A Framework for Analyzing Present Viability, Rural Livelihoods and Poverty in World Development'. *World Development* 27(12): 2021–2044.

Bernstein, Henry. 1992. 'Agrarian Structures and Change: Latin America'. In *Rural Livelihoods: Crises and Response*, edited by Bernstein, Henry; Crow, Ben and Johnson, Hazel. Oxford: Oxford University Press and The Open University, 27–50.

Brandt, Willy. 1980. *North-South: A Programme for Survival—Report of the Independent Commission on International Development Issues*. London: Pan Books.

Hentschel, Thomas; Hruschka, Felix and Priester, Michael. 2002. *Global Report on Artisanal and Small-Scale Mining*. England: International Institute for Environment and Development and World Business Council for Sustainable Development.

Huber, Maria. 2014. 'Extractivism and Transition in Africa: Opportunities and Challenges'. MPhil Thesis, University of Stellenbosch, South Africa.

Mabhena, Clifford. 2012. 'Mining with a "Vuvuzela": Reconfiguring Artisanal Mining in Southern Zimbabwe and its Implications to Rural Livelihoods'. *Journal of Contemporary African Studies* 30(2): 219–233.

Maponga, Oliver and Meck, Maideyi. 2003. 'Illegal Artisanal Gold Panning in Zimbabwe'. *Journal of Rural Studies* 23(3): 349.

Maponga, Oliver and Ngorima, Clay. 2003. 'Overcoming Problems in Gold Panning Section through Legislation and Education: The Zimbabwean Experience'. *Journal of Cleaner Production* 11(2): 147–157.

Maponga, Oliver. 1999. 'Socio-Economic and Environmental Impacts of Alluvial Gold panning in Zimbabwe'. University of Zimbabwe—A *Progress Report*.

Mawere, Munyaradzi. 2011. 'Gold Panning in Central Mozambique: A Critical Investigation of the Effects of Gold Panning in Manica'. *The International Journal of Politics and Good Governance* 2(4): 1–8.

Mawowa, Showers. 2013. 'The Political Economy of Artisanal and Small-Scale Gold Mining in Central Zimbabwe'. *Journal of Southern African Studies* 39(4): 921–936.

Nyerere, Julius. 1973. *Freedom and Development/Uhuru na Maendeleo: A Selection for Writings and Speeches 1968–1973*. Dar es Salaam/Oxford: Oxford University Press.

Rodney, Walter. 1972. *How Europe Underdeveloped Africa*. London: Bogle-L'Overture.

Sen, Amartya. 2010. 'Human Rights and Development'. In *Development as a Human Right: Legal, Political and Economic Dimensions*, edited by Bård, Andreassen and Marks, Stephen. Cambridge: Intersentia.

Conclusion

18 Towards a resilient development future in Africa

Carol Chi Ngang

Africa's development future, which from the diverse perspectives articulated in this book—entailing improvement in living standards—depends principally on how effectively and judiciously natural resources are disposed of or utilised to the exclusive benefit of the peoples on the continent as stated in article 21 of the African Charter and other international human rights instruments. Taking cognisance of this fact, if the commitments contained in various instruments on the African agenda for development are to be taken seriously, the peoples of Africa would have legitimate reasons to be optimistic about the commitment to transform, grow and industrialise the domestic economies through beneficiation and value addition of natural resources.[1] The optimism would be even more reassuring if the intention to maximise the gains that are obtained from the exploitation or free disposal of the natural resources were to be aligned with the right to development that is granted to be enjoyed by all the peoples of Africa. It is inconceivable that Africa gave birth to the concept of the right to development[2] and in spite of the assurance in the 'equal enjoyment of the common heritage' in ensuring its realisation, is yet to substantively benefit from asserting entitlement to such a right.

It is important to reiterate therefore, as Carol Ngang illustrates in Chapter 2 that the realisation of the right to development in Africa is preconditioned on equitable (re)distribution of the common African heritage, comprising of all the natural wealth and resources, which they are entitled to exclusively benefit from and may only dispose of as they may find reason to do so. By effectively benefiting from asserting entitlement to the right to development means that the outcomes of development processes should become manifest on the ground and in the lives of the peoples of Africa in terms of improved living standards and visible changes in their socio-economic and cultural circumstances. Recognition that the right to development is attainable through asserting exclusive rights over natural resources makes the right to permanent sovereignty over natural resources a vector to the right to development. It means, as explicitly stated in article 22(1) of the African Charter that the peoples of Africa must first and foremost be able to equitably benefit from the African common heritage on the basis of which enjoyment of the right to development could then be gauged.

Dealing with the complexities in natural resource sovereignty

There is no gainsaying that natural resources play a crucial role in accelerating development and of course, in advancing living standards as has been the case in many developed and industrialised countries around the world. However, it is not to be ignored that industrialisation and development in Europe and America and presently China and other emerging economic power houses are attained at the expense of natural resources pillaged from Africa. On the basis of this understanding, the law has extensively been codified at the international, regional and domestic levels to ensure not only protection but indeed also, the permanent sovereign ownership rights that all peoples are legitimately entitled to. The contributions in this book have severally highlighted this exclusive right and in accordance, the natural resource ownership factor, which the peoples of Africa are entitled to explore and take possession of as a means to actualise the right to development. The reality however, as Isaac Shai points out in Chapter 3 is one of 'loss, anguish and postponement' as he alludes to the 'paradox of waiting' endlessly for the opportune moment to explore the wealth of natural resources on the African continent while others continue to reap huge gains from it to the detriment of the right to development in Africa as De Man and Ngang point out in Chapter 7.

The natural resource ownership factor appears to have comprehensively been understood and appropriated by indigenous peoples more than any other group of peoples. This is clearly articulated in Chapters 9 and 10 where Esther Njieassam and Ebun Abolarin, respectively illustrate how some indigenous communities in Africa have successfully claimed ownership over their ancestral lands and the natural resources thereon. Besides securing the adoption of the Declaration on the Rights of Indigenous Peoples, which makes provision to the effect that indigenous peoples and their future generations are by virtue of their unique cultural lifestyles, indigenous spirituality, traditions and customs, exclusively entitled to their lands, territories and resources, and therefore forbids anyone from encroaching thereon,[3] they have proceeded in many instances to effectively vindicate those rights before a court of law and other extra-judicial enforcement mechanisms. Even though article 21 of the African Charter equally guarantees to all the peoples of Africa exclusive ownership rights over their natural wealth and resources, not many other groups of peoples on the continent have been firm in asserting that entitlement in the manner in which indigenous peoples have done probably because the latter attach superior value to land and natural resources as an indispensable factor for their survival.

In looking at conceptual issues of sovereignty over natural resources, Faith Kabata draws attention in Chapter 8 to that while it remains a matter of contestation between African state governments and their peoples, the law clearly imposes a limitation on states' sovereignty claim over natural resources and attributes that entitlement to peoples. Acknowledging this fact has had positive implications on the extent to which the peoples of Africa can genuinely exercise sovereignty over their natural resources and hence, implications too on prospects of achieving the right to development. Exacerbated by the unprecedented substantial loss of natural resources through increasing land grabbing as Jean-Claude Ashukem demonstrates in Chapter 12, it is of essence, to draw attention once again, as highlighted in Chapter 1, to the need to not only look at peoples' sovereignty over natural resources in isolation but to increasingly emphasise its mutually reinforcing connection to the right to development.

Legal commitments on the right to development in Africa

It is worth restating that in addition to the Declaration on the Right to Development, which says the human right to development also implies the inalienable right to full sovereignty over all natural wealth and resources,[4] the African Charter equally underscores the fact that the realisation of the right to development is subject to the equal enjoyment of the common heritage of mankind.[5] This is corroborated by the entitlement to natural resource ownership guaranteed to the peoples of Africa.[6] The concept of the common heritage of mankind is understood in context to mean the African patrimony (commonwealth), which Carol Ngang elucidates on in Chapter 2. The provisions on natural resource ownership and the right to development impose two important legal commitments which state parties are obligated to comply with.

The connection between the right to development and the right to sovereign ownership over natural resources amounts to a commitment to self-sufficiency, which means that for the right to development to be achieved the peoples of Africa must self-reliantly explore the benefits that their natural resources offer. Robert Hitchcock writes that the self-reliance approach has been employed in a number of countries in Southern Africa including Botswana, Zimbabwe and Namibia where local communities have been able to form community trusts engaged in the management and utilisation of natural resources for their collective benefit.[7] While self-reliance is acknowledged to be a determining factor to development in Africa as indicated in Agenda 2063,[8] domestic laws seem to, in some instances, constitute a hindrance as Bridget Antonio-Apedzi illustrates in Chapter 15 with the example of Ghana.

Notable about natural resource sovereignty and the right to development in Africa are their unique formulation as essentially collective entitlements, which entails a commitment to collective benefit sharing[9] with the aim, as contained in Agenda 2063 to achieve 'shared prosperity and well-being'.[10] Shared prosperity and well-being connotes a development context that is supposed to incorporate strategies for the sustainable eradication of poverty and therefore, basically summarises the core purpose of the right to development, which as indicated in the relevant legal instruments discussed in this book, grants entitlement for it to be enjoyed by all peoples. An accurate understanding of the right to development in Africa illustrates that only a collective of peoples can assert a legitimate claim thereof, implying that development gains accruing from the exploitation or free disposal of natural resources ought to be shared equitably to the benefit of the collective. It is within this context that Dechassa and Jalata as well as Cephas Mandirahwe provide empirical findings in Chapters 13 and 17, respectively on how local communities in Ethiopia and in Zimbabwe collectively benefit from their natural resources. The collectivism ascribed to the right to development and to natural resources ownership obligates state governments to craft policy frameworks that guarantee equitable redistribution, which however, remains elusive.

Redressing the policy gaps

The Declaration on the Right to Development stipulates in article 2(3) that '[s]tates have the right and the duty to formulate appropriate national development policies that aim at the constant improvement of the well-being of the entire population and of all individuals'. Article 10 of the Declaration equally stipulates that to ensure the full exercise and progressive enhancement of the right to development, entails the formulation, adoption and implementation of policies, legislation and other measures. Because natural resources representa lever to the end goal of development through which constant improvement in human well-being is envisaged, the crafting of national development policies ought to prioritise equitable (re)distribution of natural resources and the interest of the peoples to whom the resources legitimately belong. The fact that the right to development in Africa remains largely unattainable with resultant adverse impact on well-being and prospects for improvement in living standards across the continent means that envisaged national development policies that are supposed to facilitate the processes for development are either non-existent or generally not suitable enough.

Looking into the apparent policy gaps, Daphine Agaba makes it clear in Chapter 4 that the 'blind adoption of policies and measures that do not directly contribute to development' in Africa generally result from the failure to fully understand or engage with the situational realities on the ground. In addressing some of these realities, Aaron Salau makes the suggestion in Chapter 5 for the adoption of 'resource constitutionalism' as a means to ensure the efficient management of petroleum resources in Africa's petrostates. Considering water as a source of life and a vital component of the right to development, Joy Saungweme highlights in Chapter 11, the need for measures in protecting the right to water in tourism-impacted communities in Africa. With a keen interest in the mining sector, Mangu and Kamga make an argument in Chapter 14 in favour of an obligatory corporate social responsibility policy framework as a means of actualising the right to development for local communities in the Democratic Republic of Congo (DRC). Mofihli Teleki's interrogation of the South African Mining Charter provides illustration on how the right to development could be made practicable in communities that host mining operations.

Even though the obligation to adopt appropriate national development policies contained in the Declaration on the Right to Development is generally non-binding, in Africa, a binding obligation is enshrined in article 22(2), corroborated by article 1 of the African Charter, which compel state parties to take the relevant legislative and policy measures among others for the realisation of the right to development. The obligation contained in article 22(2) further requires state parties, in addition to their domestic efforts, to also engage in extraterritorial (collectively with other state parties) measures to actualise the right to development. Extraterritorial measures in this instance would mean for example, harmonising national development policies, including those that have to do with the exploitation and/or free disposal of natural resources so that some level of equilibrium in living standards is established across the continent.

The binding obligation factor in states' duty to formulate appropriate national development policies for the realisation of the right to development in Africa is crucial in the sense that when the states fail in this duty, recourse can be had to the law for remedy as has been discussed in the jurisprudence on the right to development that has emanated both from the African Commission and the African Court on Human and Peoples' Rights. However, note is taken in Chapter 1 on the limits of the law in appropriately redressing questions relating to sovereignty over natural resources, which in Africa is inherently political, requiring more of political action, especially with regard to policy formulation on the issues. For instance, in spite of the positive rulings in favour of Kenya's indigenous communities adjudicated by the African Commission and the African Court in the *Endorois* and *Ogiek* cases, respectively, the orders have to date, yet to be implemented. Even so, the relevance of the law in shaping policy direction cannot be overlooked. Despite the Kenyan government's intransigence in enforcing the rulings of the African Commission and the African Court, it is noted that the *Endorois* decision significantly influenced major reforms that led to the adoption of the 2010 Constitution, which enshrines in article 56, provisions in favour of minority and marginalised groups.

When states fail in their legal obligations to adopt appropriate national development policies, they basically contravene both the sovereignty rights over their natural resources and the right to development that are guaranteed to the peoples of Africa. In essence, the peoples are justified and equipped with the weaponry to have recourse to the law for remedy, which even though in the instance, may not effect immediate change, is likely to increasingly compel African state governments to, in the medium or long term, take the necessary political actions in adopting development policies that ensure equitable redistribution of natural resources for the realisation of the right to development.

Political action demands transformational leadership, as Daphine Agaba draws attention to in Chapter 4, capable of formulating fiscal reserve policies in providing adequate responses, especially when confronted with contingencies such as the COVID-19 pandemic. The unprecedented public health emergency has not only disrupted normality throughout the world; in Africa, it has exposed the weaknesses in the coping capacity of almost every state government, driving them once again (as always) into unsustainable borrowing (exorbitant debt-to-gross domestic product (GDP) ratio) from external sources. The contradiction is that while COVID-19 infection and fatality rates have been less severe in Africa compared to other parts of the world, the medium-to-long-term fall-out and adversities will drive more people on the continent into extreme levels of poverty.[11] Speaking on 'COVID-19 and the right to development' at the 45th Session of the Human Rights Council, Michelle Bachelet among other high-profile personalities underscored the fact that the response and recovery from COVID-19 provides the opportunity to reverse longstanding cycles of poverty and inequality and in the main, require recourse to 'new development models',[12] which for Africa, as Carol Ngang posits in Chapter 2, entails generating development financing from the vast reservoir of natural resources on the continent, needed for redressing not only public health emergencies but also the myriad of livelihood sustainability challenges that the peoples of Africa continue to experience.

Concluding observations

The reality that remains to be confronted is the fact that in spite of the abundance of natural resources across Africa, their exploitation is yet to contribute to or accelerate realisation of the right to development either because there are no policies to that effect or existing ones are overtly not functional. The law requires, as explicitly stated in article 1 of the African Charter, that in order to give effect to the human and peoples' rights enshrined therein, including the right to natural resource ownership and the right to development contained in articles 21 and 22, respectively, state parties undertake to adopt legislative and *other* measures. Because existing legislative measures in the instance where they have been undertaken are apparently not sufficient, the requirement of *other* measures obligates African state governments, as a practical first step, to genuinely recognise as jurisprudence the African Commission and the African Court that have set the precedence that sovereignty over natural resources actually belongs to the peoples and not the state. Such a shift in perception is likely to significantly shape policy direction with regard to ownership, control, effective utilisation and equitable (re) distribution of the natural resources across the African continent in ensuring the realisation of the right to development. States have a more fiduciary sovereignty and policy-making mandate in ensuring effective governance of natural resources to the exclusive interest of their peoples.[13] It demands assertiveness, diligence, shrewdness and transformative political leadership in doing so.

Notes

1 See for example AU Commission. 2015. 'Agenda 2063: The Africa we want'. *African Union*, para. 72(d); OAU Lagos Plan of Action for the Economic Development of Africa 1980–2000 adopted in Lagos, Nigeria on April 1980, paras 76–87 and 172 (6.3 and 6.4).

2 Ngang, Carol. 2018. 'Towards a Right-to-Development Governance in Africa'. *Journal of Human Rights* 17(1): 110–114; Adeola, Ramola. 2018. 'The Right to Development under the African Charter: Is there an Extraterritorial Reach?' In *Perspectives on the Right to Development*, edited by Ngang, Carol; Kamga, Serges and Gumede, Vusi. Pretoria: Pretoria University Law Press, 34; Ouguergouz, Fatsah. 2003. *The African Charter on Human and Peoples' Rights: A Comprehensive Agenda for Human Dignity and Sustainable Democracy in Africa*. Netherlands: Martinus Nijhoff Publishers, 298.

3 UN Declaration on the Rights of Indigenous Peoples, Resolution 61/295 adopted by the General Assembly on 13 September 2017, arts 25, 26, 28 and 32; see also arts 8(b) and 10.

4 Declaration on the Right to Development Resolution A/RES/41/128 adopted by the UN General Assembly on 4 December 1986, art 1(2).

5 African Charter on Human and Peoples' Rights adopted by the Organisation of African Unity (OAU) in Nairobi Kenya on 27 June 1981, OAU Doc CAB/LEG/67/3 rev. 5, art 22(1).

6 *Ibid*, art 21.

7 Hitchcock, Robert. 2002. 'A Road to Self-Sufficiency: Natural Resource Management'. *Cultural Survival Quarterly Magazine*; Okidi, Odidi. 1984. 'Management of Natural Resources and the Environment for Self-Reliance'. *Journal of Eastern African Research & Development* 14: 95–96.

8 AU Commission, *Agenda 2063*, note 1, para 19.

9 Torpey-Saboe, Nichole *et al.* 2015. 'Benefit Sharing Among Local Resource Users: The Role of Property Rights'. *World Development* 72: 408–418; Nkhata, Bimo *et al.* 2012. 'A Typology of Benefit Sharing Arrangements for the Governance of Social-Ecological Systems in Developing Countries'. *Ecology and Society* 17(1): 17; Nkhata, Bimo; Breen, Charles and Mosimane, Alfons. 2012. 'Engaging Common Property Theory: Implications for Benefit Sharing Research in Developing Countries'. *International Journal of the Commons* 6(1): 52–69; Brosio, Giorgio and Singh, Jan. 2014. 'Revenue Sharing of Natural Resources in Africa: Reflections from a Review of International Practices'. *The World Bank Background Paper 90252.*

10 AU Commission, *Agenda 2063*, note 1, para 6.

11 Dabalen, Andrew and Paci, Pierella. 2020. 'How Severe Will the Poverty Impacts of COVID-19 be in Africa?' *World Bank Blocs*, 5 August <https://blogs.worldbank.org/africacan/how-severe-will-poverty-impacts-covid-19-be-africa>.

12 UN Human Rights. 2020. 'COVID-19 and the Right to Development: A Call for International Solidarity'. *Office of the UN High Commissioner for Human Rights*, 18 September <www.ohchr.org/EN/NewsEvents/Pages/COVID19-Right-to-development.aspx#:~:text=Underscoring%20the%20right%20to%20development,and%20recovery%20from%20the%20pandemic.&text=%E2%80%9CBy%20adopting%20a%20human%20rights,among%20the%20most%20disadvantaged%20people.%E2%80%9D>.

13 Ngang, Carol. 2020. 'The Right to Development in Africa and the Common Heritage Entitlement'. *Journal of Juridical Science* 45(1): 34–36; Cambou, Dorothée and Smis, Stefaan. 2013. 'Permanent Sovereignty over Natural Resources from a Human Rights Perspective: Natural Resources Exploitation and Indigenous Peoples' Rights in the Arctic'. *Michigan State International Law Review* 22(1):350; Cotula, Lorenzo. 2018. 'Reconsidering Sovereignty, Ownership and Consent in Natural Resource Contracts: From Concepts to Practice'. *European Yearbook of International Economic Law* 9: 148.

References

Adeola, Ramola. 2018. 'The Right to Development under the African Charter: Is there an Extraterritorial Reach?' In *Perspectives on the Right to Development*, edited by Ngang, Carol; Kamga, Serges and Gumede, Vusi. Pretoria: Pretoria University Law Press, 34–46.

Brosio, Giorgio and Singh, Jan. 2014. 'Revenue Sharing of Natural Resources in Africa: Reflections from a Review of International Practices'. *The World Bank Background Paper 90252.*

Cambou, Dorothée and Smis, Stefaan. 2013. 'Permanent Sovereignty over Natural Resources from a Human Rights Perspective: Natural Resources Exploitation and Indigenous Peoples' Rights in the Arctic'. *Michigan State International Law Review* 22(1): 348–376.

Cotula, Lorenzo. 2018. 'Reconsidering Sovereignty, Ownership and Consent in Natural Resource Contracts: From Concepts to Practice'. *European Yearbook of International Economic* Law 9: 143–174.

Hitchcock, Robert. 2002. 'A Road to Self-Sufficiency: Natural Resource Management'. *Cultural Survival Quarterly Magazine.*

Ngang, Carol. 2018. 'Towards a Right-to-Development Governance in Africa'. *Journal of Human Rights* 17(1): 107–122

Ngang, Carol. 2020. 'The Right to Development in Africa and the Common Heritage Entitlement'. *Journal of Juridical Science* 45(1): 34–36.

Nkhata, Bimo *et al.* 2012. 'A Typology of Benefit Sharing Arrangements for the Governance of Social-Ecological Systems in Developing Countries'. *Ecology and Society* 17(1): 17–25.

Nkhata, Bimo; Breen, Charles and Mosimane, Alfons. 2012. 'Engaging Common Property Theory: Implications for Benefit Sharing Research in Developing Countries'. *International Journal of the Commons* 6(1): 52–69.

Okidi, Odidi. 1984. 'Management of Natural Resources and the Environment for Self-Reliance'. *Journal of Eastern African Research and Development* 14: 95–96.

Ouguergouz, Fatsah. 2003. *The African Charter on Human and Peoples' Rights: A Comprehen sive Agenda for Human Dignity and Sustainable Democracyin Africa.* Netherlands: Martinus Nijhoff Publishers.

Torpey-Saboe, Nichole *et al.* 2015. 'Benefit Sharing Among Local Resource Users: The Role of Property Rights'. *World Development* 72: 408–418.

Index